GLOBAL ISSUES

FRESHWATER SUPPLY

GLOBAL ISSUES

FRESHWATER SUPPLY

Frank Caso

Foreword by Aaron T. Wolf
Oregon State University

Facts On File
An imprint of Infobase Publishing

GLOBAL ISSUES: FRESHWATER SUPPLY

Facts On File, Inc.
An imprint of Infobase Publishing
132 West 31st Street
New York NY 10001

Library of Congress Cataloging-in-Publication Data

Caso, Frank.
Freshwater supply / Frank Caso ; foreword by Aaron T. Wolf.
p. cm. — (Global issues)
Includes bibliographical references and index.
ISBN 978-0-8160-7826-4
1. Water-supply—History. 2. Water-supply—Social aspects. 3. Water-supply—Political aspects. I. Title.
TD215.C37 2010
333.91—dc22 2009042982

Facts On File books are available at special discounts when purchased in bulk quantities for businesses, associations, institutions, or sales promotions. Please call our Special Sales Department in New York at (212) 967-8800 or (800) 322-8755.

You can find Facts On File on the World Wide Web at http://www.factsonfile.com

Text design by Erika K. Arroyo
Composition by Mary Susan Ryan-Flynn
Cover printed by Art Print, Taylor, Pa.
Book printed and bound Maple Press, York, Pa.
Date printed: May 2010
Printed in the United States of America

10 9 8 7 6 5 4 3 2 1

This book is printed on acid-free paper.

For Vladimir, A. J., Tariq, Ibrahim, Salma, Nevan, Sylvia, and the staff and children of the Hartford Friendship Kids Camp

CONTENTS

PART II: PRIMARY SOURCES

PART III: RESEARCH TOOLS

List of Maps, Graphs, and Tables

Foreword

The world is out of "easy" water. In the not too distant past, we treated water as if it would always be abundant and clean. We used it once, either to drink or to irrigate or to dilute our waste, and then it went "away." Then, over time, our relationship to the Earth, and to our precious water resources, began to change. Our population grew, and continues to grow, exponentially, even as the supply of water remains the same as it has been since time immemorial. Much of the world has industrialized, adding new pressures to supplies as economies grew with new demand for dams for power generation and irrigation and regularly inventing new pollutants that are added to our streams and aquifers. We are only beginning to understand the latest set of pollutants, pharmaceuticals: how to monitor for them, their impacts on human and ecosystem health, and how to mitigate for these impacts. These pressures built within nations, endangering mostly those who live on the environmental margins, and threaten to spill across international boundaries, exacerbating political tensions.

While water quantity has been the major issue during most of the last century, water quality has been neglected to the point of catastrophe: Today, more than a billion people lack access to safe water supplies; almost 3 billion do not have access to adequate sanitation; 5 to 10 million people die each year from water-related diseases or inadequate sanitation; and 20 percent of the world's irrigated lands are salt-laden, affecting crop production.

In short, water demands are increasing, groundwater levels are dropping, surface-water supplies are increasingly contaminated, ecosystems and arable land are dying, and delivery and treatment infrastructure is aging—and all of these issues are made worse by global climate change. Twelve years ago, the World Bank estimated that it would cost $600 billion to repair and improve the world's existing water delivery systems, a number that has increased significantly since then. In terms of the number of people affected, this issue has as big an impact as HIV/AIDS and malaria, and easily bigger than earthquakes, tsunamis, hurricanes, floods, or all of the wars of any given year put together.

But what allows room for optimism in this dark story is the quiet advances we collectively have made in our relationship with water resources. Our understanding of complex physical systems is ever increasing with satellite data supplementing monitoring and computer modeling systems, allowing for new efficiencies in management. We are developing new technologies for drip irrigation and desalination, for example, as well as new drought- and salt-resistant crops. On the international front, the stresses that bring threats of tension and even violence are increasingly driving neighbors who share water to the negotiating table, resulting in new agreements between even those co-riparians who have the most distrust, including Arabs and Israelis, Indians and Pakistanis, and Azeris and Armenians. Even Californians and Oregonians.

In this book in the Global Issues series, *Freshwater Supply,* Frank Caso addresses both the painful details of the current situation and the advances we have already made. The first chapter gives a detailed global overview; it clearly defines and outlines the key challenges we currently face and analyzes the drivers of these challenges, both physical and human. Extensive surveys of the globe follow, with subsequent chapters covering the United States and hot spots around the world. Each of these global case studies describes in clear, nontechnical language the issue at hand, followed by a section on the challenges, and, most important, ending with a section on counterstrategies, the approaches we can take to help overcome the daunting issues of the day.

Allowing the student to understand both the challenges and the possibilities is one of the gifts this book offers. Take the issues of violence over water or, commonly, "water wars." The prevailing wisdom is that wars over water have been common historically and are inevitable in the future. And the arguments are compelling. When surveying the world of water conflict one cannot help but notice just how similar some sets of problems facing parties locked in hydroconflict can be, regardless of the level of economic development or geographic scale involved. *Acequia* is the term used in the United States' southwestern desert and other Spanish-speaking parts of the world to denote both an irrigation ditch and the informal institution that manages it. In *Mayordomo: Chronicle of an Acequia in Northern New Mexico,* Stanley Crawford describes his period as *mayordomo* or ditch manager of an *acequia* in New Mexico. He writes of two neighbors who:

> *. . . have never been on good terms, at least in my hearing, the lower neighbor commonly accusing the upper of never letting any water pass downstream to his place and then of dumping trash into it whenever he rarely does.*

This is, in essence, the heart of water conflicts since, well, forever. In modern days, the neighbor's complaint could be Syria's of Turkey,

Pakistan's of India, or Egypt's of Ethiopia. For all our 21st-century capabilities in water management—whether in dynamic modeling, remote sensing, geographic information systems, desalination, biotechnology, or demand management—and for all our newfound concerns with globalization, privatization, and information technologies, the crux of the disputes are still often about little more than who opens a diversion gate when, and how much garbage gets put into the water as it makes its way downstream.

In fact, this universality of water problems is embedded in the very terminology we use to describe them: *Acequias* have their roots in Spain; according to tradition, the Tribunal de las Aguas (Water Court) has been meeting to resolve disputes over the *acequias* around Valencia in the same church-front square since medieval times, if not before. But the root of *acequia* is *al-saqia,* Arabic for a gear-driven water wheel, the technology that made early irrigation possible along many of the rivers of the ancient Middle East. From the Middle East to Spain to the New World, many of the problems remain.

The fortunate corollary, of course, is that so do many of the solutions. Water has also proven to be a productive pathway to confidence building, cooperation, and arguably conflict prevention, even in particularly contentious basins. Cooperative incidents outnumbered conflicts by more than two to one from 1945 to 1999. In some cases, water provides one of the few paths for dialogue in otherwise heated bilateral conflicts. In politically unsettled regions, water is often essential to regional development negotiations that serve as de facto conflict-prevention strategies.

This historical record suggests that international water disputes do get resolved, even among enemies, and even as conflicts erupt over other issues. Some of the world's most vociferous enemies have negotiated water agreements or are in the process of doing so, and the institutions they have created often prove to be resilient, even when relations are strained. The Mekong Committee, for example, established by the governments of Cambodia, Laos, Thailand, and Vietnam as an intergovernmental agency in 1957, exchanged data and information on water resources development throughout the Vietnam War. Israel and Jordan have held secret "picnic table" talks on managing the Jordan River since the unsuccessful Johnston negotiations of 1953–55, even though they were technically at war from Israel's independence in 1948 until the 1994 treaty. The Indus River Commission survived two major wars between India and Pakistan. And all 10 Nile basin riparian countries are currently involved in senior government-level negotiations to develop the basin cooperatively, despite "water wars" rhetoric between upstream and downstream states. In fact, the only actual "water war" between nations specifically over shared water resources on record occurred more than 4,500 years ago between the city-states of Lagash and Umma in the Tigris-Euphrates basin.

However, this is true mostly for international armed conflict and water as a scarce resource. Water-related violence continues to take place at the subnational level, generally between tribes, water-use sectors, or states/provinces. In fact, there are many examples of internal water conflicts ranging from interstate violence and death along the Cauvery River in India, to California farmers blowing up a pipeline meant for Los Angeles, to recent tribal violence in Kenya. In the United States, the desert-state of Arizona even commissioned a navy (one ferryboat) and sent its state militia to stop a dam and diversion on the Colorado River in 1934.

Although wars over water between countries have not occurred, there is ample evidence showing that the lack of clean freshwater has occasionally led to intense political instability and that, on a small scale, acute violence may result. Interestingly, geographic scale and intensity of conflict seem to be inversely related. As water quality degrades—or quantity diminishes—over time, the effect on the stability of a region can be unsettling. For example, for 30 years the Gaza Strip was under Israeli occupation. Water quality deteriorated steadily, saltwater intrusion degraded local wells, and water-related diseases took a rising toll on the people living there. In 1987, the intifada, or Palestinian uprising, broke out in the Gaza Strip and quickly spread throughout the West Bank. Although there was no direct causality between water quality and uprising, undoubtedly the issue was an irritant that exacerbated an already tense situation.

Freshwater Supply skillfully outlines both the threat and promise, danger and opportunity, inherent in issues surrounding freshwater. In between, there is lots and lots of nuance, with plenty of room for human ingenuity, creativity and, finally, hope. This is the story our relationship with water tells, a story that is artfully captured in these pages.

—Aaron T. Wolf
Professor of Geography and Chair
Department of Geosciences
Oregon State University

Acknowledgments

I would like to thank Claudia Schaab for shepherding this project. I also would like to thank Professor Aaron T. Wolf of Oregon State University and Professor Jeremy Carl of Stanford University for their insightful comments on the manuscript, and Mr. Robert Moore of Hartford's Metropolitan District Commission for taking the time to answer my questions concerning domestic water usage and distribution issues. Special bows go to Bronwyn Collie for her ability to sculpt the manuscript into its present shape and to Alexandra Simon for her wonderful copyediting, to Alexandra Lo Re for editorial support in seeing the book through its final stages, and to Shawna Kimber for the keen eyes needed of a proofreader. Lastly, I can't thank my wife, Katherine, enough for her support.

Metric Conversion

Most countries use the metric system, even to measure volume and flow of water. The United States does not use the metric system, and because water volume in gallons would be cumbersome the measurement acre-foot is used. Both metric and U.S. terms are used in this book. Below is a list of conversions.

1 cubic meter = 1,000 liters

1 cubic kilometer = 1 billion cubic meters = 1 trillion liters

1 liter = .264 U.S. gallons

1 U.S. gallon = 3.785 liters

1 cubic meter = 264 U.S. gallons

1 U.S. gallon = .00378 cubic meters

1 cubic kilometer = 810,713 acre-feet

1 acre-foot = 1,233 cubic meters = 325,851 U.S. gallons

1 kilometer = .621 miles

1 mile = 1.6 kilometers

1 kilogram = 2.2 pounds

1 pound = .45 kilograms

PART I

At Issue

1

Introduction

Aside from the air we breathe, freshwater is our most precious resource, something upon which all life depends. Without freshwater, humans can live for only a few days. Biologists theorize that all life evolved from the sea, so perhaps it is no accident that the human body is composed of 70 percent water—thus our need to remain hydrated.

Throughout history humans have tended to take freshwater for granted, generally assigning little value to it beyond their immediate needs. That is probably because it seemed to be in abundant supply—each rainstorm brought more water, as did the spring thaw high in the mountains. In some cases, where people moved into areas where water was scarce, such as the western United States, technological innovations enabling water to be transported in large quantities helped uphold the illusion of bounty. But it is an illusion, and one fostered by water's role in history: Easy access to freshwater is responsible for the birth of civilization—along the Tigris and Euphrates, the Nile, the Ganges, the Indus, the Yangtze, and the Yellow Rivers. Freshwater has contributed to the development of various societies to such an extent that, generally speaking, the wealthiest nations are the best-watered ones. Indeed, the Industrial Revolution, which began in the United Kingdom, would not have been possible without easy access to water.

Nevertheless, easy access to freshwater is diminishing, even in water-rich countries such as the United Kingdom, the United States, and Canada. And in the water-scarce countries the acquisition of freshwater is a daily trial, and its lack a cause of misery and death.

THE ISSUE

Water is found all over the world in one form or another, and the amount of water on the Earth is finite and fixed. Clouds do not create more water, they merely contain existing water in a different form as part of what is called the hydrologic cycle. Simply put, the hydrologic cycle is the process by which surface water, notably from the oceans, but also from lakes, ponds, rivers, wetlands, and the like, evaporates, forming clouds in the atmosphere that then return the evaporated water to Earth in the form of rain, snow, or hail. Much of this water falls back into the oceans, which cover 70 percent of the Earth's surface, or into lakes, rivers, and the like, while the rest seeps into the ground to become groundwater. Under normal conditions, groundwater assists in replenishing surface water or remains in an aquifer, a natural underground water reserve, or it makes its way back to the sea.

How much of the planet's vast, but finite, supply of water is suitable for use by humans? Some 97 percent of all the water on Earth is found in the oceans, and two-thirds of the remaining 3 percent is locked in glaciers and ice sheets or is unreachable groundwater. That leaves approximately 1 percent to sustain life on land. The picture is further complicated by the fact that that 1 percent of usable freshwater is not evenly distributed around the world; as noted, some nations are considered water rich, some are water stressed, and others struggle somewhere in between. But the situation is not so tidy as that. Experts predict that possibly by the middle of this century, certainly by the end of it, the number of water-stressed countries will increase. To put it another way, the majority of the world's population by the year 2050, which could be as high as 9 billion people, could be living under circumstances of dire per capita freshwater quantity and/or poor quality.

There is evidence that human tampering has altered the hydrologic cycle; though the amount of water remains constant, it may no longer be as available as it once was in some places. Actions such as draining wetlands and diverting rivers have caused climactic changes in certain areas, making them drier. Deforestation and urbanization have also contributed to dwindling water supplies, as has climate change, or global warming, which is exacerbated by deforestation and population increase. In terms of population, consider the following: China, a water-stressed nation, has a slightly larger supply of freshwater than Canada, one the world's water-rich nations. The reason why China is water stressed is that its population is far larger than Canada's, at more than a billion people, compared with around 33 million. Overall, approximately 3 billion people live in countries that are either water stressed or where water is in chronically short supply,[1] and even in the non-stressed areas water allocation and wastage are becoming problems.

Water quality is another important aspect of the global water crisis. Pollutants are found in freshwater sources throughout the world and include industrial toxins, pesticides and fertilizers from agricultural runoff, animal waste, and human sewage and other waste from domestic sources. Naturally occurring toxins, notably arsenic, which is found in the Earth's crust, can dissolve in slow-moving or standing water. Improper irrigation methods, or simply over-irrigation, have caused salinity, an increase in salts, in many of the world's rivers and the adjoining and irrigated land, while poor or nonexistent sanitation has turned many surface supplies of water, such as rivers, into bacterial petri dishes. Indeed, waterborne diseases are a leading cause of death in developing nations, especially among young children.

The vast majority of the world's freshwater supply, approximately 70 percent, is used for agricultural purposes. Industry accounts for 22 percent of freshwater consumption, while domestic and municipal use together consume 8 percent.[2] Again, water usage varies by nation and even by regions within nations. For example, the average daily per capita freshwater usage in the United States is 151 gallons (approximately 193 liters), while the per capita daily freshwater usage in Ethiopia amounts to just three gallons (11.36 liters).[3] The United States, despite having some remarkably arid areas, is a water-rich nation—so much so that approximately 7 billion gallons of water per day is used for landscaping—while Ethiopia, which includes the headwaters of the Blue Nile, is water stressed.

Water in Agriculture

According to *The Water Atlas*, at the turn of the 21st century, Central Asia, from Pakistan in the south to Kazakhstan in the north, had the highest per capita freshwater usage for agriculture. With the exception of Pakistan and Afghanistan, the countries in this zone were once part of the Soviet Union and many of their farming practices remain a legacy of that time. Their annual per capita rate is a minimum of 1,000 cubic meters of water (264,000 gallons), used mainly for irrigation. Other nations at the high-end range are Azerbaijan, Syria, Sudan, and Thailand. Prior to the current Iraq War, which began in 2003, Iraq was also in this category.[4] Ironically, despite the high usage of freshwater for agricultural purposes, a large percentage of the population in many of these countries is undernourished. The Central Asian countries—as well as Azerbaijan, Oman, Egypt, Iraq, North Korea, Japan, Surinam, and Chile—also fall into the highest category for percentage of arable land that is irrigated. All of the above-mentioned irrigate 51 percent or more of their arable land, with the Central Asian countries irrigating at least 76 percent, as of 2000. Salinity is common in these areas due to over-irrigation and/or poor drainage methods and ultimately renders the land useless for agriculture.[5]

Because many crops are water intensive, the production of food itself can involve large amounts of water. Rice and soybeans are two of the more water-intensive crops. It takes 1,900 liters (approximately 502 gallons) of water to produce two pounds (just under a kilogram) of rice, and a minimum of 1,650 liters (436 gallons) of water to produce two pounds of soybeans. At the other end of the scale, two pounds of wheat requires 900 liters (238 gallons) of water, and two pounds of potatoes, 500 liters (132 gallons). The beef and poultry industries consume by far the most water. It takes 15,000 liters (3,960 gallons) of water to produce two pounds of beef, and 3,500 liters (924 gallons) of water to produce the same weight in poultry. These latter figures include the amount of water used to grow the grain to feed the animals, which is the largest component in the equation.[6]

Agriculture can also lead to the pollution of the freshwater supply. Industrialized, monocultural farming practices, in particular, require more chemicals (fertilizers and pesticides), which leach into the soil or run off into rivers and streams, contaminating sources of drinking water. Fertilizers generally belong to the phosphate and nitrate families, while pesticides run a gamut of chemical compounds. The latter, especially, can cause health problems in humans; and both may undermine the health of the aquatic ecosystems into which they drain—phosphates, for example, promote the growth of algae in both seawater and freshwater. During the 1960s and 1970s, the global use of fertilizers increased dramatically: by more than threefold in the developed world and almost tenfold in developing countries. By 1981, the industrialized nations still used twice as much fertilizer as developing countries—78 million tons combined compared with a combined 39 million tons. Since that time fertilizer use in developed countries has been steadily declining, partly due to the increasing popularity of organic foods but mainly because of more stringent freshwater laws, while its use in the developing world continues to soar. By the early 1990s, the amount of fertilizer used in developing countries surpassed that of industrial nations, and in 2001 it stood at 86 million tons as opposed to 50 million tons. Many of the nations with the highest fertilizer use are water stressed or in crisis; they include South Africa, Sudan, Egypt, Jordan, and Bangladesh.

Pharmaceuticals in Freshwater

In recent years hormones, antibiotics, antidepressants, tranquilizers, painkillers, anti-seizure medications, and chemotherapy drugs are among the numerous pharmaceuticals that have been detected in the freshwater supplies of developed nations, including the United States. At present, there is some disagreement over whether a keener ability to detect pharmaceutical pollution or an increase in consumption of pharmaceuticals is the cause, although the latter seems to be gaining credence in most quarters. Pharmaceuticals enter the freshwater system when excess amounts, no longer needed, are flushed down toilets or improperly disposed of in dumpsites where they eventually leach into the groundwater. However, the most common way for pharmaceuticals to enter a municipality's freshwater system is by passing through humans. In the future this will become particularly troublesome in those communities where graywater is recycled.

While pharmaceuticals have so far been discovered in concentrations too small to do any harm to humans, the trend is nevertheless a disturbing one. Moreover, the long-term effects on biota and aquaculture more generally have yet to be determined, though short-term mutations and damage has already been observed in some species of fish and shellfish. At present there is almost no abatement of pharmaceutical pollution, and water treatment is inadequate. The issue has been of great concern in Europe since the early 1990s, but has yet to catch the public's imagination in North America and elsewhere.

Water in Industry

The use of freshwater in industry increased in the latter half of the 20th century throughout the world, and that trend is expected to continue through the first quarter of the present century. In 1950, approximately 204 cubic kilometers (approximately 54 trillion gallons) of freshwater was used for industrial purposes. By 2000, this figure had increased to 776 cubic kilometers (approximately 205 trillion gallons), and it is projected that by 2025 worldwide use will be approximately 1,170 cubic kilometers (309 trillion gallons). In this area, the developed nations have been and remain the true water guzzlers. By 2000, the highest users of water for industry were the United States, Canada, western Europe, Russia, Kazakhstan, China, and Australia. Among the less-developed nations, Egypt and Chile use the most water for industry. Pollution caused by the use of water in industry occurs in various forms and includes heavy metals such as lead and mercury, organic pollutants, and a category of toxic chemicals known as persistent organic pollutants. Organic pollutants "use up vital oxygen in the water," while persistent organic pollutants "are a group of carbon-based chemicals that share a number of characteristics: they all persist in the environment for a long time, concentrate in the food chain, travel long distances, and are linked to serious health effects."[7] Two of the most well-known persistent organic pollutants are dichlorodiphenyltrichloroethane (DDT) and polychlorinated biphenyls (PCBs).

The computer and electronics industries are two of the biggest users and polluters of freshwater, and a reason the developed nations have continued to outpace the rest of the world in industrial water use and pollution. Indeed, wastes from these industries affect the land and air, as well as water. The wastes from just the first step in semiconductor manufacturing include spent solvents, acids, alkaline cleaning solutions, and deionized water. During the four-step process, pollutants including aqueous metals, spent etchant solution, spent aqueous developing solutions, and chromium are also produced. The pollutants from the manufacture of printed wiring boards include spent acids, acid solutions, alkaline solutions, spent electroless copper baths, spent developing solutions, aqueous metals, spent etchants, spent solvents, lead, nickel, copper, and silver.[8] As for water use, it takes 3,787 gallons of water to manufacture a one-inch silicon wafer.

Home and Municipal Use of Water

There is a marked disparity between countries in terms of home and municipal water usage, and this stems from both a country's wealth and its population's access to water. Those in water-rich, wealthy nations by and large have immediate access to freshwater—that is, it is piped into homes and workplaces. Unsurprisingly, the more difficult it is to access freshwater, the less per capita is used, and where water is more difficult to access, sanitation is often poorer. Domestic water usage is highest in the United States, Canada, and the United Arab Emirates, all of which at the turn of the century were using a minimum of 200,000 liters (52,800 gallons) per capita per year. In addition to the basic uses, this included such things as watering lawns and gardens, washing cars, and filling swimming pools. Water usage per capita in developing nations often falls below international health standards, which are minimally defined at 50 liters (13.2 gallons) per person per day for all activities: drinking, cooking, washing, and sanitation. In Ethiopia, Eritrea, Uganda, Rwanda, Burundi, and Tanzania, all of which are in the Nile Basin, domestic water consumption at the turn of the century topped off at 9,000 liters (2,376 gallons) per person per year, or less than half of the 50 liters a day required to maintain health. In Bolivia, Pakistan, Bangladesh, and the Nile basin nations of Sudan and Kenya annual per capita freshwater consumption ranged from 10,000 to 49,000 liters (2,640 to nearly 13,000 gallons) and in India and Egypt it fell in the 50,000 to 99,000 liter (13,200 to 26,000 gallons) range, while Mexico was in the second highest group, with annual per capita consumption of between 100,000 and 199,000 liters (26,400 to 52,500 gallons). Of course even within a country there is a range of usage, depending on wealth and ease of access.[9] Moreover, wherever water is hard to access the job of retrieving it generally falls to women and girls. Girls old enough to carry water jars (usually on their heads) to the home may miss a good portion of each school day just performing this task, or may be pulled out of school altogether. Essentially, women and girls in these situations spend hours each day walking to and from the water source—and the return trip naturally takes longer. They must also wait in line for their turn to fill their jar and, depending on how steady and fast the water flow is, actually filling it can also take a good deal of time.

Freshwater as a Source of Energy

The second half of the 20th century may very well go down in history as the era of great dam-building. Dams were constructed to control floods, create reservoirs for irrigation and recreational purposes, and create hydroelectric power for a modern world ever in need of energy. The United States led the way with engineering marvels like the Glen Canyon, Grand Coulee, and Hoover Dams, which were all built to capture western water.[10] Their construction was spurred by a number of factors, not least of which was the rivalry between the United States Bureau of Reclamation and the United States Army Corps of Engineers for various federal dam projects, but which also included the growth of agribusiness in the U.S. West, the desire for inexpensive hydropower (which in turn created the proliferation of "cash-register dams" that were supposed to pay for their construction costs and subsequently turn a profit by selling the electricity they produced), the arid climate, and U.S. western water law wherein prior appropriation took precedence. High dams of 500 feet (150 meters) or more, of which Hoover Dam was the first in the world, were deemed modern marvels of engineering (largely because of the sheer amount of hydroelectric power they produced) and thus became the model for other such dams, particularly in China, which was home to 46.2 percent of the world's dams of all sizes by 2003. The United States was a distant second, with 13.8 percent, while India was third, with 9 percent.[11] The majority of dams built in the United States, China, and India are for flood control, irrigation, and other purposes. By the year 2000, hydropower provided just 16 percent of China's total power, in the United States the figure was 6 percent, and in India it was 14 percent.

This was not the case in many other countries that sought to make the best use of their water resources. For example, China and India's hydropower percentages are low compared with neighboring Nepal's, which stood at 98 percent at the turn of the 21st century; Egypt stood at 31 percent; Ethiopia, 97 percent; Bolivia, 50 percent; Brazil, 87 percent; Chile, 46 percent; Paraguay, 100 percent, and Peru, 81 percent. In the water-starved Middle East, Lebanon, Jordan, and Israel gleaned 6 percent, 1 percent, and 0 percent, respectively, of their power from hydropower.[12]

Whenever humans reengineer nature, it seems there is a hidden cost that is either overlooked (at first) or ignored (sometimes both), and dam-building is no exception. Over the past decade or so, dams have come under question because of their environmental costs and their toll on people in terms of displacement, as well as the consequences to the ecosystems. Furthermore, the stored water in reservoirs so large they are deemed lakes is subject to greater loss from evaporation than it would have been in rivers. And some rivers—such as the Rio Grande and Colorado in the United States and the Yellow in northern China—are so dammed that some years they no longer flow to the sea.

The widely held assumption that a dam's period of usefulness is nearly limitless as long as the dam itself is properly maintained is not correct. Unless the silt that builds up in the reservoir is removed, the dam will cease to function as such; instead, it will morph into an artificial waterfall when the buildup reaches the top of the dam and the water is forced over. If the dam produces hydropower, the silt will also have clogged the intake pipes by then. Nevertheless, the World Bank and the International Monetary Fund (IMF) are generally not averse to providing the financing for dams, especially in developing nations where the twin ideas of a reservoir and hydropower are too good to pass up. This helps to explain why 60 percent of the world's rivers have either been dammed or diverted for irrigation purposes.

THE CHALLENGES

Pollution

The pollution of the freshwater supply, aside from eutrophication and leaching which occur naturally, has occurred in tandem with the rise of civilization. Throughout most of history, however, pollution caused by humans was at low enough levels that rivers, streams, lakes, ponds, and wetlands were able to cleanse themselves (by the process of inflow and outflow, and as part of an interdependent ecosystem) and so remained healthy. Over the course of many centuries, this balance gradually changed. With the rise and spread of the Industrial Revolution beginning in the late 18th century and the technological revolution of the 20th century, pollution increased at a much faster rate, leading to much higher levels around the world. Essentially, the problem related to the disposal of waste material by washing it down rivers, dumping it into lakes or wetlands, or burying it. Though waste had been disposed of in these ways since prehistory, industry and technology create more waste at a faster rate than nature can handle. Until the mid-20th century, pollution of the freshwater supply was largely ignored. Since then, increased awareness has led to steps to correct the situation—laws have been enacted and enforced, and international agreements signed.

The pollution of the world's waterways causes many problems. First, it removes water from the available supply—which is compounded by population growth and climate change—and polluted water must be treated to return it to a safe state, often at enormous expense. Second, governments more often than polluters themselves bear the brunt of these expenditures, and restoring a natural resource is not likely to be high on the agenda during economic hard times. Third, freshwater pollution alters ecosystems. Fourth, groundwater reserves can be compromised by the pollution of surface freshwater.

Agricultural pollution, which as previously noted consists primarily of nitrates and phosphates in fertilizers, can have unintended consequences. While sometimes the chemicals seep down into the groundwater, most often they enter surface water by way of runoff. The purpose of using nitrates and phosphates in agriculture is to add nutrients to the soil and increase crop yields. Freshwater, however, does not need such nutrients; their addition to the water supply is called nutrient loading. Among the problems caused by nutrient loading is eutrophication. Eutrophication is a natural process whereby plant and algal growth is stimulated by higher amounts of nutrients in a body of water. However, when nutrient loading occurs through such things as runoff, this process speeds up to the point where the water cannot repair itself. Then it is called galloping eutrophication. Galloping eutrophication can lead to dead zones in bodies of water such as lakes, and high concentrations of nitrates in the drinking water supply can cause problems as serious as blue baby syndrome, when infants "convert nitrate into nitrite, which stops their blood from transporting oxygen and can lead to suffocation and death."[13] Once again, over the past 30 years, the use of phosphates and nitrates has declined in Europe and North America, but continues to grow in the developing world.

Industrial pollution covers a wide range of toxic chemicals, heavy metals, and other substances. Cadmium, lead, and mercury are the most common heavy metals found in freshwater,[14] and removing them often requires expensive, long-term abatement procedures. Meanwhile, organic pollutants discharged into a body of water monopolize oxygen. They include a subclass known as persistent organic pollutants (POPs), which remain in the environment a long time. Because of this, they become concentrated in the food chain, and the higher up the chain, the higher the concentration of POPs. Two of the most deadly POPs are DDT and PCBs; both present extreme hazards to all forms of life, including people.

Agricultural and industrial pollution affect mainly surface water, but groundwater can also be polluted in the same and other ways. Arsenic may naturally leach into aquifers. This has occurred in a number of places around the world including the United States, but perhaps nowhere as much as in Bangladesh. Groundwater may also be polluted by overdrawing from coastal aquifers, causing the water table to drop below sea level. The salt water eventually mixes with the freshwater, causing salinity.

Sanitation

According to a 2005 study undertaken by the World Health Organization (WHO), more than 3.4 million people die each year around the world as a result of waterborne diseases. This is overwhelmingly due to the fact that more than 1 billion people do not have sufficient access to clean drinking water to remain healthy. Young children account for most of the deaths, especially as a result of diarrhea that leads to severe dehydration. The water in many cases is polluted with coliform bacteria from human feces. The vast majority of deaths from waterborne diseases occurs in Southeast Asia, Africa, and the eastern Mediterranean. Among the most common waterborne illnesses are cholera, typhoid, polio, guinea-worm disease, and trachoma. Of these, trachoma is not caused by drinking bad water but is transmitted from person to person in unhygienic environments where people do not have sufficient water to wash. Mosquito-borne diseases such as malaria, dengue, and West Nile virus are common in developing equatorial nations. Standing water attracts mosquitoes, which lay their eggs in the water. The mosquitoes carry the virus, which infects the people they "bite." Animals may also receive the virus and become hosts, thus transmitting it to future generations of mosquitoes. These diseases are not confined to equatorial regions. West Nile Virus, for example, has become widespread in the United States. The difference is that it is seasonal in North America, the mosquitoes dying off in the late fall. In equatorial regions, it is a year-round problem.

All of this leads back to inadequate sanitation. Of the countries discussed in this book, at the turn of the century the United States, Canada, Egypt, Israel, the Palestinian territories, and Jordan were in the highest category for percentage of population with access to adequate sanitation facilities, at between 91 to 100 percent. Uganda, Kenya, Burundi, and Tanzania were in the next highest group—between 76 and 90 percent of their populations had adequate sanitation facilities. Mexico, Bolivia, Sudan, and Pakistan fell into the third level, at between 51 and 75 percent. Among the countries occupying the fourth level were India and Bangladesh, at between 26 and 50 percent. At the bottom were Ethiopia, Eritrea, and Rwanda, with 25 percent or less people having adequate sanitation facilities.[15]

Improper sanitation and the ensuing disease and death exacerbates poverty; for example, the death of a parent often results in a loss of income, forcing children to leave school and enter the workforce to help maintain the family. And the death of a child lessens the potential of the whole family.

Public versus Private Ownership of Water

The debate over who owns and who should own water is becoming increasingly complicated as freshwater scarcity expands, as multinational companies—at times at the behest of the World Bank and the IMF—inject themselves into the debate, and as global trade agreements such as the North American Free Trade Agreement (NAFTA) and the General Agreement on Tariffs and Trade (GATT) and the World Trade Organization (WTO) move closer to commodifying water.

Perhaps foremost in this debate is a problem that at first appears like a simple semantic issue, a matter of which word is best suited to describe the relationship of humans to water. On one hand are those who strive to have water declared a human right, while on the other are those who insist it is a human need. The side championing water as a human need seems, at least for the present, to have won the argument. The second World Water Forum held in The Hague in March 2000 declared water a human need. In doing so, it essentially approved the further commodification of water.[16] While it is an indisputable fact that humans need water to survive, other human needs such as food and adequate clothing, without which we would die, have been turned into for-profit commodities. The difference, of course, is that they are manufactured. Water, on the other hand, is a resource, and by proclaiming it a need instead of a right the World Water Forum turned that resource into a source of profit. If water had been proclaimed a human right, then everyone in the world would have the right to quench their thirst, cook their food, bathe, and enjoy adequate sanitation. This would place, first, governments and, second, transnational water companies on notice. The two key organizations behind the four-day World Water Forum—Global Water Project (GWP) and the World Water Council (WWC)—count among their members many water companies.[17]

Two of those companies were the multinationals Vivendi Universal (as it was known at the time) and Suez—both are French companies with roots going back to the mid-19th century, when France began privatizing water delivery during the reign of Napoléon III, Emperor of the French. Vivendi (known as Compagnie Générale des Eaux until 1998) is the larger of the two and is primarily a media conglomerate, but by the beginning of this century its water division was operating in 90 countries. In 2000, Vivendi Universal created Vivendi Environnement as the umbrella company for its water and waste management divisions, and in 2003 Vivendi Environnement became Veolia Environnement.[18] Veolia Water is now the largest water company in the world.

Suez began its corporate life a few years later and was initially known as Compagnie Universelle du Canal Maritime de Suez. As its name implies, it was the company that, under the diplomat Ferdinand de Lesseps, was largely responsible for the construction of the Suez Canal. In 1997, Suez went full force into the water delivery business when it merged with Lyonnaise des Eaux to form Suez-Lyonnaise des Eaux. The company has since undergone further restructuring: In 2008 it merged with Gaz de France to form GDF Suez. Its water division, Suez Environnement, then became a stand-alone company (in which GDF Suez owned 35 percent, making it the largest shareholder), which delivered water and sanitation services to more than 100 million people in some 130 countries by 2003. Among Suez's subsidiaries is United Water, which has customers in the United States and Australia.

Other multinational water companies that like Vivendi/Veoli and Suez operate either directly or through subsidiaries include Bouygues-SAUR, another French company with operations in approximately 80 countries; RWE-Thames, created when the German electrical company RWE purchased Thames Water; and Bechtel, the California construction company that got its start in the water business, so to speak, back in the 1930s when it was one of the companies that built the Hoover Dam on the Colorado River.

Despite some of the problems associated with the privatization of municipal water delivery and sanitation services, water activists worldwide are not unanimous in their opposition to the trend. In fact, in developing countries some favor it, seeing in privatization the opportunity for people to finally receive a service their government has failed to provide. Nevertheless, water delivery through private companies with an eye on the bottom line and shareholders' dividends, has resulted in higher prices, inefficiencies, and often poor sanitation services. In some cases, water companies have corralled a municipality's supply system to such an extent that individuals' rights to water have been superseded by the company's right to earn a profit. Most notable is the example of Cochabamba, Bolivia, discussed in chapter 3.

Water delivery companies are not the only multinationals with an interest in the water privatization debate. The numerous bottled water companies—including Coca-Cola, Danone, Nestlé, PepsiCo, and Proctor & Gamble—have seen their sales expand exponentially around the world over the past 15 years. The profits can be enormous, in part because the resource itself—water—is all but there for the taking. As a former chief executive officer of Perrier (a Nestlé subsidiary) put it: "It struck me . . . that all you had to do is take the water out of the ground and then sell it for more than the price of wine, milk, or for that matter, oil."[19]

Bottled water is often, though not always, taken from springs, and its popularity, particularly in developed countries, is in part connected with the notion that springwater has healthful benefits—for instance, it may contain more minerals. But tests on bottled water undertaken by the National Resources Defense Council (NRDC) as far back as 1999 have shown that this is not necessarily the case, and indeed sometimes bottled water, and even springwater, is less healthful than tap water.[20] This is because municipal filtration systems in the United States are usually of the highest standards. Additionally, bottled water may in fact be purified tap water, sometimes with minerals added, so the consumer may not be getting what he or she expected. In developing countries, however, where delivery systems may be old or nonexistent, bottled water may be a viable, though expensive, alternative to what might otherwise be available. In these instances, the bottled water is not the half-liter bottles, equal to slightly more than a pint, one commonly finds for sale, but larger, often five-liter (1.32 gallons) bottles. Whether spring, tap, or designer water, all bottled water has one thing in common: It is a commodity that earns companies a profit. Essentially, the debate over private versus public ownership of water and whether water is a human need or right boils down to a struggle in which, in the words of the journalist and author Cynthia Barnett, "Those who control the water control the destiny of a place and its people."[21]

Conflict over Water Resources

War as a resolution to the world's growing water crisis is something everyone fears. While this fear ought not to be taken lightly—wars over water have been fought all the way back to ancient times[22] and more recently water has been an underlying cause of some of the conflicts between Israel and its Arab neighbors, Syria and Jordan—the evidence shows that there has been more international cooperation over water than there has been conflict. A study undertaken by Oregon State University "documented more than 1,800 interactions between two or more nations" over water during the latter half of the 20th century and showed that there were slightly more than four times as many water treaties signed during this period than there were acute disputes.[23] Still, writers, politicians, bureaucrats, and activists have posited various scenarios that might lead to water wars in the near future. For example, Dr. Ismail Serageldin, an Egyptian senator and director of the Bibliotheca Alexandrina and a well-connected person among water technocrats whose previous positions included vice president for environmentally and socially sustainable development at the World Bank, chairman of the Global Water Partnership, and chairman of the World Commission for Water in the 21st Century, declared in 1995 that "the wars of the twenty-first century will be fought over water."[24] While this prediction has often been repeated, because of Serageldin's high-profile position in the world's water bureaucracy at the time he made the statement, it can be seen less as an assessment of the contemporary water situation in the world than as a rationalization for the further globalization of water delivery and sanitation services.

Wherever water is scarce or of poor quality, there is the potential for some sort of conflict. An upstream nation may decide to dam a river, which can lessen the supply to its downstream neighbor, or a more powerful downstream neighbor may cast its eyes toward a plentiful water basin upstream. Furthermore, problems related to water can exacerbate other, long-standing political problems, as has occurred between Israel and Syria, Egypt and Ethiopia, and Pakistan and India. Nevertheless, with the exception of the 1967 Six-Day War, which had some, but by no means all, of its roots in an earlier water conflict, and post-partition skirmishes in the Indus basin, in the divided state of Punjab, water wars between nations have not occurred.

The above examples pinpoint some of the globe's political hot spots. Less obvious is the potential for a reemergence of political conflict between

Mexico and the United States over Colorado River water, a point of contention during the 20th century. Since the end of World War II, Mexico has legitimized its claim to water from the Colorado River through a series of treaties with the United States. But Mexico's drier climate and increasing population in the states that claim rights to water from the Colorado could in time make those treaties worthless. Historically, there has also been conflict between U.S. states—especially California and Arizona—as to how the water from the Colorado River should be divided up. About the only claimants who currently appear to be fully satisfied are the Southwest U.S. Native American tribes, whose water rights have been affirmed by the United States Supreme Court.

Some Canadians, too, are concerned about U.S. designs on their freshwater supplies, particularly their western rivers and the Great Lakes. Largely overshadowed by the September 11, 2001, terrorist attacks and the subsequent wars in Afghanistan and Iraq was President George W. Bush's pronouncement "in July 2001—just before the famous G-8 meeting in Genoa, Italy—that he saw Canadian water as an extension of Canada's energy reserves, to be shared with the U.S. by pipeline in the near future."[25] This, among other things, has made Canadians somewhat wary of U.S. intentions regarding their freshwater, especially since the signing of NAFTA has heightened the potential for water commodification.

COUNTERSTRATEGIES

As noted, when it comes to water, cooperation among nations and among states and provinces within nations has been the rule rather than the exception. This suggests a faith in the ability of the law to protect a precious resource, and an acknowledgement of the universal value of that resource. Nevertheless, like all human systems, laws, declarations, and treaties related to water can be manipulated to benefit one group over another, which is why watchdog organizations have formed. These groups publicize water inequities, law and treaty violations, and bad water policies and give voice to those who might not otherwise be heard. They may also propose new counterstrategies or comment on those already in existence to help resolve the water crisis.

Conservation

Conservation is perhaps the most obvious of all the strategies to help counter the global water crisis, but it does not apply equally to all areas of the world. Clearly, it can be applied in countries that are rich in water, such as the

United States, Canada, and most of western Europe, more easily than in those countries that suffer a scarcity of water. Indeed, in countries and regions where freshwater is scarce, conservation tends to be ingrained—to many, it is second nature. It is not so much conservation, as a matter of survival.

There are many ways of conserving water. It can be conserved particularly in agriculture by replacing poor practices that over the centuries have resulted not only in high levels of water wastage, but in soil degradation. In water-scarce countries such as Israel and Ethiopia, the long-term prospects for agriculture depend on conservation. Conservation may also come through technological advances that in the past 60 years have reduced the amount of water used in industry and agriculture, made municipal water delivery more efficient, and have enabled wastewater to be treated and recycled. New attitudes toward water have caused lawmakers and the public to rethink some previously accepted ways of managing water, such as dams accompanied by huge reservoirs, which have high evaporation rates.

Water Reclamation Projects

Given that the vast majority of the Earth is covered by non-potable salt water, why not reclaim it for our use? In fact, desalination plants are in use around the world, and their numbers are increasing. Furthermore, the technology improves with the construction of each new plant (desalination technology has improved at a better rate over the past 30 years than solar technology). The two basic processes employed in desalination are distillation and reverse osmosis, with some plants using a combination of the two. Distillation requires salt water to be heated. The water rises as steam, while the salts and other impurities remain behind. The steam is captured, then cooled to become purified freshwater. Distilled water is so pure that, with the exception of fossil glacier water, it is the purest freshwater available. In the reverse osmosis process, electric or diesel pumps force the salt water through a semipermeable membrane to separate out the salts and other impurities. Because the energy required for both processes is very high—2003 estimates range from 25 to 200 kilowatt hours (kWh) of electricity to produce 1,000 liters (264 gallons) of freshwater by the distillation process, and six kWh of electricity for every 1,000 liters of freshwater produced by reverse osmosis[26]—as is the cost itself, it is no surprise that desalination plants are more prevalent in wealthy countries, particularly the oil-rich states of the Middle East. Saudi Arabia is one of the world leaders in developing and improving desalination processes. In the United States, there are major desalination plants in southern California, Florida, and West Texas—the last purifies the brackish Colorado River water that goes to Mexico. Most desalination plants use electric or diesel pumps, however, and so are large emitters of greenhouse gases. There are several alternatives, two of which are already in operation. Japan has seven nuclear-powered desalination plants. India has two such plants, while Kazakhstan and Pakistan have one each. Nuclear-powered plants are gaining credence in the Middle East and North Africa: In 2007, Libya signed a contract with a French firm for the construction of a nuclear-powered plant. A desalination plant near Perth, Australia, that supplies 17 percent of the city's water supply is powered by a wind farm. Solar power is another option for desalination plants, though the technology still lags behind the other methods. Nevertheless, a small solar-powered desalination plant exists in Saudi Arabia, and another is planned in Spain.[27]

While most developing countries cannot afford to build and run desalination plants, even wealthy countries, if landlocked, may have no means of drawing water from the sea in the first place. Countries that do have desalination plants face the problem of disposing of the briny water the process leaves behind. Often it is dumped back into the sea, where, potentially, the higher percentage of salt in the water may upset the ecology of the area of outflow.

Among the other methods proposed for acquiring freshwater are melting icebergs. This would be an extremely expensive and complicated process; it would require, for example, at least two ships (one pushing, the other pulling) to tow the iceberg to the desired destination. Furthermore, crossing sea lanes with an iceberg would require careful logistical planning. Building the infrastructure to deliver the iceberg's water would be no easy task either, but likely the least complicated part of the process. Regardless, the ecological effects of an iceberg sitting even temporarily in a new environment would have undesirable effects on the marine life of the area, since the water is bound to cool down. In the same way, the area from which the iceberg was taken would suffer negative ecological consequences.

In times of drought some countries import freshwater by tanker, but the technology to do this is not well developed and the expense too great. New methods include the use of giant plastic bags, connected together and floating in a convoy towed by a ship, the water in the bags acting in concert with the ocean's waves so as not to rip apart from the motion. Technological problems aside, the main stumbling blocks are often political and environmental. Leaders of water-scarce countries are generally averse to having another country control their lifeblood. Conversely, water-rich areas may not want to release their supplies of freshwater because of the possible ecological damage caused by the continuous draining of rivers or lakes. The precedence for such fears lies in one of the most grievous examples of environmental destruction due to water drainage—the plight of the Aral Sea in Kazakhstan, formerly part of the Soviet Union. In 50 years, since the beginning of the irrigation projects that diverted water from the Amu Darya and Syr Darya Rivers, both of which fed the Aral Sea, the Aral has shrunk by more than 50 percent and lost two-thirds of its volume. What water remains in the sea is saltier, and salts from the dry areas of the seabed are picked up by the winds, creating a health hazard for humans and further despoiling the land. Fishing villages that once thrived by the Aral Sea no longer exist.[28]

But the news is not all bad when it comes to freshwater and ecology. For example, in the past few decades scientists have gained an understanding of the true value of wetlands in the hydrologic cycle. No longer are they seen as swamps or marshes that ought to be filled in to prevent disease or whose land can be developed for commercial or domestic use. Wetlands, in addition to being the habitat of numerous species of plants, fish, and animals, also act as stopover points for birds and other migratory animals and enhance the hydrologic cycle by serving as filters for the water that flows into them. In some cases, this includes seawater. One recent reclamation effort involves the marshes in southern Iraq, which were drained during the reign of Saddam Hussein. This all but eliminated the thriving culture of the Marsh Arabs, who had lived in the area for centuries. Whether reclamation efforts will invigorate their culture remains to be seen; at the very least, the ecological balance in the area may be restored. The restoration of the Everglades in Florida, the largest wetlands area in the United States, is another reclamation project that has gained momentum in recent years.

---

[1] Robin Clarke and Jannet King. *The Water Atlas: A Unique Visual Analysis of the World's Most Critical Resource.* New York and London: The New Press, 2004, p. 22.

[2] These figures were given as part of the 2008 exhibition Water: H_2O-Life, at the American Museum of Natural History in New York City.

[3] Water: H_2O-Life.

[4] Clarke and King, pp. 32–33.

[5] Clarke and King, pp. 34–35.

[6] Clarke and King, pp. 34–35.

[7] Clarke and King, p. 40.

[8] "EPA Office of Compliance Sector Notebook Project: Profile of the Electronics and Computer Industry." Available online. URL: http://74.125.95.132/search?q=cache:uV1BPa1L7cIJ:www.epa.gov/compliance/resources/publ ications/assistance/sectors/notebooks/elec mpsnpt1.pdf+Computer+industry+and+water+pollution&cd=9&hl=en&ct=clnk&gl=us. Accessed May 27, 2009.

[9] Clarke and King, pp. 30–31.

[10] For a lucid account of the era of dam-building in the United States and the rivalry between the Bureau of Reclamation and the Army Corps of Engineers, see Marc Reisner. *Cadillac Desert: The American West and Its Disappearing Water.* New York: Penguin Books, 1993.

[11] Clarke and King, p. 44.

[12] The percentages in this and the preceding paragraph are from Clarke and King, pp. 102–109.

[13] Clarke and King, p. 56.

[14] They are also found in seawater, witness the problem with mercury in tuna that has plagued the world for more than a quarter of a century.

[15] Clarke and King, pp. 50–51.

[16] Numerous sources discuss this debate and the role of the World Water Forum. One of the best is Maude Barlow and Tony Clarke. *Blue Gold: The Fight to Stop the Corporate Theft of the World's Water.* New York: The New Press, 2002.

[17] The World Water Forum is held every three years. The first was held in Marrakech, Morocco, in 1997; the third in Kyoto, Osaka, and Shiga, Japan, in 2003; the fourth in Mexico City in 2006; the fifth in Istanbul in March 2009.

[18] In 2006 Vivendi Universal became Vivendi SA.

[19] Quoted in Anita Roddick with Brooke Shelby Biggs. *Troubled Water: Saints, Sinners, Truths and Lies about the Global Water Crisis.* West Sussex, U.K.: Anita Roddick Books, 2004.

[20] For further information, see "Summary of NRDC Test Results: Bottled Water Contaminants Found." Available online. URL: http://www.nrdc.org/water/drinking/bw/appa.asp. Accessed May 28, 2009.

[21] Cynthia Barnett. *Mirage: Florida and the Vanishing Water of the Eastern U.S.* Ann Arbor: University of Michigan Press, 2007.

[22] In c. 2500 B.C.E., two city-states in ancient Mesopotamia, Lagash and Umma, fought a war over irrigation water from the Tigris River.

[23] Aaron T. Wolf. "'Water Wars' and Other Tales of Hydromythology." In Bernadette McDonald and Douglas Jehl, eds. *Whose Water Is It?: The Unquenchable Thirst of a Water-Hungry World.* Washington, D.C.: National Geographic Society, 2003.

[24] Quoted in Marq de Villiers. *Water: The Fate of Our Most Precious Resource.* New York: Houghton Mifflin, 2000, p. 13.

[25] Maude Barlow and Tony Clarke. *Blue Gold: The Fight to Stop the Corporate Theft of the World's Water,* p. 71.

[26] Clarke and King, p. 69.

[27] Derek Sands. "Analysis: Mideast Turns to Nukes for Water." *Energy Daily* (9/14/07). Available online. URL: http://www.energy-daily.com/reports/Analysis_Mideast_turns_to_nukes_for_water_999.html. Accessed May 29, 2009.

[28] There is an even worse disaster waiting to happen. For years, the Soviet military stored its spent nuclear fuel on an island in the Aral Sea. Since the island no longer exists because of the shrinking of the Aral Sea, the radioactive material is a danger to the environment and vulnerable to thievery by criminal gangs.

2

Focus on the United States

The freshwater crisis in the United States, particularly in the West and the Midwest, is entwined with issues related to water in both Mexico and Canada and, some argue, more so since the passage of the North American Free Trade Agreement (NAFTA) in January 1994. Both the United States and Canada are considered water-rich nations, although the United States is one of the few countries in the world that is geographically big enough to be water rich and suffer water crises at the same time. For nearly 150 years, Americans have taken insufficient account of the landscape in their desire to "overspread the continent," as an early proponent of the doctrine of Manifest Destiny put it, and having done so they continue to migrate to the country's driest areas—first remaking them, then attempting to maintain them as verdant paradises of agricultural bounty or suburban gardens. And all of that takes water. The hotter and drier the climate, the more water is needed. And the more people arriving in an area, the more water is needed.

Thus it was in the U.S. West that the 20th-century worldwide phenomenon known as the great era of dam-building was kicked off. Rivers were tamed, their water sent along artificial canals and aqueducts to settlements large and small, achieving results not dreamed of by the earliest settlers. All this required varying degrees of wheeling and dealing, arm-twisting, and political pressure, as well as compacts and treaties between states vying for a scarce resource, between cities and rural areas, between states and the federal government, and between the United States and its neighbors to the north and south, Canada and Mexico. And that merely covered the quantity of water needed. Once the large-scale irrigation of agricultural crops came into play, as well as industrialization and population growth, quality also became an important issue.

In short, settlement came at a cost. Over-irrigation and irrigating land with poor drainage eventually led to a buildup of salts in the soil, as well as the runoff of salts, along with fertilizers and pesticides, into rivers and streams. Today, excessive salt in the soil, called soil salinity, has led to the retirement of thousands of acres of once-prosperous farmland. Some in the western states are now looking elsewhere for water, notably toward Canada, whose western provinces are one of the world's last refuges of wild rivers, and toward the Great Lakes, which are the largest continuous source of freshwater in the world.

The American West is not the only part of the United States experiencing water strain, nor is it the only part that could draw water from the Great Lakes. Much closer to the lakes are the Great Plains, originally designated on U.S. maps as the Great Desert due to the dearth of surface water. During the 20th century, this barren land was transformed into the great wheat, corn, cotton, and alfalfa farms that have an almost-mythic hold on the American imagination. The land was transformed through the drawing of water from what at the time was one of the world's most extensive aquifers. The Ogallala Aquifer stretches from South Dakota, a little north of the Nebraska state line and west of the Missouri River to west Texas, below the Colorado River.[1] As well as South Dakota, Nebraska, and Texas, the Ogallala Aquifer lies beneath parts of Kansas, Wyoming, Colorado, Oklahoma, and New Mexico. Nebraska receives the most water from the aquifer, since most of the state sits atop it. While the aquifer once held about as much water as Lake Huron, the third-largest of the Great Lakes,[2] after World War II the increased use of electric and diesel pumps meant not only that more water was drawn faster, but that farmers could pump deeper as the amount of water in the aquifer began to decline.

The main problem with drawing water from the Ogallala Aquifer is that it is primarily a fossil aquifer with a very low refill rate; the water it contains took millions of years to accumulate. As the water level recedes, more energy and thus more money are required to draw water to the surface. At some point, possibly within the next 50 years, it will likely become too expensive to pump the remaining water. If the cost of pumping goes down and the process continues, the aquifer will be drained in less than a century. Farmers in western Texas and New Mexico whose land is situated on the edges of the aquifer are already experiencing difficulty obtaining water from it.

Reduced access to water from the Ogallala Aquifer is one of the reasons that the states and provinces in the Great Lakes region have come together to protect their natural resource. These areas do not have a water crisis in terms of quantity, but seek to maintain the status quo. The state and provincial governments are concerned not only with the designs of neighboring states, but also with those of corporations that seek to profit from their resource by bottling and selling water.

Aside from the American West and the Great Plains, the Southeast of the United States is, ironically, experiencing water problems. That area's annual rainfall and numerous rivers, lakes, and aquifers ought to be sufficient to sustain the population and indeed were until the final decades of the 20th century, when rapid urban growth, unregulated industrialization, poor farming practices, and the drainage of wetlands led to shortages. Florida is particularly badly affected, as is Georgia, especially in the Atlanta area. Alabama, North Carolina, South Carolina, and even Virginia have also experienced water problems, such as coastal freshwater salination, in recent years.

Excluding Hawaii, the country's wettest regions in terms of annual rainfall are New England, the Mid-Atlantic, upper midwestern states, and northwestern states. These areas are not entirely without problems, however. For example, the Ipswich River in northeastern Massachusetts has dried up for several years due to the diversion of its water to suburban towns. In 1997, the Ipswich became the only river on the East Coast to be named among the "20 most threatened rivers" by the environmental organization American Rivers.[3] Maine and Connecticut, which like Massachusetts receive more than 45 inches of precipitation annually, are also not immune to freshwater problems. Irrigation of Maine's blueberry farms and fertilizer and pesticide runoff have interfered with river flows and threatened the Atlantic salmon in the area.[4] In April 2009, Hartford, Connecticut, and its surrounding towns were placed under an advisory to boil water after copepods and rotifers, two microscopic organisms, were discovered in the drinking water.[5]

In the Northwest, the Cascade Mountains provide a barrier for the moist Pacific air, blocking the storm clouds from moving east and making the western halves of Washington and Oregon verdant paradises, while the eastern portions are arid. Eastern Oregon is beginning to suffer the effects of an overdrawn aquifer. Meanwhile, the freshwater of the eastern United States suffered from the Industrial Revolution. It was the publicity surrounding the pollution of Lake Erie and the Cuyahoga River in Ohio that led the media and the public to draw attention to similar occurrences elsewhere in the country, helping to spur the environmental movement and the cleanup of rivers, lakes, bays, and harbors.

In any discussion of freshwater problems in the United States, it must be borne in mind that there is a philosophical difference that translates into a major legal variation concerning water use. In the East, riparian law is recognized as the basis for water allotment and use. Riparian rights, borrowed from English law by the first settlers, guaranteed water to those who owned the land abutting rivers, lakes, etc. As America moved from a primarily agrarian to a primarily industrial society, riparian water rights were altered to take into account the needs of new industries. The law of priority that protected early entrepreneurs evolved into a doctrine of reasonable use. In the West, the doctrine of prior appropriation governs water use. Formulated in California during the gold rush era, prior appropriation is similar to, but more forceful than, the law of priority. It involves a seniority system, so that those with second or third rights may receive little, if any, water during dry periods. Eventually, prior appropriation was enshrined in the law for farmers, as agriculture replaced mining as the primary economic activity of the region, and affected relations between communities and even states. The appropriation of water devolved into a competition between states and cities to stake a claim to water even before it was needed; in other words, they aimed to get in first and thereby ensure an adequate supply. This situation persists today. Marc Reisner summed it up in *Cadillac Desert,* his history of water use in the American West: "In the East, to 'waste' water is to consume it needlessly or excessively. In the West, to waste water is *not* to consume it—to let it flow unimpeded and undiverted down rivers" (emphasis in the original).[6]

THE AMERICAN WEST

As late as the mid-19th century, water was not a problem for the inhabitants of the territory that became the western states of the United States of America. They lived in areas of abundant rainfall such as the Pacific Northwest or were nomadic, such as the Plains tribes, or had settled along rivers and lakes that were able to sustain their small populations. Even the earliest U.S. settlers, such as the Mormons, managed to carve out a successful farming and ranching civilization in the Utah desert (and elsewhere) through the judicious use of water. However, when the region's population began to explode, first in California during the gold rush of 1848–1855, then in other areas where the mining of precious metals had created boomtowns and big cities, the race for freshwater began.

In the early history of the mapping of rivers and rainfall in the American West one name stands out—Major John Wesley Powell. A Civil War hero who had lost an arm in the Battle of Shiloh, Powell taught geology at Illinois Wesleyan University and was one of the founders of the Illinois Museum of Natural History. In 1869, after exploring the Rocky Mountains, Powell led a 10-man, three-month expedition to explore the Green and Colorado Rivers (the Green being a tributary of the Colorado), including the Grand Canyon. Two years later, he led a second expedition, though on this trip Powell himself spent much of the time exploring the land around the Colorado River. Powell served as director of the United States Geological Survey (USGS) from 1881 to 1894, and the year after he retired he published *Canyons of the*

Colorado (later retitled *The Exploration of the Colorado River and Its Canyons*). Powell was one of the first to understand that the 100th meridian was the dividing line that separated the water-abundant East from the arid West, realizing that only irrigation farming would work in the West.

Los Angeles and Owens Valley

More has likely been written about Los Angeles's quest for water than that of any other city in the United States. The water issue in Los Angeles has even made it into the movies.[7] This is because of a larger-than-life civil engineer named William Mulholland who, as head of the Los Angeles Department of Water and Power in the early 20th century, oversaw the construction of the 233-mile Los Angeles Aqueduct that transported water from the Owens River to the city. The aqueduct took six years to construct and was completed on November 5, 1913. While Mulholland's foresight led to the birth of metropolitan Los Angeles, his plan also caused severe environmental degradation, ruined the valley's agricultural economy, and led millions to migrate to what had once been a desert:[8] At the beginning of the 20th century, Los Angeles's population numbered just 100,000, a figure that doubled within four years. Mulholland, of course, was not the only one responsible for the acquisition of water from Owens Lake; he was assisted by former Los Angeles mayor Fred Eaton and Joseph B. Lippincott, an engineer for the United States Reclamation Service (later the Bureau of Reclamation).[9]

The tragedy was not that the aqueduct was built, or that Los Angeles acquired Owens Valley water in a legal, though somewhat manipulative manner, or indeed that Los Angeles, despite its rapid population growth, did not make much use of the water for years. The real tragedy is that *all* of the water was taken—Owens Lake dried up, as did the economy of the valley in which it lay. In time, the dried lake bed became a health hazard when winds blew the salts and dust into the air, not unlike what is currently happening in the area surrounding the shrunken Aral Sea. The situation continued for decades until the City of Los Angeles rectified it somewhat by returning enough water to Owens Lake to keep the lake bed moist and the salts and the dust down.

But if Los Angeles did not at first use the Owens River water, who did? That is the second part of the tragedy. The water was redirected to the San Fernando Valley, where, as a result, farming acreage grew exponentially and which therefore used more water from the Los Angeles Aqueduct than did Los Angeles itself right up until the Great Depression. A great deal of San Fernando acreage was owned by Los Angeles movers and shakers, at least one of whom, Moses Sherman, sat on the board of the city's water commission. The tragedy of the Owens River diversion was twofold. It helped attract more people to an arid region than it could naturally support, and it set the precedent for cheap water that led to high amounts of water wastage that persist to this day. The story of the Los Angeles Aqueduct illustrates an important trend in the history of U.S. water use: Water tends to flow toward money.

The Colorado River

How many ways, and how many times, can a river be divided up? That is the story of the Colorado in the past century. Various compacts and treaties and dams have so apportioned the river's water that it no longer flows all the way to the Gulf of California. Although various groups had been tampering with the Colorado River in both its upper and lower basins (including in Mexico) for some time, the wholesale apportionment of its water began in the early decades of the 20th century, with the settlement of southern California, and particularly Los Angeles. In the early 1920s, the farmers of California's Imperial Valley, just north of the Mexican border, sought relief from the repeated flooding of the Colorado River.[10] What they wanted was a canal. The Bureau of Reclamation declared that a canal would be useless without a dam upriver "to control flooding and to capture silt."[11] The supply of electricity to the fast-growing city of Los Angeles also played into the equation, and the idea of a dam that would also produce hydropower became more feasible.

At that point, the other states in the Colorado Basin realized they had better get involved in river negotiations in order to prevent California from gaining the upper hand. Negotiations for dividing up the river took place in Santa Fe, New Mexico, with Secretary of Commerce Herbert Hoover serving as chairman. Delegates from Wyoming, Colorado, New Mexico, Utah, California, Nevada, and Arizona attended this "river convention" and signed the Colorado River Compact on November 24, 1922. Nevertheless, arguing among those states still to ratify the compact forced Congress to make the authorization of the canal and the dam provisional on ratification by six of the seven states—and six did ratify, with Arizona holding out. The compact was the first in a series of Colorado River agreements involving various states and the federal government that, collectively, became known as the Laws of the River. At Herbert Hoover's urging, the compact divided the river's water between the upper basin states of Wyoming, Colorado, New Mexico, and Utah and the lower basin states of California, Nevada, and Arizona. The dividing point between the two basins was Lee's Ferry in Arizona. Each basin was to receive an average of 7.5 million acre-feet of water,[12] but the allotments for each state were determined in 1928 at: California, 4.4 million acre-feet; Colorado, 3,881,250 acre-feet; Arizona, 2.8 million acre-feet; Utah, 1.725 million acre-feet; Wyoming, 1.05 million acre-feet; New Mexico, 843,750 million acre-feet; and Nevada, 300,000 acre-feet. (An acre-foot is the amount of water it takes to cover an acre of land with one foot of water, or approximately 325,851 gallons.) Mexico was mentioned in the third section of the compact, but its claim to water from the Colorado River would not be an issue for decades.

Having divvied up the river's water, it was time to contain the flooding. On December 21, 1928, the federal government authorized the Boulder Canyon Project, which included the construction of the All-American Canal, so named because it would lie entirely north of the Mexican border, unlike its predecessor the Alamo Canal, which was largely in Mexico. After a last-ditch effort by Arizona to halt the project ended in defeat in the Supreme Court, construction began in 1931. The All-American Canal was completed in 1942 and remains the world's largest irrigation canal.[13] The Boulder Canyon Project included the construction of Boulder Dam (formally renamed the Hoover Dam in 1947). Six Companies, Inc., a primarily western consortium of contractors—among whom was the partnership of Warren Bechtel and Henry Kaiser—won the right to construct the dam with a bid of $48,890,955—just "$24,000 more than the cost calculated by Bureau of Reclamation engineers."[14]

The depression-era construction of the dam is legendary. It took slightly more than four years to complete and was dedicated on September 30, 1935, by President Franklin D. Roosevelt. The dam itself is a majestic beauty, as is Lake Mead, the reservoir created by the dam and named for Elwood C. Mead, the man who oversaw the dam's construction as head of the Bureau of Reclamation. At the time, the Hoover Dam was the largest concrete structure in the world and generated the most hydroelectric power in the world.[15] The federal government had parceled out the electricity generated by the dam long before its completion. In 1929, Secretary of the Interior Ray Lyman Wilbur awarded a contract for 36 percent of the 3.6 billion kilowatt hours (kWh) to the recently formed Metropolitan Water District (MWD) of Southern California (a consortium of Los Angeles and surrounding cities). The states of Arizona and Nevada were each allotted 18 percent, the City of Los Angeles was awarded 13 percent, the Southern California Edison Company was allotted 9 percent, and various other Southern California cities that were not part of the MWD received the remaining 6 percent.[16] The money received for these electricity contracts was used to pay for the construction of the dam.

The construction of Hoover Dam and Lake Mead led to the transformation of not only the American West, but many other places east of the 100th meridian. Many of the dams that followed were cash register dams. That is, they were hydropower producers, and the sale of the electricity they generated was intended to offset construction and maintenance costs. The fact was that that often was not the case, although this did not stop the Bureau of Reclamation or the Army Corps of Engineers from building more dams (generally, but not always, the latter's projects were in the East).

Perhaps the most interesting fact about the Hoover Dam's hydropower is that the largest percentage went to the MWD. It received more than one-third of the dam's hydropower because it needed it for its own public works project, the Colorado River Aqueduct. The aqueduct took eight years to construct (1933–41) and runs 242 miles over the California desert and over and through the San Jacinto Mountains, just west of Palm Springs. It requires five pumping stations to deliver approximately 1.3 million acre-feet of water per year, which is more than 25 percent of California's annual entitlement under the Colorado River Compact.

The construction of the Colorado River Aqueduct called for the construction of another dam on the river, Parker Dam (completed in 1938), and its accompanying reservoir, Lake Havasu. In this case, once it was settled that Arizona would be receiving water from Lake Havasu, the acrimony between California and Arizona abated somewhat. Smaller dams, reservoirs, and diversions were also built on the Colorado River and its tributaries, further reducing the river's flow. Naturally, the brunt of the decreased flow was felt downriver, and downriver on the Colorado means Mexico. In 1944, the United States and Mexico signed a treaty that guaranteed "1.5 million acre-feet a year of the Colorado's flow to Mexico, without cutting allocations to the seven states, and created the International Boundary Water Commission to administer it."[17] U.S. reasons for signing the treaty were largely pragmatic: A wartime ally was mollified, and the country's southern flank was secured.

The next big public works project built on the Colorado River was the Glen Canyon Dam, the construction of which began in 1956 as part of the Colorado River Storage Project. By 1963, the dam was already impeding the Colorado's flow, although it was another three years before it was formally dedicated. Furthermore, Glen Canyon was flooded to create the dam's reservoir, Lake Powell—named for John Wesley Powell, who had led the first expedition to explore the river and its canyons. Environmentalists and others view the flooding of Glen Canyon as a great tragedy. They include David Brower, the former head of the environmental group the Sierra Club, who excoriated himself in print for acceding to the construction of the dam and the loss of the canyon.[18]

The loss of the canyon was not the only problem on the Colorado. Even before the Glen Canyon Dam began functioning, things had worsened for the inhabitants of the Mexicali Valley in Mexico. By the early 1960s, the water delivered to the area was extremely saline, containing approximately "2,500 parts per million of salinity."[19] In contrast, freshwater has a salinity level of 500 parts per million or less. The level of salts in the water flowing to Mexico put that water in the brackish category. The Bureau of Reclamation had dug more than 60 wells in the Wellton-Mohawk Valley of Arizona to pump out salty water caused by irrigation. The water had been ruining crops in Arizona and was dumped into the Gila River, close to its confluence with the Colorado. In 1963, when Glen Canyon Dam began impeding the Colorado's flow, the problem worsened because now even less water was making its way downstream to dilute the salts. Saline water, besides being undrinkable, is bad for most crops. At best plants are stunted, and at worst they do not grow at all.

On the Mexican side, the Presa Morelos Dam, just south of the border, had been constructed in 1948, along with a canal to carry the water to the Mexicali Valley. Thus, south of the Presa Morelos system, the Colorado's flow was further impeded. During the 1950s and 1960s, the population of the Mexicali Valley was booming, and throughout the latter decade and into the 1970s Mexican officials conducted negotiations with their U.S. counterparts over the quality of water being sent to Mexico. During the years of these negotiations, the water quality did not improve, especially after the mid-1960s when more farmers in southern Arizona were discharging salty water directly into the Colorado. For some time, negotiations were stuck on whether or not the water pumped from the water table even qualified as Colorado River water: In a technical sense—the U.S. point—it was part of the Colorado basin. While the Mexican argument was that it was not part of the river itself, but unusable water diverted into the river.

Mexican negotiators eventually invoked the naming of their country in the Colorado River Compact, contending that Mexico was a partner in the river along with the seven U.S. signatory states. (Arizona had finally signed the Compact in 1944.) This may have had some impact on the agreement that was eventually reached, as some, such as Jose Trava, have noted that "sympathy for Mexico was building" in the United States during the presidential election year of 1972.[20] Other writers are more cynical, tying the 1973 agreement between the two countries to the fact that oil had recently been discovered in Mexico.[21] Whatever the reason, the result was the construction of a massive desalination plant near Yuma, Arizona. Some have criticized the building of the plant, pointing out that it would have been far less expensive simply to retire farm acreage in Arizona and further upstream—acreage that is bound to be retired eventually anyway because of soil salinity—than to construct a massive, power-consuming desalination plant. (The construction of which, incidentally, was completed in 1992, but after nine months of operation a flood washed away a diversion canal, and the plant remained offline for 15 years.) A further unintended consequence of the damming and diversion of the Colorado River and its tributaries is that the river has ceased to flow all the way to the Gulf of California and therefore no longer carries silt to its delta. As a result, an important wetland has been destroyed and the ecology of the area, including the Gulf of California itself, has been seriously compromised: Wildlife has died off and fresh and salt water no longer mix along the gulf coast, which harms coastal marine life. Likewise, the water in the giant reservoirs built on the river, like water everywhere else on earth, is subject to the hydrologic cycle. And since these reservoirs are situated in some of the driest parts of the United States and Mexico, their rates of evaporation are very high. Decades ago, when these dams were constructed, they were almost universally viewed as a boon for society for their flood control, hydropower, water storage, and recreational possibilities. As it turns out, that was far too narrow a view.

While most of the public works projects on the Colorado River have benefited farmers, they have also created new tourist industries, particularly around the reservoirs, improved sanitation, and provided drinking water and water for industrial use. Nevertheless, the improved water supply has also brought more people to the region, creating a need for even more water. California was the fastest-growing state in the union during the latter half of the 20th century and as a result is bisected by canals and aqueducts, its rivers dammed and diverted, and yet its water crisis continues.

The Central Arizona Project

It wasn't until February 3, 1944, that Arizona ratified the Colorado River Compact. Prior to becoming a signatory, the state had been drawing water from the river, but not its full allotment.[22] Prompted by California's increasing thirst, leaders in Arizona decided they had better make their state's right to the water legitimate before their more populous neighbor laid claim to it. Once California, or any other signatory state of the compact for that matter, began using Arizona's share of the Colorado's water, the state's leaders reasoned, tradition and pragmatism would make it difficult to redirect the flow.

ARIZONA V. CALIFORNIA

The animosity between Arizona and California came to a head in 1934, in what came to be known as the Great Colorado River War. Specifically, the trouble had to do with California's construction of the Parker Dam (located on a section of the Colorado River that serves as the border between California and Arizona), which Arizona interpreted as an attempt to hoard water. The governor of Arizona, Benjamin Moeur, sent a small number of National Guard to "repel the threatened invasion of the sovereignty and territory of the state of Arizona," which was considered more a political ploy than anything else.[23] Indeed, when Congress authorized the dam's construction a few months later, there was little Moeur or anyone else could do to stop it. However, Congress could not legislate away Arizona's fear of California's need for water. A second issue in the Great Colorado River War related to Arizona's construction of dams east of Phoenix—the Salt River Project—in order to conserve water that would otherwise evaporate before it ever reached the Colorado River. The argument between the two states raged over the amount of conserved water that should be charged against Arizona's Colorado River Compact allotment. California vowed to block construction of the massive Central Arizona Project (CAP), the purpose of which was to water the central part of the state of Arizona using some of the state's allotment of the Colorado River.

In 1948, the upper basin states began drawing their full allotments from the Colorado River, hoping that their tradition of prior usage laws would protect their allotments in the face of California's expanding population. This essentially upped the ante for the two remaining lower basin states, Arizona and Nevada, both of which would experience their own population booms in the decades following World War II. But the dispute over the Salt River Project dragged on, so that by 1952 Arizona turned to the United States Supreme Court to decide the matter. That decision did not come until June 3, 1963 (the final arguments were not made before the Court until 1962), when, in a majority opinion written by Justice Hugo Black, the court awarded Arizona "almost all the advantages it sought in the 1922 compact."[24] But the decision did more than that: It confirmed each states' allotment of Colorado River water, and it acknowledged Native American water rights, which the states were disinclined to do. At any rate, the court's decision was a green light for Arizona to go ahead with the CAP. Nevertheless, it took another 10 years and the passage in Congress of the Colorado River Basin Project Act of 1968 for the CAP to get off the drawing board.

Historically, Arizona had underused its annual allotment of Colorado River water because the areas where it was most needed—Phoenix, Tucson, and approximately 1 million acres of the state's agricultural district—were too far away from the river and, unlike California, the state had neither the money nor the political will to correct the situation. Also unlike California, Arizona had traditionally had enough groundwater to satisfy its needs. After World War II, however, it was clear that that would not always be the case. CAP was therefore a way for Arizona to draw its full share of river water and to ease the drawdown from the aquifer. In Arizona lore, the hero of CAP is Senator Carl Hayden (1877–1972), who along with fellow Arizona senator Ernest McFarland first introduced the CAP legislation in Congress in 1946—thus the money to pay for the project would come from the federal government. During the course of the next 22 years, Hayden continued pushing for CAP legislation, finally succeeding in 1968.[25] Three years later, the Central Arizona Water Conservation District was created as the administrative agency for CAP responsible for reimbursement to the federal government for construction costs. Construction began in 1973, and the project was completed in 1993 at a cost of approximately $4 billion. At 336 miles, it is the largest aqueduct system in the United States.

CAP begins at Lake Havasu and ends south of Tucson. In 2004, President George W. Bush signed the Arizona Water Settlements Act, which allotted 650,724 acre-feet of CAP water either directly to Arizona's Native American tribes (primarily, the Gila River Indian Community and the Tohono O'odham Nation) or to the Secretary of the Interior for contract to the tribes. The remaining 764,276 acre-feet was allotted for "non-Indian municipal and industrial entities, the Arizona Department of Water Resources, and non-Indian agricultural entities."[26] Of that amount, CAP delivers 620,678 acre-feet of Colorado River water to 55 cities, towns, and water companies—nearly one-quarter of Arizona's allotment under the Colorado River Compact.[27]

Like California's many water projects, once CAP became a reality it contributed to Arizona's population boom, which in turn increased the region's water use. That CAP has provided an economic boost to the state is undeniable, but many environmentalists and water experts fear the costs will be high. During and after construction, numerous steps were taken to mitigate the negative effects of the canal on the environment. Nevertheless, the Colorado no longer flows to the sea. Likewise, the rate of evaporation is higher from the canal than it would have been had the water remained in the river. Still, groundwater depletion has slowed down, buying the state more time in its ongoing battle with the desert.

Las Vegas

Since the last decade of the 20th century, Las Vegas, Nevada, like Phoenix, Arizona, has been among the fastest growing cities in the American West. But Nevada, under the terms drawn up after the signing of the Colorado River Compact, receives only 300,000 acre-feet of freshwater per year. The state uses its entire allotment each year and needs more. This pits Las Vegas, which receives an annual average of 3.8 inches of rainfall,[28] against Phoenix and Southern California, and in that three-way race Las Vegas is at a geographical advantage because it is upriver and has the money to compete successfully for water, at least against Phoenix. But the city faces an image problem: It is commonly perceived to be wasteful in terms of water. Indeed, according to the author and journalist Diane Raines Ward, the fastest-growing metropolitan region in the United States has the potential to transform the "state's rural landscape" into "a dustbowl."[29] Since many of the freshwater sources that were once close to the city have dried up, such fears are very real.[30] Las Vegas proper now has a population of more than

500,000, but its greater metropolitan area includes more than 2 million people and this figure is expected to be well over 3 million by the year 2025. Presently, the majority of the city's water, approximately 80 percent, comes from Lake Mead, but the reservoir, though still bounteous, has been under stress for decades and may not be able to support a continuously growing population.

The Upper Colorado River Basin

Use it or lose it. Initially, it was the law of prior appropriation that drove the upper basin states to draw their full annual allotments from the Colorado River even before they needed them. The thirstier lower basin states were sure to infringe on their neighbors' allotments—as California had been doing for years—and the upper basin feared that once its water became part of the lower basin's traditional use, prior appropriation would take precedence, and they might never be able to recover what was accorded to them under the Colorado River Compact and as per the 1928 agreement that initiated the Boulder Canyon Project.

As a result, while diversions in the lower Colorado Basin are larger and better known, those in the upper basin are more numerous. Two important reasons for this are that the upper basin is not as dry as the lower basin (making large-scale projects less necessary) and the upper basin states have been less prone to conflict among themselves than their counterparts to the south. Nevertheless, one of the larger upper basin projects, the Central Utah Project (CUP), which is neither as colossal nor as controversial as CAP, was designed to divert water to the areas of Utah where the population was densest and which also had the best farmland. CUP legislation was introduced in Congress by Senator Abe Murdock the same year as CAP legislation, 1946. Dealing with both projects created a tangle, and Congress made CUP legislation contingent on water management in the upper basin, including guaranteeing the 7.5 million acre-feet of Colorado River water to the lower basin, as called for by the 1922 compact. The result was a new agreement among Colorado, New Mexico, Utah, Wyoming, and Arizona, called the Upper Colorado Basin Compact, which was signed on October 11, 1948.[31] With the signing of this compact, the Colorado River Storage Project (CRSP), a Bureau of Reclamation project that included Arizona and of which the CUP was a subproject, took a step closer to becoming reality. It took another eight years, however, before Congress appropriated the money—$760 million—for the Utah portion of the CRSP.[32]

The CRSP was not without controversy. One of its earliest projects was the construction of a dam at Echo Park, Colorado, on the Green River and another at Split Dam Mountain (also in Colorado) that together would have fed into the CUP. These dams were intended to be cash register dams whose hydropower would have offset CRSP's costs.[33] However, the dams would have flooded Dinosaur National Monument on the Utah-Colorado border, and when knowledge of the project became public it was met with a great deal of opposition from environmentalists, economists and others, including historian Bernard de Voto. The Sierra Club, led by David Brower, was in the forefront of the fight against the dams and indeed managed to halt their construction through a compromise that allowed the Colorado to be dammed at Glen Canyon. As previously noted, Brower later regretted this because it resulted in the flooding of Glen Canyon.

The CUP also faced challenges from other environmental groups and was on President Jimmy Carter's "hit list" of economically unfeasible water projects.[34] But it was the slow pace of construction (because of challenges and rising costs) that prompted Congress, in 1992, to enact the Central Utah Project Completion Act. Essentially, this formed a partnership between the U.S. Department of the Interior and the Central Utah Water Conservancy District, which had administrative control of the project. A third group, the Utah Reclamation Mitigation and Conservation Commission, was created as another partner.

While concerns about the CUP lingered, the CRSP moved ahead with the project in the other upper basin states. Among the larger projects, built on tributaries of the Colorado River, were the Flaming Gorge Project on the Green River in Utah and Wyoming, the Navaho Project in New Mexico and Colorado on the San Juan River, and what was originally called the Curecanti Project on the Gunnison River in Colorado, but was later renamed the Wayne N. Aspinall Storage Unit in honor of the Colorado congressman who advocated for public water projects in Colorado and throughout the West.

While all these projects have been the driving engine behind the phenomenal growth of the American West since the end of World War II, the states involved have in turn become victims to the projects' successes. Just as the availability of freshwater in the lower basin states has attracted more people and industry and made farming more feasible, so the same can be said for the upper basin states, though growth rates there are not as rapid. Now having to continue the search for freshwater, the states must also contend with a reduction in water quality. Nevertheless, the West has far exceeded the estimations of John Wesley Powell, who declared in his 1878 *Report on the Lands of the Arid Region of the United States* that only about 20 percent of the area was fit for large numbers of people. Indeed, testifying in 1893 before the National Irrigation Congress, Powell declared, "I tell you gentlemen, you are piling up a heritage of conflict and litigation of water rights, for there is not sufficient water to supply the land."[35] Now, in the early 21st century, the major rivers and their tributaries in the American West have been dammed to the point where there are no major waterways left to develop, yet the area's water problems linger, and in fact the region's water crisis has worsened over the past quarter century.

Current Crises

There are so many causes of the Southwest's water problems that it is hard to pinpoint where to begin. The Colorado River Compact, the initial "law of the river," was drawn up at a time when rainfall and snowmelt were at unprecedented highs.[36] The Colorado's average annual flow is actually about 14 to 14.5 million acre-feet, as opposed to the 17.5 million acre-feet assumed in the compact, thus there is actually less water to divvy up. So how have the seven states and Mexico managed to get their share of freshwater? First, most were not taking their full allotments throughout the more than 80 years of the compact's life, and in the upper basin water was drawn from tributaries such as the Green and Gunnison Rivers, while in the lower basin it was drawn from the Gila River.

Second, the success of Hoover Dam gave the U.S. Bureau of Reclamation (and to a lesser extent in the American West the U.S. Army Corps of Engineers) its raison d'être, that being the construction of dams and reservoirs, not to mention canals and aqueducts to carry the water and hydroelectric power plants. The many reservoirs dotting the western landscape have done three things: provided water storage, provided recreation areas, and affected river flow. Now the era of dam building in the United States is all but over. Dams, and especially the high ones, are no longer universally viewed as beneficial. The buildup of silt behind the dams has reduced flow and therefore the amount of hydropower. Moreover, the silt not making its way downriver has caused ecological damage to the delta.

The slower flow of the Colorado and its tributaries has meant a poorer quality of water is being drawn and/or diverted. This is especially true where irrigation farming (combined with poor drainage) has added tons of salts to the river, as well as fertilizers and pesticides. Most affected, as noted, is Mexico, but everywhere downriver remedial steps have been, or need to be, taken. Another factor that contributes to the water crisis in the American West is climate change. The area is growing warmer and drier. One effect of this is a reduction in the snowpacks in California's Sierra range and the Rocky Mountains. In 2007, "the snowpack in the Sierra Nevada, which provides most of the water for Northern California, was at its lowest level in 20 years."[37] As mentioned, there is a high rate of evaporation from reservoirs, and on the canals and aqueducts that carry diverted water, and the rivers move more slowly due to the dams and diversions. By late 2007, Lake Powell was at approximately 50 percent capacity, while Lake Mead was at 49 percent capacity.[38]

Since agriculture replaced mining as the prime industry of the region, it has been the largest consumer of freshwater in the area. While evaporation and leakage have taken their toll, perhaps the greatest waste occurs because water is relatively inexpensive due to direct or indirect government subsidies. Farmers, and more specifically agribusinesses, have few incentives to conserve water. As noted, in terms of quality, farm runoff has severely degraded freshwater in the American West.

All of these issues will have to be addressed by municipalities, state governments, and Congress in the coming decades. Unfortunately, the issue of western water rights and use is highly politicized, with compromise difficult to reach, and sometimes when compromise is reached the cost-to-benefit ratio (whether economic or environmental) is unbalanced. Nevertheless, new and old solutions have been and are continuing to be put into use to stave off disaster.

Counterstrategies

First and foremost among countermeasures to the water crisis is conservation, although using less water per capita does not necessarily affect overall consumption in the face of rising population. While conservation efforts have reduced water use in the United States in the past 20 years, and even California and Las Vegas have made inroads in this area, many consider efforts to remake the desert landscape no longer viable.

One scheme to conserve water involves graduated pricing for consumers, from households to agribusiness. Under this system, the more water that is used, the more it will cost. Essentially, it is a use tax designed to promote conservation. A version of this is already in practice in Los Angeles, but critics contend that the base price of $2.85 per day for the first 885 gallons and the increase to $3.40 for the next 885 gallons are both too low to inspire conservation and merely cover the costs of delivery and the system's infrastructure as was intended. Average daily domestic use in Los Angeles is 125 gallons.[39] Another method of conserving water involves reusing treated wastewater, or gray water. One of the leading municipalities in this regard is Aurora, Colorado, a suburb of Denver but a growing city in its own right, with a population of more than 300,000. Aurora's problem is threefold. It is just west of the Ogallala Aquifer and east of the Colorado River Basin, and Denver has a prior appropriation claim to most of the nearby mountain water. Thus, with its population estimated "to surpass 500,000 by 2035," conservation and reuse are not choices, but necessities. The city has a plan, called Prairie Waters, to recycle treated wastewater emptied into the South Platte River, but this has yet to be authorized by taxpayers. Aurora currently relies on purchasing water from neighboring towns and from local farmers who fallow their fields and lease their water rights.[40]

As noted, a desalination plant was built in Arizona to purify water from the Colorado River guaranteed to Mexico. Another plant, to desalinate water taken from the Pacific Ocean for San Diego County, has been approved by various governmental commissions after more than five years; freshwater delivery contracts with nine public agencies were signed as of late 2008. When completed in 2011, the plant, at Carlsbad, California, will deliver 50 million gallons per day (mgd), or 56,000 acre-feet per year. However, it will take 100 mgd of seawater to produce 50 million gallons of freshwater. The remaining brine will be diluted with seawater before being released into the ocean. The plant's developer has developed a climate action plan for the Carlsbad Plant that when implemented will recover at least 15 percent of the energy used in the desalination process, thereby reducing emissions. The estimated construction cost of the plant is $300 million, and no doubt other desalination plants will follow along the California coast.

One of them might be paid for by the city of Las Vegas. With money to spare but never enough water, Las Vegas, which currently pays Arizona hundreds of millions of dollars for emergency-use water, "might pay for a desalination plant on the Pacific Coast that would transform seawater into potable water for use in California and Mexico. In exchange, Nevada could get a portion of their Colorado River water in Lake Mead."[41] Complicated deals like this, involving the marketing of water and the transferral of water rights, have been increasing in the American West since the 1970s.[42] They are a testament to the stubbornness of inhabitants when it comes to maintaining their lifestyles, but also show a willingness to seek creative, nonlitigious means of recognizing and resolving a regional problem.

Not all problems find easy solutions. One plan, on the drawing board since the early 1950s, involves taking water from western Canada and piping it down to Southern California and the southwestern United States. The North American Water and Power Alliance (NAWAPA) was the brainchild of Donald McCord Baker, an engineer in the Los Angeles Department of Water and Power. As developed by the Ralph M. Parsons Company of Pasadena, California, the plan is so extraordinary and its engineering feats such marvels that it would dwarf what has already been accomplished in the western United States. NAWAPA involves hundreds of smaller projects that would dam, store, and move massive amounts of water—approximately 120 million acre-feet annually—from British Columbia eastward into Alberta and Saskatchewan, which together would receive between 19.5 million and 22 million acre-feet of freshwater for irrigation. The United States would receive a total of 78 million acre-feet, including: Idaho, 2.3 million acre-feet; Montana, 4.6 million acre-feet; Texas, 11.7 million acre-feet; and California, 13.9 million acre-feet. The plan includes a connector canal, via the Peace River to the Great Lakes and the Mississippi River system, that would eventually also deliver water from British Columbia to New Orleans and the Gulf of Mexico. In fact, NAWAPA would deliver 20 million acre-feet of British Columbia freshwater to Mexico. And that is just the water end of it. The plan also calls for dams on most of the major rivers in western Canada in order to create hydropower. The cost of all of this, spread out over the estimated 30 years it would take to complete the project, was estimated at well over $500 billion in 2007.[43]

Over the decades, some Canadian entrepreneurs and politicians have presented NAWAPA as farsighted and a boon to the western economy as have their American counterparts. But for many, including western Canadian stakeholders and environmentalists, the plan is a nightmare that has engendered fear and paranoia north of the U.S.-Canadian border, as it would not only alter desert and prairie environments but the rivers and wetlands of British Columbia. Environmental and conservation movements have been able to sway opinion away from the plan. Since the early 1990s, conservation has been in the forefront of most western hydrologic plans, yet NAWAPA has never been entirely scrapped. Many Canadians fear that NAFTA, combined with expanding water commodification and the seemingly unquenchable thirst of the U.S. Southwest, Mexico, and their own prairie provinces, will revive interest in NAWAPA, or at least some lesser version of it.

THE GREAT LAKES REGION

With approximately one-fifth of the world's surface freshwater, the situation of the Great Lakes region is nearly opposite that of anywhere else on earth—British Columbia being an exception. The five lakes—Superior, Michigan, Huron, Erie, and Ontario—together contain approximately 5,474 cubic *miles* of freshwater, and that does not take into account the rivers in the basin, such as the St. Lawrence and the Niagara. The total area of the five lakes and their combined individual drainage basins is 291,080 square miles. Public works projects in the 20th century, such as the Soo Locks (Sault Ste. Marie) and the St. Lawrence Seaway, have made the lakes navigable for ocean-bound commercial ships. Other connector canals in Illinois and New York serve smaller commercial vessels and recreation boats. The distance "from Duluth,

Minnesota [on the western shore of Lake Superior] to the eastern end of Lake Ontario is 1,156 miles."[44]

Since the 1960s, the eight U.S. states—Minnesota, Wisconsin, Illinois, Indiana, Michigan, Ohio, Pennsylvania, and New York—and two Canadian provinces—Ontario and Québec—that border the basin have been firmer in their oversight duties.[45] The main problem for the states and provinces, as well as the federal governments of Canada and the United States, is pollution of the lakes. Centuries of habitation and more than a century of agriculture and industry in the vicinity of the lakes and their tributaries have not only contributed to the destruction of wetlands but created numerous problems that Canada and the United States did not begin to effectively correct until the latter half of the 20th century.

That is not to say that the region was previously ignored, although not until the 1871 Treaty of Washington was the boundary between Canada and the United States finally settled, including demarcation through the Great Lakes. The treaty also laid out navigation rules for the St. Lawrence River. Thus, a major precedent was set for the two countries to work together to control river traffic and resolve any problems that might emerge in the future. In 1905, the two countries created the International Waterways Commission to advise the respective governments "about levels and flows in the Great Lakes, especially in relation to the generation of electricity by hydropower." However, "its limited advisory powers proved inadequate for problems related to pollution and environmental damage. One of its first recommendations was for a stronger institution with the authority for study of broader boundary water issues and the power to make binding decisions."[46] This led to the Boundary Waters Treaty of 1909, which in turn created the International Joint Commission (IJC).[47]

Aside from boundary issues, the Treaty of Washington was concerned with navigation rights. In the 1950s, Canada and the United States authorized and completed the St. Lawrence Seaway (1957), which allowed commercial oceangoing vessels to navigate the lakes (although today many of these vessels are too big for the locks). This contributed to some of the problems, including pollution, destruction of wetlands (as harbors were expanded) and, later, the entry of invasive species of marine life. Nevertheless, as noted by Professor Mark Sproule-Jones, the alteration of the Great Lakes ecosystem cannot be attributed solely to navigation: "[G]iven diminishing shipping and fishing uses as well as the early exploitation of most hydroelectric power sites, the primary use of the Great Lakes nowadays is for waste disposal."[48] In fact, waste disposal and unchecked runoff into the lakes and tributaries had been occurring for decades. Two major events in the 1960s finally caused the United States and Canada to examine the state of the region's ecology and take remedial action.

Lake Erie and the Cuyahoga River

Erie is the smallest of the Great Lakes. It is the shallowest and holds the least amount of water, though its surface area is larger than that of Lake Ontario. It is bordered by the states of New York, Pennsylvania, Ohio, and Michigan and by the province of Ontario. A good deal of the land within Erie's basin is farmland, and fertilizer runoff into the lake's tributaries in the decades after World War II increased the levels of phosphorous and nitrates in the water and sediment. (At this time, phosphates were also used in other items, such as laundry detergents.) The increased level of phosphorous resulted in eutrophication, or algal blooms in the lake.[49] Algae began to cover the lake surface, absorbing the available oxygen to the detriment of the aquaculture, and also making it unsuitable for use by humans. Lake Erie, "by the 1960s, was often characterized as 'dead,' which meant that massive algal blooms were occurring and that several near-shore areas were largely devoid of aquatic life."[50]

Then, in 1969, the Cuyahoga River, a tributary of Lake Erie that runs through Cleveland, Ohio, caught fire. It is believed that sparks from a passing train ignited an oil slick on its surface. The oil and assorted debris burned for about half an hour and caused approximately $50,000 worth of damage. The irony was that Cleveland "city officials had authorized 100 million dollars to improve the Cuyahoga River's water before the fire occurred." In fact, the fire was not the first on the Cuyahoga, it was the *tenth*—they had occurred as far back as 1868. The previous river fire in 1952 had caused more than $1.5 million dollars in damage.[51] The difference was that this one, coming on the heels of Lake Erie's problems, caught the attention of the national media. The Lake Erie and Cuyahoga River episodes gave impetus to the expanding environmental movement in North America and particularly the Clean Water Act of 1972 and the Great Lakes Water Quality Agreements.

The Clean Water and Safe Drinking Water Acts

The Clean Water Act of 1972 was one of a spate of environmental laws passed by Congress in the late 1960s and throughout the 1970s. Three others were the National Environmental Policy Act (1969), the Safe Drinking Water Act (1974), and the Endangered Species Act (1973). The earliest measure, the National Environmental Policy Act (NEPA), "was one of the first pieces of legislation in the United States to specifically address protection of the whole environment—air, land, and water—and the organisms living in them, including humans. . . . The purpose of NEPA is to evaluate the environmental impacts of all activities sponsored by the federal government," that is, receive federal funding.[52] This law and the Endangered Species Act passed four years later became the bane of water developers (and developers in general) who sought to build in areas that were found to harbor endangered species or whose environmental impact statements failed to adequately remedy any destruction a project might cause.

The Clean Water Act (CWA) had a historical precedent in the Rivers and Harbors Act of 1899, the first U.S. federal environmental law. That law, which is still in force though by and large superseded by the CWA, made it a crime to discharge refuse into navigable waters and tributaries without a permit. Even so, many of the nation's waters, including areas of the Great Lakes, were highly polluted by the late 20th century. The first sentence of the CWA, Title 1, Section 101 (a) states: "The objective of this Act is to restore and maintain the chemical, physical, and biological integrity of the Nation's waters."[53] In other words, the law intends to make U.S. freshwater potable, protect aquaculture, and ensure the cleanliness of the water for recreation purposes. Title 3 of the Act describes standards of water quality and programs "for developing effluent limitations, reviewing water quality conditions, preventing the discharge of oil and hazardous substances, and maintaining clean lakes."[54] It also contains measures for enforcement of the law.

In 1974, the Safe Drinking Water Act became law. It "defines the maximum concentrations of contaminants allowed in . . . drinking water. It defines the maximum contaminant levels (MCLs) for inorganic substances, organic substances, and microorganisms."[55] These levels are set by the Environmental Protection Agency (EPA). By 1986, there were 23 "regulated contaminants" but amendments to the Safe Drinking Water Act passed by Congress and signed into law by President Ronald Reagan that year required the EPA to increase that number to 112 by 1995. These laws primarily affected the U.S. watersheds in the Great Lakes region and tributaries feeding the lakes, though they did help protect the Great Lakes themselves. However, 1972 also saw the beginning of a series of stronger efforts by the United States and Canada to clean up the lakes and the surrounding environment.

The Great Lakes Quality Water Agreements

The eutrophication of Lake Erie was not the only environmental problem related to the Great Lakes, it was merely the one best known to the public. In the early 1970s, the time had clearly come for better international cooperation and statewide and provincial programs to clean up the lakes and the rivers in the area. More than 100 years of dumping industrial and agricultural wastes into the lakes had sent them into a downward spiral. The St. Lawrence, Niagara, Detroit, St. Clair, and St. Marys Rivers, as well as the Cuyahoga River, all suffered varying levels of pollution. The IJC had begun looking into the problem as far back as 1912, completing several studies and making recommendations to both the United States and Canada, including one in 1965 during the Lake Erie crisis in which it suggested "that measures be taken to limit phosphorous inputs."[56] Neither the province of Ontario, nor the states of Michigan, Ohio, Pennsylvania, and New York, nor the federal governments of either country acted on that recommendation at the time. By the early 1970s, however, attitudes had changed, and the governments were willing to shake off their torpor.

The United States and Canada signed the first Great Lakes Water Quality Agreement (GLWQA) on April 15, 1972, six months prior to President Nixon signing the Clean Water Act into law. The purpose of the agreement was of course to control pollution, though at that time the focus was on phosphates and nitrates, and the dumping of both did decline. One of the treaty's weaknesses was that its oversight was confined to Lakes Erie and Ontario. Thus, after a few years the 1972 GLWQA was deemed inadequate.

On November 22, 1978, the two countries signed a more comprehensive agreement, which sought to control the dumping of toxic substances into all five lakes and their tributaries. "Joint water quality objectives and standards were set, commitments to implement control programs were made and the IJC mandate of monitoring progress was revamped and continued. Additionally, more stringent standards were set across the board with regard to water quality. These measures included implementation of municipal and industrial abatement programs, reductions in nutrient loadings, and a decrease in toxic chemical discharge. And a new concept was introduced: the Great Lakes basin ecosystem."[57]

In addition, the 1978 GLWQA included abatement programs for pollution from shipping, dredging, onshore and offshore facilities, contaminated sediments, and polluted groundwater and subsurface sources, as well as airborne pollutants and toxic substances. It also made provisions for surveillance and monitoring. The 1978 agreement was further concerned with management, remedial action, and contingency plans; it also spelled out the powers and responsibilities of the IJC and various protocols between the United States and Canada. In 1987, the agreement was further amended by protocol to add new programs and initiatives. Thus, it remains the most important instrument for preserving water quality in the Great Lakes region.

Love Canal

Environmental treaties and laws cannot undo the past. In the Great Lakes region, this point was driven home with cruel irony: The most infamous incident of toxic waste dumping in the United States was being uncovered even as lawmakers were hashing out the final details of the 1978 GLWQA. It occurred at Love Canal in Niagara Falls, New York.

The Love Canal, named for William T. Love, the man who proposed the project in 1892, was never really a canal. It was a giant ditch, approximately 3,200 feet long, that was supposed to be a canal of between six and seven miles long connecting the upper and lower Niagara River and harnessing the water to create electricity for domestic and industrial use. However, the project was abandoned during the financial crisis of the 1890s. In the 1920s, the ditch, still known as the Love Canal, became a dump site for toxic chemicals. By 1953, the canal was owned by the Hooker Chemical Corporation, a subsidiary of Occidental Petroleum. Hooker Chemical had purchased the site in 1947, but had been using it as a toxic dump since 1942. It disposed of the toxic chemicals in drums. But Hooker wasn't alone in using the site. "The City of Niagara Falls and the United States Army used the site as well, with the city dumping garbage and the army dumping possible chemical warfare material and parts of the Manhattan Project [which had developed the first atomic bomb]."[58] In 1953, Hooker Chemical closed the site, filled in the ditch with dirt, and sold the property to the City of Niagara Falls for a dollar. Later a school and a residential neighborhood were built on and near the site of the Love Canal, and by 1978 health problems, especially in children attending the school, were evident. It was later discovered that "82 different compounds, 11 of them suspected carcinogens, [had] been percolating upward through the soil, their drum containers rotting and leaching their contents. . . ."[59] How much of the toxic waste found its way into the Niagara River and subsequently Lake Ontario is unknown. The upshot was that Congress created the Superfund program to clean up Love Canal and other toxic waste sites throughout the United States, of which there were thousands. As for the GLWQA, the Love Canal incident stiffened the resolve of the environmental groups on both sides of the border and that of stakeholders in the Great Lakes ecosystem to protect the basin.

The Great Lakes Charter

On February 11, 1985, the eight governors and two premiers of the Great Lakes basin states and provinces signed an unprecedented agreement to work together on the state and provincial level to clean up the basin and monitor and preserve it in the future. This agreement, the Great Lakes Charter, draws from and supplements the GLWQA of 1978. In one very important aspect, it differs from the GLWQA and, for that matter, from every other such agreement in the world at that time. The charter recognizes the cohesiveness of the Great Lakes region as a whole—not simply that it cuts across artificial political boundaries, but that it has its own natural boundaries. It also recognizes that anything that affects one section of the region may affect other sections. Thus, the charter makes it clear that major diversions or uses of water in any state or province require the consent of all the states and provinces of an area. Such consensus offers added protection to the lakes and the regional ecosystem.

Current Crisis

Depending on who is doing the formulating, there are either very many crises in the lakes' ecosystem or one overarching crisis with more than 40 areas of concern. Either way, it comes down to pollution. Toxic chemicals, nutrient loading, and heavy metals harm marine life, birds, and other wildlife in the ecosystem, cause beaches to be closed, affect drinking water, and cost offending industries and agricultural concerns a great deal. Unfortunately, the majority of the areas of concern (AOC) that were listed in the 1990s have yet to be delisted. Many are in the tributary rivers and bays of the Great Lakes, some are in harbors and estuaries, while the connecting rivers account for seven AOCs: one each on the St. Marys, St. Clair, and Detroit Rivers, and two each on the Niagara and St. Lawrence Rivers. As might be expected, metropolitan Toronto is one of the region's worst polluters.

Drinking water is generally not at risk in the Great Lakes region, though there have been specific AOCs in this matter. These include seasonal increases in nitrate levels in municipal water supplies from the Maumee River in Ohio, which runs into southwestern Lake Erie, and bad taste and odor due to algae in Saginaw Bay on Lake Huron in Michigan. Much of the ecosystem's problems have to do with degradation of the environment, which in turn proves harmful to marine and other wildlife. In some areas there are restrictions on fish (and wildlife) consumption due to pollution.[60] For example, more than 159,000 kilograms of polychlorinated biphenyls (PCBs) were discovered in the sediment of Michigan's Kalamazoo River.[61] In other areas, the concern is the degradation of the benthos, or organisms that live in the sediment. And despite the attention given to eutrophication in the 1960s and 1970s, that remains a problem in some of the AOCs, as does the loss of wetlands.

A second crisis is the lower levels of the Great Lakes. In 2007, vessels transported approximately 35 million tons of coal across the Great Lakes, an 8 percent decrease from the previous year and a 7.6 percent decrease over the previous five years. Operators of the vessels blamed "unusually low water levels for reduced loads, especially in the final months of 2007."[62] The water levels of the lakes are part of another ongoing concern of many in the region: the protection of their resource from diversion. States in the arid U.S. West have coveted Great Lakes water for decades—NAWAPA being only one of their giant engineering options—and the High Plains states look to their midwestern and eastern neighbors as predictions for the future of the Ogallala Aquifer become increasingly dire. However, since the Great Lakes Charter requires the unanimity of the region's leaders before any major diversion can take place, the water itself seems protected.

Nevertheless, there is cause for concern, especially on the Canadian side. If water should be considered a commodity, while disregarding its life-giving, life-preserving, and life-protecting qualities, international treaties such as NAFTA may supersede the Great Lakes Charter. This regional fear extends to the corporations that bottle drinking water for sale throughout the world.

Counterstrategies

Water conservation and recycling are both being implemented in the Great Lakes region. One notable example is the city of Chicago, on the south shore of Lake Michigan, which in 2008 began metering its municipal water. Tracking water use and tying it to a payment schedule inevitably leads to conservation, especially among the heaviest users, such as industry. Conservation and recycling may yet have to be enacted on a much broader, regional scale, depending on whether the lower water levels in the lakes are the result of cyclical fluctuations or climate change. Any permanent drop in the lakes' water levels would further hamper outside regions' attempts to tap into the Great Lakes and/or their tributaries to satisfy their own freshwater needs. It would also affect the ecosystem as a whole.

The fear of Great Lakes water being diverted in large amounts was lessened somewhat in 2008 with the congressional passage of the Great Lakes Compact. Nevertheless, a provision known as the bottled-water loophole in the legislation "allows water to be diverted from the basin if it is in containers holding less than 5.7 gallons."[63] Accordingly, in August 2008, "Nestlé Waters North America was granted permits for a new well and pipeline at its Ice Mountain facility in Mecosta County, Michigan, where it bottles 700,000 gallons a day. Nestlé also recently renewed permits for its plant in Guelph, Ontario."[64] As the University of Chicago law professor Anu Bradford noted, "Under the WTO and NAFTA, there is no obligation for a state to extract its natural resources. The difference comes when it makes the decision to allow an entity to commercialize it and they do commercialize it. Then it is a product and you can't ban the export."[65] The Nestlé operations are not directly diverting water from any of the Great Lakes, but rather from creeks and lakes within the ecosystem that feed the Great Lakes. Grassroots campaigns have begun on both sides of the border to close the bottled-water loophole, while Nestlé Waters North America feels its operations have been targeted unfairly.[66]

Controversies aside, pollution remains the number one problem affecting the Great Lakes. To address the AOCs, various remedial action plans (RAPs) have been put into effect, with varying degrees of success. By the early 21st century, only one RAP had been successful enough that the AOC was delisted—that is, the problem was deemed remedied.[67] Activating remedial plans is complicated where AOCs fall under the jurisdiction of more than one state/province or even of both Canada and the United States. Likewise, the number and variety of stakeholders (some of whom may be the polluters themselves) are complicating factors. Since there are strict guidelines for listing and delisting AOCs, each remedial action plan works toward a specific goal or goals, depending on the type and amount of pollution that needs to be cleaned up, who the polluter is or was, the environmental effect of the pollution, the cost of the abatement, the governmental entities involved, and public input. While bureaucratic processes can be slow, action plans are in place and the situation is far from hopeless.

SOUTHEASTERN UNITED STATES

If John Wesley Powell could not have conceived of the population growth in the U.S. West 150 years ago, neither could he have foreseen the freshwater problems that have begun to occur in the southeastern portion of the country. After all, the southeastern United States lies east of the 100th meridian and receives more than enough rainfall every year to support its population—or does it? In the early years of the 21st century, the state of Georgia has experienced a prolonged drought, particularly in the sprawling Atlanta area. But Atlanta's water problems transcend the drought; indeed, they touch on some of the crises faced not only in the U.S. West and the Great Lakes region, but in developing nations. Even so, Atlanta's freshwater problems are dwarfed by those of Florida. One of the fastest growing states in terms of population although this trend may now be in reverse, Florida spent a good deal of the 20th century doing two things: "developing" its wetlands—that is, draining them—and attracting people to occupy the developed land. Additionally, the north of the state receives far more annual rainfall than the south.

In North Carolina, thousands of households in the counties of Dare, Pamlico, and Pender reported poor water quality even in the late 1990s and early 2000s, according to a report issued by the U.S. Department of Agriculture (USDA). Furthermore, North Carolina was involved in a "water war" with neighboring Virginia over diversion rights to Lake Gaston, which straddles both states. Virginia had planned to pipe water from the lake to Virginia Beach, the state's largest city (which lacks a municipal water supply) more than 70 miles away. However, North Carolina sought a federal injunction against the pipeline on environmental grounds. The USDA report also noted that 32 communities in Virginia and 377 in Kentucky had poor water quality. Additionally, it listed 37 communities in South Carolina that did not even have running water.[68] For the most part, this was due to infrastructure problems rather than diminished supply; nevertheless, people's health and quality of life were negatively affected.

Atlanta's Water Problems

Like the West and Florida, population growth is behind Atlanta's water crisis. But there is more to the story than this. For one thing, metropolitan Atlanta relies solely on surface water, primarily from Lake Lanier, a reservoir created in 1953 by the damming of the Chattahoochee River. The river begins in the Blue Ridge Mountains, northeast of Atlanta, and runs through metropolitan Atlanta and continues to the southwest, where it serves as the boundary between Alabama and Georgia; it then turns eastward until it reaches Lake Seminole on the Florida border (at the panhandle).[69] In Florida, the Chattahoochee, after it merges with the Flint River, is known as the Apalachicola River. Thus Atlanta, one of the largest metropolitan areas in the United States, sits upriver on the Chattahoochee, the smallest river in the country that provides water to a metropolitan area.[70]

Atlanta's problems arise from a number of conditions: rapid population growth; the area's reliance on the Chattahoochee River as its source of freshwater; aging infrastructure; drought; and a lack of good water management. Since the population of the city of Atlanta, as of 2007, is estimated by the United States Census Bureau to be greater than 500,000 and that of its metropolitan area to be in excess of 5.6 million, Atlanta in the past few years has become a microcosm of the global water crisis caused by more people sharing a finite resource.[71] Increased population combined with drought, including higher rates of evaporation in Lake Lanier and the Chattahoochee River, have helped pitch metropolitan Atlanta into a water crisis.

Before the drought intensified in the early years of the 21st century, Atlanta's metropolitan water delivery infrastructure brought home some of the problems related to water commodification and globalization to people used to simply turning on their faucets to receive clean freshwater. In 1999, the City of Atlanta and United Water Resources, a subsidiary of the French transnational water giant Suez Lyonnaise, brokered what became "the largest water-privatization deal in U.S. history" when the two parties signed a

20-year contract for United Water to run the city's water system at a cost of $22 million dollars a year.[72] City leaders turned to United Water rather than raise the water rates. The deal lasted all of four years, in the end both sides blaming the other for bad faith.

United Water claimed Atlanta was less than forthright in explaining the condition of the city's water pipes and treatment plants, while the city pointed to the poor service and even poorer water quality since United Water took over delivery and treatment services. Certainly, the city's water infrastructure was at least 100 years old at the time of the privatization deal and in a state of disrepair, but it seems unlikely that the company would have entered into such a deal without prior knowledge of the system, especially its age. The issue reached crisis point when some city residents had to boil the water from their faucets because of health concerns, while others at times received no water at all because of infrastructure repairs. In the end, the city's water system reverted to municipal oversight and rates were raised; "the City of Atlanta judged that it was better to deal with the problem itself, rather than depend on a private company to make the necessary upgrades."[73]

Atlanta's greater withdrawals of water from the Chattahoochee River via Lake Lanier mirror circumstances in many areas of the world, where urban dwellers are in conflict with farmers over water. In this case, the farmers are downstate and downriver in Georgia. Thus, in the first decade of the 21st century, the downstate Georgia farmers faced a twofold water crisis: drought compounded by greater diversion upriver.

Florida's Worsening Crisis

It seems almost impossible to believe that Florida faces a water crisis with the Everglades, Lake Okeechobee (the third largest lake in the United States), and the state's numerous rivers, including the Apalachicola. Moreover, the state receives some 50 inches of rain per year and sits atop the Floridian aquifer, the largest and deepest aquifer in the southeastern United States. Yet Florida does indeed face a water crisis.

Throughout the 20th century and into the 21st century, Florida has focused on development. Its population increased exponentially in the latter half of the 20th century, from approximately 2.7 million people in 1950 to more than 17.7 million in 2005, making it the fourth most populous state in the country. Population expansion was accompanied by urban growth and sprawl, increased agriculture and industry, and numerous major public works projects, especially water projects that affected the state's topography and environment. The best known is the draining of the Everglades, the largest wetlands in the United States. In 1948, the U.S. economy was entering a boom phase after nearly 20 years of depression and war, and with the disastrous floods of 1928 still on the minds of many Floridians, Congress authorized the U.S. Army Corps of Engineers "to replumb the entire bottom half of the state to provide flood protection and freshwater for urban and agricultural lands."[74] The massive program was known as the Central and South Florida Project and was administered by the Central and South Florida Flood Control District, later the South Florida Water Management District.

The "replumbing" of the area not only provided more water for agriculture, but also more land. The project "designed a large area of the northern Everglades, south of Lake Okeechobee, to be managed for agriculture.... Called the Everglades Agricultural Area, it encompassed about 27 percent of the historic Everglades and was a major factor in the economic justification of the [Central and South Florida] Project."[75] Among the unforeseen problems caused by the project were a reduction in the quality of freshwater, caused by runoff from the Everglades Agricultural Area. This led to an increase in cattails, a reedlike marsh plant, which in turn adversely affected the remaining area of wetlands as a habitat for wading birds. Second, the drainage canals built to prevent flooding of the land reclaimed for agriculture instead caused problems, including water shortages and a reduction in water quality caused by higher concentrations of salts, among other things, in the water. Furthermore, the canals did not always prevent flooding. An increase in mercury was also noted. It is theorized that mercury occurs naturally in Everglades soil in small amounts and that draining so much of the wetlands led to concentrations of the highly toxic element, which entered the food chain "with predators containing the highest levels."[76] Such predators include raccoons and alligators; fisher-folk have also been warned of the danger.

The filling in of the Everglades and associated swamps and marshes of South Florida has contributed to at least one other major negative effect: There is less rainfall because there is less water evaporating, though it is unclear whether this is a contributing factor for the drought plaguing other parts of the Southeast, particularly Georgia. Nevertheless, the domino effect on the environment is clear: Lower rainfall amounts contribute to less surface water (whose reduction is increased by human tampering), less surface water requires more groundwater pumping, which then reduces the water table enabling saltwater incursion. The latter has occurred to the freshwater supply of some coastal cities of South Florida.[77] Another side effect of a lower water table is subsidence, the lowering of the ground above the depleted aquifer and, in many cases, a concurrent result—a sinkhole. In Florida, sinkholes have swallowed cars and damaged homes.

The Florida peninsula has its own rainfall line, much like the 100th meridian, dividing the usually wet northern part of the state from the drier southern half. Thus, the destruction of approximately 50 percent of the Everglades ultimately exacerbated a situation it had intended to solve. Scientists have concluded that the destruction of the wetlands has also led to "more severe freezes in the winter," which has been disastrous for Florida's citrus growers.[78] The state also has been locked in a battle with Georgia and Alabama over those states' diversions of the Chattahoochee River (and in Georgia's case also the Flint River), which ultimately affect the Apalachicola River, which bisects Florida's panhandle. The Apalachicola, formed by the merging of the Chattahoochee and Flint Rivers at the border of Georgia and Florida, flows to Apalachicola Bay in the Gulf of Mexico, which is home to a huge shellfish industry, now threatened by the decrease in river flow. In 1997, the states embarked on negotiations to fulfill the terms of the Apalachicola-Chattahoochee-Flint Compact to devise an allocation formula for the basins waters, but failed to reach an equitable agreement by the deadline of August 31, 2003.[79] The compact then became void. Since that time the states have mostly engaged in a water war in the courts. However, more and more the issue of water conservation is taking into account how much water is needed for the rivers themselves, not just human use. The 2003 "agreement" would have allowed Atlanta and the Metropolitan North Georgia Water Planning District to "set aside up to 50 percent more water in Lanier for the region," which obviously meant less Chattahoochee water for Alabama and Florida. In 2008, though, a federal appeals court rebuffed Georgia's plan when it ruled that diverting more water from Lake Lanier required congressional approval. Thereupon, Georgia requested that the U.S. Supreme Court review the decision, but in January 2009 the Supreme Court denied the request, sending the dispute back to federal court in Jacksonville, Florida.[80] The Supreme Court's decision not to review the case was interesting in that many observers of this ongoing "water war" have long believed it will only be resolved by a decision from the Court.

Counterstrategies

While population growth and drought continue to plague water managers in Georgia and Florida (and to some extent Alabama and the rest of the Southeast), federal and state governments have developed strategies for water conservation and sharing.

COOPERATION IN GEORGIA

The USGS's Cooperative Water Program (CWP) is assisting water managers in Georgia on issues such as water conservation and use, contamination, and information on managing surface and groundwater.[81] The information it provides covers the water resource potential in northern Georgia, salinity problems in the coastal Upper Floridian aquifer, groundwater resources in other parts of the state, nonpoint sources of pollution that affect the metropolitan Atlanta system, and statewide water sampling. The information is used in certain parts of the state to develop resources, while studying the "mechanisms of groundwater flow and nitrate contamination."[82]

Meanwhile, Atlanta's water problems continue. In 2008, the drought abated somewhat, though by year's end Lake Lanier was still far below its normal level. Near-continuous rainfall in early January 2009 caused its level to rise three feet, but it was still 12 feet below its normal January level.[83] By September 2009, Atlanta and a good deal of the Southeast faced the opposite problem. Prolonged rainfall caused massive flooding in the regions, with parts of Atlanta itself under water. Needless to say the flooding will have adverse effects on the city and region's water and sewer systems. Meanwhile, the Metropolitan North Georgia Water Planning District (whose 15 counties and more than 90 cities includes Atlanta) has developed three integrated plans for dealing with its water crisis. The plans cover water supply and conservation, long-term wastewater management, and watershed management; all three were still in the draft stage in early 2009, though earlier measures dealing with these issues have been adopted and implemented.

The newer water supply and conservation plan calls for the construction of three new reservoirs and investigation into three additional reservoirs. The plan also calls for the construction of six new water treatment plants and the expansion of the existing 28 plants. Water reuse is also addressed, as the 10 percent goal outlined by the state has not yet been met.[84] Finally, the watershed plan includes "six model ordinances which provide for post-development stormwater management, floodplain management, conservation/open space development, illicit discharge and illegal connection controls, litter control and stream buffer protection."[85] The plan is not without its detractors, especially environmental advocates who claim it relies on expensive 20th-century technology to conserve water.

EVERGLADES RESTORATION

One problem that environmentalists, water managers, the Army Corps of Engineers, politicians, and citizens need to resolve is the reduction of the Everglades. Many of the negative effects of the last half-century on the ecosystem could yet be abated. In 2000, President Bill Clinton signed into law the Water Resources Development Act, which in turn promulgated the Comprehensive Everglades Restoration Plan (CERP). "CERP provides a framework and guide to restore, protect and preserve the water resources of central and southern Florida, including the Everglades. . . . The goal of CERP is to capture fresh water that now flows unused to the ocean and the gulf and redirect it to areas that need it most. The majority of the water will be devoted to environmental restoration, reviving a dying ecosystem."[86] It is estimated that it will take around 30 years and $60 billion to complete CERP. The plan has 60 components, the 13 most important being:

- surface water storage reservoirs
- water preserve areas
- management of Lake Okeechobee as an ecological resource
- improved water deliveries to the estuaries
- underground water storage
- wetlands treatment
- improved water deliveries to the Everglades
- removal of barriers to sheet flow (shallow surface water flow not concentrated into channels)
- storage of water in existing quarries
- reuse of wastewater
- pilot projects
- improved water conservation
- additional feasibility studies.[87]

In November 2004, USGS issued a report on underground water storage and the Floridian aquifer recovery in southern Florida. The report outlined the use of aquifer storage and recovery (ASR) wells and provided percentages of recovery of various wells, though not all of them. At that time, the percentages ranged from 0.5 percent recovery in Lee County on the Gulf of Mexico to 78 percent recovery in southern Palm Beach County. The purpose of the wells is "to store excess freshwater available during the wet season in an aquifer and recover this water during the dry season when needed for supplemental drinking water supply."[88] The thinking behind all of this is that as more of the Everglades are restored, they will contribute to increased local rainfall (and less severe freezes), which in turn will increase the water table in the aquifer, as well as strengthen the Everglades environment.

CANADA AND MEXICO

Throughout this chapter it has been pointed out how U.S. water policy can affect its northern and southern neighbors and in some instances is promulgated in partnership with them. It is worthwhile, then, to examine briefly water problems and issues within Canada and Mexico as part of the interconnectedness of the overall issue.

Over the past quarter-century, Canadians have become wary about motives concerning their water resources because of all the nations in the Western Hemisphere, Canada has the most to lose. While the federal and provincial governments have taken some steps to ensure the viability of the country's freshwater ecosystems, Canada lacks a national water law, and Environment Canada claims that protective measures taken so far are not adequate. In a report issued in 2005, the department warned of a "national water crisis" spurred by "pollution and overextraction."[89]

Canadian activists and politicians have halted the privatization of municipal water delivery in some of the country's biggest cities, including Halifax, Québec City, Toronto, and Vancouver. In varying degrees, municipal privatization occurs in Hamilton, Ontario, and in Moncton and Sackville, New Brunswick. Activists remain concerned about NAFTA and the commodification of water, which they fear may lead to depletion of resources by for-profit transnationals, particularly U.S. companies or their subsidiaries. Ever vigilant, they "have successfully stopped the commercial export of water from the Great Lakes, British Columbia, and Newfoundland."[90] But the threat of globalization looms over Canada's water, as does the ethical obligation to assist those who are literally dying of thirst or have access to only poor-quality water.

Mexico faces a different set of problems. Much of the country, like the western United States, is water deficient. The problem of low flow at the lower end of the Colorado River heightens the overall problem of over-diversion of the river. More serious is the crisis faced by the Mexico City metropolitan area.[91] The groundwater supply to the most populous city in North America is running so low because of overuse that sections of the city have subsided due to over-pumping. In fact, Mexico City's need for water is so great and the aquifer so depleted that water must be pumped "from as far as 120 miles away, which escalates costs about 55 percent."[92] For Mexico City and many other municipalities, a major problem is infrastructure. Just as Atlanta's water delivery system had to be overhauled because of age, so too does Mexico City's, which "loses an estimated one-third of its water through leakage before it reaches consumers."[93] This is not a total loss because the water eventually seeps back into the aquifer, but it adds to the overall cost, especially that of pumping more water.

Unlike Canada, and perhaps because of the scarcity of the resource, Mexico does have a national water law. Its original law was the Federal Water Law of 1971, followed by the Federal Water Law of 1992. The latter was amended in December 2003. In 1983, the Mexican Supreme Court declared groundwater a national property, and in 1989 the government established the National Water Commission (Comisión Nacionel del Agua, CONAGUA).

In addition to its rapidly depleting groundwater source, Mexico City receives water from the Lerma-Balas and Cutzamala River systems.[94] Diverting water from these rivers places water managers at odds with the indigenous peoples of the area, especially the Mazahua, who have organized protests against pollution of the water supplies of indigenous communities, crop destruction, and the obliging of people "living in the western reaches of the state of Mexico to pay the costs of providing water for an ever more thirsty Mexico City." The government signed agreements in 2004 "that promised to indemnify landowners for their losses, promote investment in infrastructure for the provision of potable water for the community, and reforest the area" and in 2005 "to implement a sustainable development plan," but has yet to live up to its end of the bargain.[95]

Finally, Mexico has its own worries when it comes to NAFTA. Although the border region has benefited somewhat from an influx of jobs, the agreement and globalization have tended to concentrate Mexico's resources, including water, in fewer hands, as evidenced by the protests of the indigenous, who have organized to make their issues known and their voices heard. Like Canadians, they hope to reverse, or at least limit, the growing trends of globalism over resources.

---

[1] This is a separate Colorado River that runs exclusively in Texas. Austin, Texas's capital, is situated on it.

[2] It is estimated that Lake Huron holds approximately 850 cubic miles of freshwater. See *The World Almanac and Book of Facts, 2007.* New York: World Almanac Books, 2007, p. 707.

[3] Robert Glennon. *Water Follies: Groundwater Pumping and the Fate of America's Fresh Waters.* Washington, D.C.: Island Press, 2002, pp. 99–111.

[4] Glennon, pp. 127–141.

[5] The organisms are harmless and are found in all untreated freshwater. The concern of the Metropolitan District Commission, the regional water company serving the Hartford area, was that smaller organisms such as giardia might also have passed through the filters.

[6] Marc Reisner. *Cadillac Desert: The American West and Its Disappearing Water.* Revised edition. New York: Penguin Books, 1993, p. 12.

[7] *Chinatown,* directed by John Huston.

[8] For two competing viewpoints on William Mulholland's career, see Catherine Mulholland, *William Mulholland and the Rise of Los Angeles.* Berkeley and Los Angeles: University of California Press, 2000, and Reisner, *Cadillac Desert.*

[9] For more on the history of the Los Angeles Aqueduct and the Owens Valley story, see Reisner, pp. 52–103, and Mulholland, pp. 112–264.

[10] Earlier flood control assistance came in 1907, when the Southern Pacific Railroad dumped a massive amount of rock to control flooding and push the river back to its original channel.

[11] Jose Trava. "Sharing Water with the Colossus of the North." In *Western Water Made Simple,* ed. Ed Marston, Washington, D.C.: Island Press, p. 176. I am indebted to Trava for much of what follows about U.S.-Mexico relations regarding the Colorado River.

[12] That is, the upper basin had to ensure the delivery of 75 million acre-feet of water over a 10-year period. The 10-year period clause was added to offset dry years when the river would flow at a lower rate.

[13] The canal is part of the All-American Canal System, which is part of the Boulder Canyon Project. The All-American Canal System includes the Imperial Dam and the Coachella Canal, among other public works. See "Boulder Canyon Project: All-American Canal System." U.S. Department of Interior, Bureau of Reclamation. Available online. URL: http://www.usbr.gov/dataweb/html/allamcanal.html. Accessed September 30, 2008.

[14] Joseph E. Stevens. *Hoover Dam: An American Adventure.* Norman: University of Oklahoma Press, p. 46.

[15] The Grand Coulee Dam, built in 1945 on the Columbia River in Washington, has surpassed the Hoover Dam in both these respects, as have subsequent dams around the world.

[16] Stevens, pp. 31–32.

[17] Trava, p. 177.

[18] To be fair, the construction of the Glen Canyon Dam was a tradeoff. Brower and others managed to keep the Bureau of Reclamation from building a dam further upriver on the Colorado.

[19] Trava, p. 178.

[20] Trava, p. 179.

[21] See, for example, Reisner, pp. 463–465.

[22] Most of the states were drawing less than their allotted amounts at this time.

[23] Quoted in Glennon, p. 193.

[24] Joe Gelt. "Sharing Colorado River Water: History, Public Policy and the Colorado River Compact." Water Resources Research Center, University of Arizona. Available online. URL: http://ag.arizona.edu/AZWATER/arroyo/101comm.html. Accessed October 5, 2008.

[25] Initially opposed to such public works projects, Hayden was also one of the congressional leaders to pass the legislation for construction of the Grand Coulee Dam in Washington.

[26] Arizona Water Settlements Act, 2004. Title I, Sec. 104, subsection (c), 1A. If the reader will pardon the pun, the quoted material is merely a drop in the bucket of what this law covers, as it is considered the most complex water law in U.S. history. Available online. URL: http://www.azwater.gov/dwr/Content/Hot_Topics/AZ_Water_Settlements/GRIC_files/Cong_R ecord_S437%20-108-360.pdf. Accessed October 10, 2008.

[27] "CAP Subcontracting Status Report, October 5, 2009: CAP Non-Indian Municipal and Industrial Subcontracts." Central Arizona Project. Available online. URL: http://www.cap-az.com/includes/media/docs/SubcontractStatusReport-10-05-09.pdf. Accessed January 12, 2010. These status reports are updated.

[28] This makes it one of the driest metropolitan areas of the world, comparable to the western Sahara and the Arabian desert. See Diane Raines Ward, *Water Wars: Drought, Flood, Folly, and the Politics of Thirst.* New York: Riverhead Books, 2002, p. 67.

[29] Ward, pp. 67–68.

[30] Ward, p. 68.

[31] Arizona and New Mexico retained rights as upper and lower basin states.

[32] Craig Fuller. "Central Utah Project." In *Utah History Encyclopedia.* Available online. URL: http://historytogo.utah.gov/utah_chapters/utah_today/centralutahproject.html. Accessed October 15, 2008.

[33] Fuller. "Central Utah Project."

[34] Such is the ongoing power of the water interests in the U.S. West that some contend it was the "hit list" and not the hostage crisis in Iran that most contributed to President Carter's defeat in the 1980 election. It is perhaps no coincidence that his opponent, Ronald Reagan, was the former governor of California—the thirstiest state in the Union.

[35] Quoted in Ward, p. 68.

[36] Another reason for the incorrect figure is that it "was based on about eighteen years of streamflow measurement with instruments that, by today's standards, were rather imprecise." (Reisner, p. 262).

[37] Jon Gertner. "The Future Is Drying Up." *New York Times Magazine* (10/21/07).

[38] Gertner.

[39] Gertner.

[40] Gertner.

[41] Gertner.

[42] See Charles W. Howe. "Increasing Efficiency in Water Markets: Examples from the Western United States." In Terry L. Anderson and Peter J. Hill, eds. *Water Marketing—The Next Generation.* Lanham, Md., and London: Rowman & Littlefield, 1997.

[43] Reisner, pp. 486–495. Also "The North American Water and Power Alliance (NAWAPA)." San José State University Department of Economics. Available online. URL: http://www.sjsu.edu/faculty/watkins/NAWAPA.htm. Accessed October 22, 2008.

[44] All figures taken from "The Great Lakes" in *The World Almanac and Book of Facts, 2007,* p. 707.

[45] The province of Quebec does not border any of the Great Lakes, but the St. Lawrence River runs through it to the Gulf of St. Lawrence.

[46] "Chapter Five: Joint Management of the Great Lakes." The Great Lakes Atlas, U.S. Environmental Protection Agency. Available online. URL: http://www.epa.gov/glnpo/atlas/glat-ch5.html. Accessed October 30, 2008.

[47] Both the Treaty of Washington and the Boundary Waters Treaty were signed between the United States and Great Britain, acting on behalf of the dominion of Canada. However, Canada did have its own representatives at both negotiations.

[48] Mark Sproule-Jones. *Restoration of the Great Lakes: Promises, Practices, Performances.* Vancouver: University of British Columbia Press, 2002, p. 107.

[49] Eutrophication in lakes is actually a natural process. When caused by human intervention, which always speeds up the process, it is called cultural eutrophication.

[50] Sproule-Jones, pp. 31–32.

[51] "Cuyahoga River Fire." Ohio History Central: An Online Encyclopedia of Ohio History. URL: http://www.ohiohistorycentral.org/entry.php?rec=1642. Accessed November 1, 2008.

[52] Kenneth M. Vigil. *Clean Water: An Introduction to Water Quality and Water Pollution Control,* 2nd ed. Corvallis: Oregon State University Press, p. 88.

[53] "Federal Water Pollution Control Act, as Amended by the Clean Water Act of 1977." Clean Water Act (CWA), U.S. Environmental Protection Agency. Available online. URL: http://www.epa.gov/npdes/pubs/cwatxt.txt. Accessed November 1, 2008.

[54] Vigil, p. 83.

[55] Vigil, p. 85.

[56] Sproule-Jones, p. 39.

[57] Sproule-Jones, p. 41.

[58] Lois Marie Gibbs. *Love Canal: The Story Continues . . .* 20th anniversary edition. Gabriola Island, British Columbia, and Stony Creek, Conn.: New Society Publishers, 1998, p. 21.

[59] Eckardt C. Beck. "The Love Canal Tragedy," *New York Times* (8/1/78), and quoted in *EPA Journal.* U.S. Environmental Protection Agency. Available online. URL: http://www.epa.gov/history/topics/lovecanal/01.htm. Accessed October 15, 2008.

[60] Wildlife, such as birds, eat the fish. Studies have shown the higher up the food chain, the more toxins remain in an organism.

[61] Sproule-Jones, first page of Appendix B (unnumbered).

[62] "Great Lakes Tonnage Tumbles Due to Lower Water Levels." *Platts Coal Outlook* (1/14/08).

[63] Kari Lydersen. "Bottled Water at Issue in Great Lakes: Conservation and Commerce Clash." *Washington Post* (9/29/08): p. A7.

[64] Lydersen, p. A7.

[65] Quoted in Lydersen, p. A7.

[66] Lydersen, p. A7.

[67] The AOC was Collingwood Harbour on the Ontario side of Lake Huron, on the southeast tip of the Georgian Bay.

[68] Cited in Paul Simon. *Tapped Out: The Coming World Crisis in Water and What We Can Do about It.* New York: Welcome Rain Publishers, 2001.

[69] Lake Seminole is an artificial reservoir created by the Jim Woodruff Dam, which impedes the flow of the Chattahoochee and Flint Rivers.

[70] Cynthia Barnett. *Mirage: Florida and the Vanishing Water of the Eastern U.S.* Ann Arbor: University of Michigan Press, 2007, p. 115.

[71] Actually, the population of the metropolitan area has nearly doubled since 1990.

[72] Barnett, pp. 164–165.

[73] Ken Midkiff. *Not a Drop to Drink: America's Water Crisis (and What You can Do)*. Novato, Calif.: New World Library, 2007, p. 96.

[74] Barnett, p. 49.

[75] "Everglades Agricultural Area." Duke University Wetlands Center. Available online. URL: http://www.nicholas.duke.edu/wetland/eaa.htm. Accessed November 30, 2008.

[76] "Current Problems Facing the Region." Duke University Wetlands Center. Available online. URL: http://www.nicholas.duke.edu/wetland/current.htm. Accessed November 30, 2008.

[77] Barnett, p. 7.

[78] Barnett, pp. 60–61.

[79] The compact became federal law in November 1997; its original deadline was December 31, 1998, which was extended to the 2003 deadline. See Josh Clemons. "Water-Sharing Compact Dissolves." Mississippi-Alabama Sea Grant Legal Program. Available online. URL: http://www.olemiss.edu/orgs/SGLC/MS-AL/Water%20Log/23.3watershare.htm. Accessed January 14, 2009.

[80] Stacey Shelton. "Georgia Loses Round in Fight over Lanier Water." *Atlanta Journal-Constitution* (1/12/09). Available online. URL: http://www.ajc.com/services/content/metro/stories/2009/01/12/lanier_water_fight.html. Accessed January 17, 2009.

[81] CWP is a national program directed by the USGS in partnership with states, municipalities, and Native American tribes.

[82] "Helping Solve Georgia's Water Problems—The USGS Cooperative Water Program." USGS. Available online. URL: http://pubs.usgs.gov/fs/2006/3032. Accessed December 17, 2008.

[83] Shelton. "Lake Lanier No Longer Rising." *Atlanta Journal-Constitution* (1/15/09). Available online. URL: http://www.ajc.com/services/content/metro/stories/2009/01/15/lake_lanier_drought.html?cxtype=rss&cxsvc=7&cxcat=13. Accessed January 16, 2009.

[84] "Water Supply and Water Conservation Management Plan" (draft). Metropolitan North Georgia Water Planning District. Available online. URL: http://www.northgeorgiawater.com/files/2008-12-12_WaterSupply_Conservation_Public_Comment_DRAFT.pdf. Accessed January 16, 2009.

[85] "District-wide Watershed Management Plan." Metropolitan North Georgia Water Planning District. Available online. URL: http://www.northgeorgiawater.com/html/253.htm. Accessed January 16, 2009.

[86] "About CERP: Brief Overview." Comprehensive Everglades Restoration Plan. Available online. URL: http://www.evergladesplan.org/about/about_cerp_brief.aspx. Accessed January 14, 2009.

[87] "CERP: The Plan in Depth—Part 1." Comprehensive Everglades Restoration Plan. Available online. URL: http://www.evergladesplan.org/about/rest_plan_pt_01.aspx. Accessed January 14, 2009.

[88] "Review of Aquifer Storage and Recovery in the Floridian Aquifer System of Southern Florida." USGS. Available online. URL: http://pubs.usgs.gov/fs/2004/3128. Accessed December 17, 2008.

[89] Maude Barlow. *Blue Covenant: The Global Water Crisis and the Coming Battle for the*

Right to Water. New York and London: The New Press, 2007, p. 30.

[90] Barlow, p. 120.

[91] The population of the Mexico City metropolitan area is approximately 20 million.

[92] Simon, p. 62.

[93] Simon, p. 128.

[94] See Cecilia Tortajada. "Water Management in Mexico City Metropolitan Area." *Water Resources Development* 22, no. 2 (June 2006): 353–356. Available online. URL: http://www.adb.org/Documents/Books/AWDO/2007/br03.pdf. Accessed January 21, 2009.

[95] Stefanie Wickstrom. "Cultural Politics and the Essence of Life: Who Controls the Water?" In *Environmental Justice in Latin America: Problems, Promise, and Practice,* ed. by David V. Carruthers. Cambridge, Mass.: MIT Press, 2008, p. 307.

3

Global Perspectives

The nations examined in this chapter—Bolivia; Egypt, Ethiopia, and Sudan; Israel, Jordan, and the Palestinian territories; and India, Pakistan, and Bangladesh—all face enormous challenges when it comes to freshwater. These challenges include not only the usual ones to do with quantity, quality, and delivery, but also political and ethical issues related to nationalism, regionalism, globalism, and environmentalism.

Of course, these are not the only countries struggling under the strain of water crises. Water scarcity in sub-Saharan Africa strains intraregional relations; the nations of the Mahgreb (Saharan Africa) all battle an extremely dry climate (except along their coastlines);[1] while South Africa has proved to be a regional innovator in terms of water policy. The Middle East—Israel, the Palestinian territories, and Jordan, aside—faces enormous water problems. As the upriver country for both the Tigris and Euphrates Rivers, Turkey's dam projects have serious consequences for Syria and Iraq. The countries on the Arabian Peninsula, having long overdrawn their meager supplies of groundwater, use their oil profits to import water, while perfecting the technology of desalination. China, as mentioned in chapter 1, struggles with its own water crises, which in turn can affect such countries as Vietnam, Laos, Thailand, Cambodia, and Myanmar, all of which the Mekong River either runs through or borders, as well as India, Bangladesh, Nepal, and Bhutan. The Mekong River originates in China, but that country (along with Myanmar) is not a member of the Mekong River Commission that manages the river, thus China's upriver status—combined with its economic and military power—poses a constant threat to the other nations of the Mekong Basin. And while Europe is by and large a wet continent, abuses such as pollution, over-pumping of groundwater, privatization, and profligacy have created some serious problems there. The continent's southernmost regions, particularly Spain but also Sicily and Crete, suffer from a lack of quality freshwater. Finally, in Australia, which is the driest continent on Earth and whose interior is largely desert, past abuses have led to water problems particularly in the southeast of the country. In that region, problems including overdrawing from and pollution of the Murray–Darling River system began to be corrected in 1997 through a multistate river basin commission, and in 2003 a river restoration program known as the Living Murray Initiative commenced.

No continent with the exception of Antarctica is immune to the global water crisis, and indeed no country is immune either.[2] Just as a drop of water can travel anywhere on or above the earth via the hydrologic cycle, so the freshwater crisis is one we all share. Thus, while the following discussions examine the individual crises in various nations and regions, the reader should keep in mind that the problem is a global one.

BOLIVIA

In the 21st century, Bolivia, the poorest country in South America, has emerged as a leader in turning back some of the harmful environmental and economic trends of the latter half of the previous century. In two important episodes, Bolivia's water dealings allowed for nothing short of a social and political revolution in that country.

Dating back to 1935, Bolivia is a signatory nation to numerous freshwater treaties with its neighbors. In the 1950s, it signed three important treaties with Peru concerning the use of the waters of Lake Titicaca, which is on the border between the two countries. Bolivia's capital, La Paz, is located to the east of the lake. The last of the trio of agreements, signed in 1957, concerned the use of the lake's waters for hydropower. In 1993, the two countries created the Autonomous Binational Authority of the Lake Titicaca Basin, the Desguadero River, Lake Poopó, and the Coipasa Salt Pan System, whose mandate was to resolve differences between Bolivia and Peru in the region.

Aside from the important issues of glacier melting, the quality of freshwater, and equitable water distribution, Bolivia's water concerns were international in scope. Among them was the pollution of the Pilcomayo River, a 1,600-mile tributary of the Paraguay River, which originates in the Andean foothills in Bolivia and runs through Argentina and Paraguay. More than 90 percent of the Pilcomayo is in Bolivia, and runoff and dumping from that country's silver, tin, lead, zinc, antimony, and arsenic mines, as well as other untreated waste, have polluted the river. Downstream in Argentina and Paraguay, the Mataco and Guaraní tribes have relied on the river for centuries to sustain their cultures. In addition to drawing drinking water from and fishing the river, they use the water for irrigation. Their pleas to the respective national governments to do something about the worsening water quality went unheeded until an incident in 1996 brought things to a head. An accident in a Bolivian mine caused a dike to burst, sending "300 tons of tailings [residue from the ore] into a tributary of the Pilcomayo. By the time the Pilcomayo reached Argentina and Paraguay . . . the river was covered with metal sediment, including high concentrations of poisonous lead and arsenic that were well beyond international water quality guidelines for human consumption."[3] The incident stirred the downriver nations out of their laissez-faire attitude toward Bolivian mining pollution and nearly involved them in a shooting war with their northern neighbor. Bolivia initially agreed to clean the river and regulate its mines but could really do neither properly due to a lack of funds and instead invoked a right of "ownership" of the Pilcomayo (as the upriver country and the country within whose borders most of the river ran).

As the three countries edged closer to open warfare in 1997, the international environmental agency Green Cross stepped in to mediate.[4] Fortunately, there was precedent for Green Cross and the countries' negotiators to work with: In February 1995, the three countries had signed a treaty calling for joint management of the Pilcomayo River. Four months later, the countries signed another joint-management treaty for the Bermejo and Grande de Tarija Rivers. The agreement Green Cross brokered among the countries called for all three to step back from their warlike stance, identify the various toxins in the river, and draw up a cleanup schedule in which Bolivia would take the lead, with assistance from Argentina and Paraguay, and with each country contributing an equal amount of money to the project, in addition to funds from the World Bank and environmental organizations.[5] Essentially, Bolivia reversed its position, while acknowledging its responsibility toward the river. Simultaneously, both Argentina and Paraguay acknowledged Bolivia's financial dilemma and to a lesser extent their own responsibilities toward the river. Cleanup plans were finally implemented in 2008.

The Cochabamba Water War

While the Bolivian government was willing to negotiate with international partners, it was unwilling to do so with its own citizens. In early 2000, the city of Cochabamba, Bolivia's third largest, located approximately 120 miles southeast of La Paz on the Rocha River, experienced civil unrest over the privatization of its water supply by the national government. The strife became known as the Cochabamba Water War and marked the beginning of a new era for Bolivians by helping to ensure the 2005 election of Evo Morales as president.

The passing of the law signified as DS 21060 by President Víctor Paz Estenssoro, many contend, "set in motion the process of neoliberal structural adjustment."[6] Neoliberalism describes economic and social policies that have metamorphosed into globalism, its main features including deregulation, an international free market, and privatization.[7] In fact, Bolivia's flirtation with neoliberalism may have expedited the Pilcomayo River agreement brokered by Green Cross, as the World Bank, one of the driving institutions of globalism, was also involved. The World Bank made a loan to Bolivia contingent on the privatization of Cochabamba's water system.

That the water system, Servicios Municipales de Agua Potable y Alcantarillado (SEMAPA), served only about 50 percent of Cochabambinos seemingly made it easier to privatize, as well as to upgrade and expand as part of that process. Meanwhile, the majority of the remainder of Cochabamba's population, that is those not served by SEMAPA, received water from private (often cooperative) wells and rooftop cisterns. Furthermore, Cochabamba's water quality was compromised. In 1992, for example, 500 people died in a cholera epidemic. When on October 29, 1999, the government passed Law 2029, this not only paved the way for water privatization but strengthened the privatizing companies' hands—that is, it granted them monopolies by declaring traditional water systems illegal "within the territory covered by a privatization contract."[8] For example, the 40-year privatization contract the Bolivian government signed with Aguas del Tunari made it illegal for Cochabambinos to use even private wells and cisterns. Thus, Aguas del Tunari effectively "owned" the rainfall, which it asserted by capping wells and cisterns.

Aguas del Tunari, registered in the Cayman Islands, took control of Cochabamba's water supply three days after the passing of Law 2029. In 1999, the consortium that owned Aguas del Tunari consisted of four Bolivian companies, as well as Abengoa of Spain and International Water, the latter a subsidiary of the U.S. company Bechtel.[9] International Water's stake amounted to 27.5 percent, which made Bechtel the largest shareholder. At any rate, water rates immediately increased, drawing modest protest from Cochabambinos.[10] These protests—which initially took the form of roadblocks—led to large-scale activism on the part of Cochabambinos, as well as peasants in outlying areas and *cocaleros* (coca growers), whose main spokesperson was Evo Morales.

The activism arose out of the factory workers' union (Federación de Fabriles), which had been struggling against the vicissitudes of neoliberal policy in Bolivia for more than 10 years. Union leadership lent its organizing expertise to the socially divergent but quickly coalescing activists, and the Coordinadora de Defensa del Agua y de la Vida (Coalition in Defense of Water and Life) was born. The main spokesperson for the Coordinadora was Oscar Olivera, who was also a union leader. In early December 1999, the Coordinadora began to step up the pressure on the government and Aguas del Tunari, and in early January 2000, amid further protest that led to an open town meeting (*cabildo*), the Coordinadora "gave the government until January 11 to tear up the contract with Aguas del Tunari, to repeal the water law, and to reverse the rate hikes."[11] When the deadline came and went without these things occurring, more roadblocks were set up in protest, and in response the police used tear gas on the crowd.

The Coordinadora was determined to continue the fight and scheduled massive protests in Cochabamba for February 4, 2000. For its part, the government, in need of World Bank funds and determined to remain a player in the globalized economy, decided to send in the army to assist the police in breaking up the protest. The army blockaded Cochabamba and began firing tear gas on the protest marchers. In turn, the protesters armed themselves with bricks and stones and began setting up barricades. By the next day, they had managed to reach the city center, the goal of the four main columns of protest marchers, and a truce was signed by the government on February 6, 2000, that called for a freeze on rate increases. The Coordinadora gave the government two months to enact the new policy, but within six weeks it was understood that not only would the government renege on its promise, but that Aguas del Tunari was to remain Cochabamba's water company.[12]

During the two-month interval between the February protests and the Coordinadora-imposed deadline for implementing the truce, the archbishop of Cochabamba tried unsuccessfully to mediate the situation. On March 22, 2000, some 96 percent of the approximately 50,000 voters in a Coordinadora-sponsored referendum voiced their disapproval of the Aguas del Tunari contract but, as expected, the referendum had no effect on the government's water policy in Cochabamba. Coordinadora leaders therefore continued making plans for the next round of protests, so that when the deadline passed and nothing had been done to alter the situation, people were ready to renew their opposition to Aguas del Tunari. The Coordinadora was well organized, having divided itself into sectors that held assemblies, whose decisions were taken to town meetings for validation. It has been estimated that 50,000 to 70,000 people attended the open-air meetings. Eventually the decision was made to take over the offices of Aguas del Tunari as a symbolic gesture.[13] The government had anticipated further protests once the April 4 deadline passed, and the police were armed with tear gas and rubber bullets. The Cochabambinos and others armed themselves with Molotov cocktails as well as rocks and bricks. Such was the situation that on April 6, 2000, Oscar Olivera and other Coordinadora leaders were arrested and briefly detained. Then, on April 8, with the unrest beginning to spread to other parts of the country, the Bolivian president Hugo Banzer declared a 90-day state of siege, which is legal under the Bolivian constitution. Cochabamba remained the epicenter of the protests, but the day after Banzer's declaration violence broke out near the city of Achacachi and striking police officers and soldiers clashed in the capital city of La Paz. In Cochabamba, on the day President Banzer declared the state of siege, 17-year-old Victor Hugo Daza was killed by an army captain who fired into a crowd.[14] In all six people, not all of them civilians, died in Cochabamba and elsewhere as a result of the conflict.

Daza's death, the spreading unrest, the escalation of violence, and the realization that Cochabambinos were prepared for the long haul loosened the grip of government hard-liners during negotiations. On April 10, an agreement was reached between Coordinadora and the government that promised the nullification of the Aguas del Tunari contract, as well as repeal of the water privatization provision of Law 2029. It was also no longer legal to charge peasants for well water.[15] President Banzer resigned from office for health reasons and Vice President Jorge Quiroga, a former IBM executive and native of Cochabamba, was sworn in on August 7, 2001.

As for Aguas del Tunari, it sought compensation for the cancelled contract, asserting that the Bolivian government had violated a trade agreement with the Netherlands, which is where the Bechtel subsidiary, International Water, is incorporated.[16] In February 2002, the consortium and Bechtel took their case to the International Centre for the Settlement of Investment Disputes (ICSID); it was heard as *Aguas del Tunari v. Republic of Bolivia.*[17] The consortium sought $25 million in compensation. Since ICSID is under the auspices of the World Bank, few observers, especially in Bolivia, thought Bechtel would lose the case, especially when in February 2003 ICSID announced it would not allow public or media participation in the lawsuit.[18] Nevertheless, after nearly four years of proceedings, both parties agreed to settle before a decision was handed down. In January 2006, Bechtel announced in an official company news bulletin: "The Government of Bolivia and the international shareholders of Aguas del Tunari declare that the concession was terminated only because of the civil unrest and the state of emergency in Cochabamba and not because of any act done or not done by the international shareholders of Aguas del Tunari (Bechtel, Befesa, Abengoa, and Edison). As a result of the settlement, the claims against Bolivia currently before the International Centre for Settlement of Investment Disputes will be withdrawn."[19] Thus, the company was absolved of all blame in exchange for dropping its indemnity to Bolivia.

A second water war occurred in La Paz and in one of its poor suburbs, El Altook, in 2004. Aguas de Illimani, a subsidiary of the French water company Suez, had signed a 30-year water and sewage contract with the Bolivian government. Protests and clashes led the government to once again cancel a contract that would have privatized water, claiming Aguas de Illimani had failed to live up to the contract.

In 2005, Evo Morales won Bolivia's presidential election as the country's indigenous majority and nonindigenous progressives, spurred by years of political action, backed one of their own. It was the Morales government that ended the Bechtel lawsuit, and in 2006 President Morales also instituted a Ministry of Water. Thus, it appears unlikely a water transnational or a consortium will successfully contract with the Bolivian government in the near future.

Shrinking Glaciers

Since the beginning of the 21st century, scientists have studied the phenomenon of melting glaciers, not only at the poles, Greenland, the Swiss Alps, and the Himalayas, but also in the Andes. "According to the Byrd Polar Research Center . . . Andean glaciers have retreated by as much as 25 percent in the last three decades of the 20th century."[20] Global warming is behind this phenomenon. A global increase in temperature of about one degree centigrade, though seemingly slight, has had a dramatic effect on the glaciers in the Bolivian Andes. As Cochabambinos and others were fighting over access to rainwater and piped ground and surface water, the supply in the form of runoff from glaciers was beginning to dwindle. In fact the "Zongo glacier on the . . . mountain of Huayana Potosi is retreating by 10 yards per year," while "the 18,000-foot Condoriri, the glacier that supplies the largest reservoir in the Bolivian highlands, is shriveling so fast that scientists fear a scarcity of drinking water in the decades to come."[21] Furthermore, there has been less rain and snowfall—at least in comparison to the period from the 1980s to the 1990s. Measurements at the Milluni Lagoon, which lies at the foot of Huaya Potosi, averaged 22.4 inches of precipitation per year throughout the 1980s, but this dropped to 17.88 in the following decade—a decline of more than 20 percent. The main factor in this decline is the more frequent occurrence of the dry El Niño winds coming off the Pacific Ocean.[22] The Bolivian glaciers aren't the only ones in the Andes that are shrinking, nor is Bolivia the only nation in the region to experience less precipitation—Venezuela and Peru, in particular, have similar problems.

Counterstrategies

As always, the solutions to the problems concerning the quantity and quality of freshwater are regional and interdependent. And as always, the first word is conservation. This means not just using less, but wasting less because of old and leaky urban delivery infrastructure and outmoded irrigation systems. The establishment of Bolivia's Ministry of Water in 2006 was a major step in addressing these problems, and the new ministry has begun making its voice heard internationally. During the Fourth World Water Forum held in Mexico City in 2006, ministry delegates declared their basic position that water is a human right and not a need, contrary to how delegates had voted at the second forum.

The first positive steps toward cleaning the Pilcomayo River were finally taken in 2007. In April that year, the Bolivian government shut down 19 ore-processing plants in the department of Potosí that had been polluting the river. Contrary to the Green Cross–brokered agreement, the pollution had continued unabated until that time. Indeed, it had reached the point where "experts consider[ed] the [toxic waste] sediment production process to be beyond human control: 60 million cubic meters per year that led to abrupt changes in the flow, from 3,000 or more cubic meters per second to lows of just three cubic meters."[23] The same year Bolivia, Argentina, and Paraguay, with support from the European Union, began designing a cleanup plan for the river. It involved the construction of five dams, feasibility studies into irrigation, the rehabilitation of field stations to measure sediment and water levels, the recovery of farmland, and the recovery of wetlands in Argentina and Paraguay.[24]

Bolivia had begun taking new steps to manage water use even before the election of Evo Morales in 2005. In October 2004, the government promulgated Law 2878, which "establish[ed] the norms regulating the sustainable use of water resources in the activities of irrigation for the agricultural and livestock and forestal production."[25] Basically it takes into account traditional and indigenous uses, thereby securing them as law. Law 2878 is seen by Bolivia's lawmakers and water experts as the first step toward a national water law.

In December 2007, the Water Ministry organized the first national water meeting, at which was declared the motto: "Water, belonging to all and for all."[26] This slogan seemed to ratify the premise of the Cochabamba water war. As for Cochabamba itself, the city still had water distribution and quality problems in 2009, despite SEMAPA's efforts. As when Aguas del Tunari had taken over the water system, only half the city was connected to it. Nevertheless, by May 2009 things had begun to turn around with an infusion of $13 million from Japan, which had carried out a feasibility study and spent $420,000 to design a strategic development plan. That plan calls for the expansion of the water treatment facility, the installation of pumps, the creation of a main distribution network for the southeastern portion of Cochabamba, and improved sanitation services throughout the city. It is expected that by 2012 83 percent of Cochabambinos will receive potable water through SEMAPA, and that this figure will rise to 95 percent by 2027.

EGYPT, ETHIOPIA, AND SUDAN

No other river is as synonymous with a nation as the Nile is with Egypt. Yet the Nile does not begin in Egypt and, in fact, it flows northward through nine nations. In all, 10 nations make up the Nile River Basin.[27] The Nile has two major tributaries: the White Nile, whose source is generally considered to be the Kagera River in Burundi, which flows into Lake Victoria; and the Blue Nile, whose headwaters are Lake Tana in Ethiopia. The two meet at Khartoum, the capital of Sudan, where they form the Nile proper. The three northernmost countries of the basin have the major claims on the Nile's water, with Egypt historically predominant. The Nile is responsible for the rise of Egyptian civilization, and Egyptians have been drawing its water for irrigation farming for nearly five millennia.

In the 20th century, expanding populations and new technologies to extract water brought changes to the Nile Basin and to the Nile itself, especially in Egypt and Sudan, where various dams, barrages, and means of diversion were constructed. In 1959, the two countries signed an agreement that paved the way for the construction of the Aswan High Dam. Because of that dam and other diversions, the Nile's flow to the Mediterranean Sea has decreased, and the coastline itself is receding. Furthermore, the historic flooding of the Nile, which carried fertile soil from the Ethiopian highlands, no longer occurs, and indeed was one of the reasons for the Aswan High Dam's construction, and soil salinity has increased.

Water Use in the Nile Basin

Egypt has remained the most prominent user of Nile water down through the modern era. Prior to the 19th century, the upriver areas (at that time still tribal and under European colonial rule) drew little water from the White Nile—and even today the equatorial nations of the Nile Basin draw a much smaller percentage of its water than do Egypt or Sudan. During the region's colonial period, Great Britain recognized and protected Egypt's historical water rights through various treaties and agreements, to the detriment of the other countries. In the early 20th century, Italy, which had acquired a colonial empire in East Africa, along with France, another colonial power in the region, sought no such protection for Ethiopia, thus no precedent was set and tensions over water have increased in the late 20th and early 21st centuries.

By the middle of the 20th century, most of the Nile Basin states had come into existence as independent nations (though Eritrea did not gain final independence from Ethiopia until 1993), with implications for water use that, as their populations expanded, threatened regional stability. Nevertheless, Egypt remained the primary Nile water user, though it ceded more water to

Sudan in 1959. Ethiopia, where approximately 80 percent of the Nile's water originates, was using only about 1 percent as of the late 1990s. As for the six equatorial states, their combined use was negligible as far as the downriver nations were concerned because of the huge amounts of annual rainfall they received. Nevertheless, by the end of the 20th century, population growth and development required more water even in these countries. In 1990, the region's population was approximately 251 million people, and by July 2008 estimates placed it at nearly 404 million.[28] Furthermore, some parts of the region experienced reductions in annual rainfall in the early 21st century.

Some of the upstream nations began to assert their rights to Nile water with the signing of the Kagera Basin Agreement in 1977. The agreement went into effect in 1978, and original signatory nations were Burundi, Rwanda, and Tanzania, with Uganda joining in 1981. Its purpose was to develop the Kagera Basin for "hydropower, agriculture, trade, tourism, and fisheries,"[29] but its most important effect was its independent stance from Egyptian domination. The Kagera Basin Agreement effectually abrogated a 1929 treaty whose signatories were Egypt and Great Britain (on behalf of Sudan, Kenya, Tanganyika, and Uganda) that required Egyptian approval of development projects on the White Nile and the equatorial lakes. The 1977 agreement also nullified a 1934 treaty signed by Great Britain, on Tanganyika's behalf, and Belgium, for Burundi and Rwanda, in which the client states agreed not to dam the Kagera River. (The effect of the two earlier treaties was to protect Egypt's water rights, while the last also benefited Sudan.) However, 30 years after the signing of the Kagera Basin Agreement, many of its initiatives had still not been acted on and the four countries were still negotiating a framework that would allow for joint action and management of the basin.[30]

Taking a cue from the upriver states, Ethiopian president Mengistu Haile Mariam (r. 1974–91) openly challenged Egypt's historical water rights. On February 16, 1978, he declared, "If [Egyptian president Anwar] Sadat wants to protect the Nile Basin because water is life to his people, he must know that the Nile has one of its sources in the Ethiopia he wants to destroy. It is from here that comes the dark blue alluvial soil so dear to the Egyptian fellahin. Furthermore, Ethiopia has a head of state who cares for the lives of his people." It was three months before President Sadat (r. 1970–81) responded to Mengistu, and when he did he reiterated Egypt's position in no uncertain terms: "We depend upon the Nile 100 percent in our life, so if anyone, at any moment thinks to deprive us of our life we shall never hesitate to go to war because it is a matter of life or death."[31] While this proclamation was directed at Ethiopia, it also served as a general warning to all of the upriver nations. Meanwhile, tensions between Egypt and Ethiopia remained high and the regional war of words threatened to spill over into the Middle East when

Egypt reiterated its threat to Ethiopia and to Israel, the latter hoping to aid the former through its water development projects. The hostility between Egypt and Ethiopia only began to dissipate after Sadat was assassinated in 1981 and his vice president, Hosni Mubarak (r. 1981–), had taken over as president. Under Mubarak, Egypt offered Ethiopia 2 billion cubic meters (approximately 528 billion gallons) of water to help alleviate the effects of its devastating drought during the 1980s and promised to divide future stored water added through development with Ethiopia and Sudan.

As the major economic and military power in the Nile Basin, Egypt was playing its power card against its weaker—though according to 2008 estimates slightly more populous—regional partner to the south, while relying on colonial treaties to legitimate its claim to Nile water. While the Blue Nile, about half of which is in Ethiopia, contributes the vast majority of water to the Nile proper, there is no tributary to the river north of Atbara, Sudan (itself north of Khartoum). In effect, when it comes to Nile water, Egypt, a desert but for the river, is solely a receiver nation. For its part, Ethiopia historically had high annual rainfall, but during the "1970s and 1980s there were climate changes in Ethiopia due to drought. . . . Following the change plentiful rain falls in the southern and central areas of the country, while the northern area receives less."[32] This marked a reversal of Ethiopia's rainfall pattern and is what has led to tensions between Ethiopia and Egypt. Lake Tana, the source of the Blue Nile, is in northern Ethiopia, and if that lake is receiving less rainfall, the river will receive less, and so will Egypt, downriver.

Despite climate change and the ensuing drought and famine, Ethiopia's population has grown the fastest of any of the Nile Basin countries and is now estimated to be larger than that of Egypt. This alone would make the need for water development urgent on the part of the nation's leaders. One study, completed by the mid-1980s, determined 71 sites for dams in Ethiopia, 19 of which were "suitable also for electricity production at a rate of about 8,700 megawatts." The study also identified three reservoir sites with "carrying capacities of 85 billion cubic meters (22.44 trillion gallons)."[33] However, internal troubles, more than Egyptian threats, have been responsible for Ethiopia's failure to carry out any of these water development plans. These troubles included a military coup that deposed Emperor Haile Selassie; the Eritrean War of Independence, sparked by the emperor's disbanding of the federation and Ethiopia's annexation of Eritrea; drought and famine; and civil war, which in 1991 finally brought down the ruling military junta. Essentially, Ethiopia lost decades when it came to developing its water resources or negotiating with Egypt (and to a lesser extent, Sudan) in relation to water development.

Like its neighbor to the southeast, Sudan has been, and continues to be, wracked by civil war. Nevertheless, dating back to the colonial period, Sudan has taken steps toward water development within its territory and has come to various agreements with Egypt on Nile water use and storage. By the early 21st century, there were four major dams operating in Sudan: the Sinnar (constructed in 1926) and the Roseires (1950) on the Blue Nile, the Jabal Awlia (1937) on the White Nile just below Khartoum, and the Khashm Al-Gerba (1964) on the Atbara River, west of the Eritrean border. As might be expected, the Blue Nile dams are extremely important to Sudan's economy and development. Irrigation water captured from the Sinnar Dam is responsible for "up to 60 percent of the country's agricultural production" while the hydroelectric power plants at the Roseires Dam supply "nearly half of Sudan's power output."[34] Presently under construction is the Merowe High Dam (named for the nearby city of Merowe), which is also known as the Merowe Multipurpose Hydro Project or the Hamdab Dam. It is located on the Nile about 210 miles north of Khartoum. When completed, Merowe is expected to be the largest hydropower project on the continent.

The dam is not without its critics, who note that "the project will destroy archeological sites both directly through engineering and construction works, and indirectly through environmental changes."[35] The dam also calls for the relocation of people. The original plan was to move the thousands of people who will be affected into the Nubian Desert, which the Sudanese government, after much protest by the Hamdab of that region, largely decided against (some of the displaced may yet wind up in the desert). However, the settlement areas to which people were moved in the early 2000s have no groundwater supply, and many returned to their original homes. Perhaps seeing no alternative to resettlement, the people themselves supplied three different locations as options: the bank of the Nile in north Sudan, an agricultural area in central Sudan, and on the edge of the reservoir that the Merowe dam will create.[36] Other critics point out health risks, noting that a high incidence of disease is expected to occur as a result of the project. Finally, the environmental impact of the project has not been adequately studied, nor has the rate of evaporation in the proposed reservoir been analyzed.

The Merowe Dam is not Sudan's only pending controversial water development project. In the south, the White Nile flows into the Sudd Swamps, which lose a vast amount of water to evaporation. The Jonglei Diversion Project, designed to divert some of that water before it evaporates, has been an on-again, off-again project since the late 1970s, with work being interrupted by civil war. Discussion for the project dates back to the 1920s and reflects how far Egypt's influence in the Nile Basin extends. Originally, Egypt favored the project because as the downriver state it would enjoy more discharge. Sudan opposed it because the project not only offered little toward Sudanese development but also threatened the hundreds of thousands of Nilotics who lived in the Sudd Swamp region and who relied on the swamp for fishing and cattle grazing.[37] In 1954, Sudan and Egypt reached a compromise on the Jonglei Canal, but it took another 20 years to even approve the plan. Under the revised plan, the canal would extend eastward 225 miles and average 122 feet in width and 16 to 26 feet in depth. Its multiple purpose would be to reduce flooding in the Sudd area, irrigate land along the canal's path, increase shipping for 188 miles of the canal's length, and increase the discharge of the White Nile by 4.7 to 5 billion cubic meters annually.[38] While about two-thirds of the canal was dug by the early 1980s, civil war halted the process and destroyed much of the work that had already been undertaken. In fact, the canal continued to be opposed by people in southern Sudan for developmental, cultural, and environmental reasons and was possibly among the causes for their taking up arms against the central government.

But not everything has gone poorly for the people of Sudan. Another of the conditions of the Jonglei Canal agreement was that Sudan and Egypt would equally divide the excess discharge. In fact, this was per a historic 1959 water agreement between the two countries.[39] Because of that agreement Egypt was able to push ahead with its project to replace the dam at Aswan with an even higher one and create a massive reservoir for water storage in the desert. The dam became the crowning public works project of the regime of Egyptian president Gamal Abdel Nasser (r. 1954–70), but the treaty that made it possible also benefited Sudan, whose annual allotment of Nile water increased from 4 billion cubic meters (slightly more than a trillion gallons) to 18.5 billion cubic meters (4.88 trillion gallons). Meanwhile, Egypt's share increased from 48 billion cubic meters (12.67 trillion gallons) to 55.5 billion cubic meters (14.65 trillion gallons).

The Aswan High Dam

The Egyptian government first began discussing construction of a new dam at Aswan in 1952, soon after the military coup that ousted King Faruq. The motivation behind the Aswan High Dam was to create a reservoir large enough to make Egypt independent of the vagaries of the Nile's flooding cycle and of the upriver nations, which Egypt had historically feared would reduce its water supply. In this, perhaps, Egypt's leaders anticipated the independence of the upriver countries (some, like Ethiopia, had already achieved it) from European interventionists (notably Great Britain), making Egyptian dominance of the Nile Basin more difficult. But the plan to replace the existing dam, which had been constructed in 1902, languished for seven years while politics—the creation of the short-lived United Arab Republic—and crises—notably the nationalization of the Suez Canal in 1956 and the ensuing British-French-Israeli military intervention—took precedence.

Two years after the Suez crisis, during which Great Britain and the United States backed out of a deal to finance the dam's construction, the Soviet Union came to Nasser's rescue with an offer to finance the dam. The USSR also provided engineers and construction machinery. Construction began in 1960, and the first stage was completed in 1964. That same year the reservoir, known as Lake Nasser, began filling. (Sudan's portion of the reservoir, 17 percent of the total, is known as Lake Nubia.) The project was completed on July 21, 1970, and Lake Nasser first achieved its highest capacity in 1976. Since then the dam has become a debating point among politicians and hydrologists. Like its counterparts in the American West (especially along the Colorado River), the Aswan High Dam has provided drinking water to the population, year-round irrigation water to farmers, and hydroelectric power to cities, but the cost has been enormous and is yet to be fully understood.

PROS AND CONS OF THE ASWAN HIGH DAM

When the Aswan High Dam was built the world was in the midst of an epoch of dam construction. Toward the end of the 20th century, however, hydrologists, environmentalists, climatologists, sociologists, and others began to look more closely at the efficacy of dams, and especially high dams. As a result, some have been decommissioned and others have not been funded. Nevertheless, dams continue to be built around the world, though at a slower pace than 40 or 50 years ago.[40]

In the decades after the Aswan High Dam was completed, Egypt did indeed seem to be self-reliant when it came to water. During the drought of the 1980s, Lake Nasser provided enough water for Egyptian farmers, while allowing Egypt to agree to the increase in Ethiopia's annual Nile diversion. Nevertheless, if the drought had lasted any longer—heavy rainfall in northern Sudan in 1988 rescued the area—disaster would have also struck Egypt. For one thing, Lake Nasser had reached such a low point just prior to the monsoons that hydroelectric power generation would temporarily have ceased had the drought lasted much longer. "In 1978 the volume of water in the basin [Lake Nasser] was 110 billion cubic meters (29 trillion gallons), the volume required to satisfy Egyptian water needs . . . In 1984 the discharge reached an unprecedented low, 35 billion cubic meters (9.24 trillion gallons)." In other words, much less water was coming from the White Nile and especially the Blue Nile Rivers. In 1970, the flow into Lake Nasser was 77.2 billion cubic meters (20.38 trillion gallons), in 1974 the rate was 69 billion cubic meters

(18.22 trillion gallons), and in 1978 it was 62.1 billion cubic (16.39 trillion gallons). Thus, even in the 1970s the rainfall was sporadically less and the discharge into the reservoir was lower as the decade progressed.

In the 1980s, the water situation in the reservoir got worse, before it improved. In 1984, when discharge into Lake Nasser reached a low of 35 billion cubic meters (9.24 trillion gallons), the volume of water in the reservoir stood at approximately 72.9 billion cubic meters (19.25 trillion gallons). Four years later, the volume was down to 42 billion cubic meters (11.09 trillion gallons). In 1989, however, after the drought broke, Lake Nasser's volume increased to 75.3 cubic meters (19.88 trillion gallons), and by 1991 it was at 88 cubic meters (232.3 trillion gallons).[41] Egypt teetered on the brink of disaster in the 1980s (while Ethiopia and Eritrea fell over the brink). It is uncertain whether the Aswan High Dam could rescue Egypt in the face of another major drought. After the 1980s drought, Egypt did take practical steps to insure there would be a supply of drinking water for its citizens: "The first was not to rely on the dam alone for electricity production. . . . The second . . . was to match the amount of discharge from the lake with the amount of discharge into the lake."[42]

But even as the Aswan High Dam and Lake Nasser saved Egypt from famine in the 1980s, it was creating other, long-term problems. One dilemma is that approximately 10 billion cubic meters (2.64 trillion gallons) of water per year is lost to evaporation from Lake Nasser, which is in the desert. A second problem concerns the annual flooding of the Nile River. The floods sustained Egyptian civilization for more than 5,000 years and had always been highly predictable in terms of when they would occur and their duration. The trouble had been that some years the floods would bring too much water and other years not enough, and under these scenarios crops were destroyed. By controlling the floods and thereby releasing the water year-round, Egyptian water authorities were able to increase agricultural production. But in addition to water, the Nile floods deposited nutrient-rich silt on farmers' fields. With no flooding, farmers have become more reliant on fertilizers (and therefore have to cope with the problem of nitrates in the runoff) because the fields have become less productive—a good number of acres of farmland have actually been retired because the soil is becoming more saline. Another problem is that the silt and other buildup gathers behind the dam and is gradually reducing the volume of Lake Nasser. This will, in time, negatively affect hydropower production either from lower water levels or from too much silt in the intake pipes.

The buildup of soil behind the dam has another serious consequence downriver. As much as 88 percent of the soil moved by the Nile during its discharge, primarily from the Ethiopian highlands, was historically not deposited in the fields during the floods, but downriver at the delta. But, like the fields, the Nile delta now receives only a small amount of this soil, and as a result is being reclaimed by the Mediterranean Sea. The erosion of the delta is particularly problematic from an environmental standpoint, as it affects the region's marine and other wildlife and also its fishing industry.

As noted, the displacement of the Nubians of northern Sudan, whose territory lay in the path of the Aswan High Dam's reservoir, was one of the sticking points for the dam's construction until the 1959 agreement between Egypt and Sudan. While the agreement worked out to the benefits of both countries, the Nubians were nevertheless displaced, exacting a social toll that has never been fully analyzed. But the Aswan project also required the removal of cultural structures and artifacts and the submergence beneath Lake Nasser of many others. United Nations Educational, Scientific and Cultural Organization (UNESCO) archaeologists saved the ancient temples of Abu Simbel, which date from the reign of the 19th-dynasty pharaoh, Ramses II (r. 1279–1213 B.C.E.), by relocating them to higher ground on the western bank of the reservoir. Other temples were reconstructed in museums around the world, including in Cairo and Aswan, while the Addar and Amda temples, as well as the fortress of Kasr Ibrim on the eastern side of the river, facing Abu Simbel, were lost to the reservoir's rising waters.

Other problems caused by the Aswan High Dam, which are associated with particularly high dams and their enormous reservoirs more generally, include a higher incidence of waterborne diseases and the need to secure against natural and human-caused disasters. Finally, the maintenance of the complex must be taken into account. Nevertheless, it is reckoned that the Aswan High Dam will last for another 500 years, though it is expected that long before then the effects of soil salination will have changed Egypt's agricultural practices or even forced the government to decommission the dam.

The Toshka and Sinai Projects

In 1997, the Egyptian government came up with a new plan to feed its ever-increasing population: transform the desert of western Egypt into an agricultural paradise—a "new valley." The Toshka, or New Valley, Project had its origin in the late 1970s when Lake Nasser was at its highest.[43] Since too much water threatened the structural integrity of the dam, Egypt built the Sadat Canal leading from the reservoir to Wadi Toshka in the western desert to drain off the excess water. By the end of the 20th century, the Toshka Lakes had formed. The Toshka plan is to divert even more Nile water, from Lake Nasser, using a system of canals to irrigate nearly 600,000 acres of desert land. In March 2005, the Mubarak Pumping Station went online to pump reservoir water into the Sheikh Zayed Canal. As of March 2009, it was pumping approximately 14.5 million cubic meters (3.83 trillion gallons) of water per day from Lake Nasser, though its 24 vertical speed pumps have the capacity to discharge 1.2 million cubic meters (316.8 billion gallons) of water per hour.[44] The project is expected to be completed by 2020, by which time 10 percent of the Nile's waters will be redirected to the western desert.

Periodically during the first half of the 20th century, Egypt built barrages, or diversion dams, along the Nile north of Aswan. These were the Assyut and Ziftas Barrages erected in 1902, the Isna and Nag Hammadi Barrages built in 1909, the Mohammed Ali Barrage built in 1951 (on the remains of the failed 19th-century Delta Barrage), and the Adfina Barrage, also built in 1951. Most of these also underwent improvements from time to time. Of these six, the Adfina, Mohanned Ali, and Zifta Barrages were in the delta, while the others were located on the river. The water diverted by the three delta barrages, however, had limited range as far as the Sinai Desert was concerned.

In 1994, the Egyptian government began the Northern Sinai Agricultural Development Project (NSADP) to develop approximately 415,000 acres for farming along the Mediterranean Coast of the Sinai Desert in northeast Egypt,[45] where it borders Israel and Palestinian Gaza, and increase the region's population tenfold, from approximately 50,000 people to 500,000. In order to achieve this goal, a lot of water was needed. There was a precedent, at least theoretically, for such a canal project. Back in 1978, long before the NSADP became a reality, President Sadat in a historic peace overture pledged to construct a canal across the Sinai Desert to the Negev Desert in southern Israel. At that time, the Sinai was still under Israeli occupation, but within two years Israel began to gradually withdraw its troops from the region. Sadat later extended his offer to send Nile water all the way to Jerusalem—a total of 1 percent of Egypt's allotment of Nile water would have gone to Israel. Since Egypt's share as per the 1959 agreement with Sudan amounts to 55.5 billion cubic meters (14.65 trillion gallons) of water per year, Israel's allotment would be 555 million cubic meters (146.5 billion gallons). The canal by which this water was to travel across the Sinai Desert was named the Al Salam Canal (Peace Canal).[46]

The water never reached Israel because of political reasons on both sides. First, then Israeli prime minister Menachim Begin politely turned down the offer because the quid pro quo would have required Israel ceding East Jerusalem to the Palestinians. Israeli nationalists were also wary of having Egypt's "hand on the tap." Meanwhile, Sadat had also hoped the overture would render a solution to the West Bank problem in the Palestinians' favor. For their part, other Arab nations opposed the deal with Israel, while Nile Basin nations, notably Ethiopia, opposed sending Nile water out of the basin, despite Israel's offer to assist Ethiopia in its water development plans.

Rumors of a military coup in Egypt, with the water deal presumably one of the reasons for the discontent, was the straw that broke the camel's back, and Sadat ended the water-for-Israel proposal. Nevertheless, construction of the Al Salam Canal was begun prior to his assassination in 1981.[47]

With Israel withdrawing from the Sinai, the Al Salam Canal would now make part of the desert bloom by irrigating nearly 650,000 acres. The Al Salam Canal Project consists of two phases: Phase One is the canal itself, which extends 162.5 miles and irrigates 228,000 acres of desert. The project's second phase is much more complicated: The construction of an underground pipeline has been completed—beneath the Suez Canal—to channel Nile water into the Sinai. This will water 415,000 acres. There are also branch canals to convey water.[48]

Water Crises

For the downriver nations—Egypt, Sudan, Ethiopia, and Eritrea—water scarcity, or the threat of it, is an ongoing concern. Eritrea is a special case, in that it has no riparian rights to the Nile River and lies only partially within the basin. Its case for water vis-à-vis the other basin states is a weak one, yet it suffered through the same drought as Ethiopia, with far less surface water to fall back on. In any case, the majority of the Nile Basin countries are contending with enormous water challenges that are projected to worsen by the middle of this century. Egypt and Eritrea face water scarcity, defined as having an annual internal renewable rate of less than 1,000 cubic meters (264,000 gallons) per person.[49] The equatorial nations of Rwanda and Burundi also fall into this category (as do their nonbasin neighbors, Djibouti, Somalia, and Libya), while Kenya has already crossed over into water scarcity. By late 2009, a prolonged drought had affected agriculture and cattle (as well as the nation's elephant herds) to the point of near social breakdown as famine threatened the country. Sudan and Uganda fall into the water-stressed category, with an internal renewable rate of between 1,000 and 1,699 cubic meters (264,000 to 448,536 gallons) per person. Classified as having insufficient water—between 1,700 and 2,900 cubic meters (448,800 to 765,600) of internal renewable water—are Ethiopia and Tanzania. Thus, of the Nile Basin countries, only the Democratic Republic of Congo enjoys a surfeit of renewable water—its per capita amount is higher than that of the United States. Based on their 2000 renewal rates, it is projected that the nine Nile Basin countries and their neighbors will be experiencing chronic water shortages by 2050.

While quantity is always the most pressing water problem in the basin, quality is also a concern. Ethiopia's water development projects have come in the form of micro-dams, but these "have led to a sevenfold increase in the incidence of malaria."[50] Industrial pollution is also a concern in the Nile

Basin, as four of the upriver countries—Tanzania, Burundi, Kenya, and Ethiopia—had daily average emissions (as of 2000) of at least 200 grams of industrial organic water pollutants per worker.[51] By the end of the 20th century, phosphorous concentrations from agricultural runoff were high in four of the basin countries, with Sudan leading the way with between 1 and 1.75 milligrams of phosphorous per liter of water. (By comparison, the United States had at the most .09 milligram per liter.) Egypt, Kenya, and Burundi were in the next highest category: 0.5 to 0.99 milligrams per liter of water, while the rest of the basin countries, with the exception of Eritrea for which there was no data available, had between 0.1 and 0.49 milligrams per liter.[52]

Counterstrategies

By 1992, it was evident that more intrabasin cooperation was needed to handle the growing demand for the Nile's water. That year the Council of Ministers of Water Affairs of the Nile riparian countries (known as NILE-COM) embarked on an initiative to promote development and cooperation. Originally Egypt, Sudan, Uganda, Tanzania, Rwanda, and the Democratic Republic of the Congo formed the Technical Cooperation Committee for the Promotion of the Development and Environmental Protection of the Nile Basin (TECCONILE). The other four basin countries were observers to the proceedings.[53] (Eritrea, being a nonriparian basin country, has remained an observer and nonvoting participant throughout the evolution of the regional organization and the ongoing meetings.) In 1995, the Nile River Basin Action Plan (NRBAP) was developed and endorsed by NILE-COM, which also requested coordinating assistance from the World Bank. It finally accepted the offer in June 1997, but stipulated that the United Nations Development Program (UNDP) and the Canadian International Development Agency (CIDA) also come in as external partners and that the NRBAP undergo a review, which it did in November 1997 in the United States. Further reviews, discussions, and plans led to the establishment of the Nile Technical Advisory Committee (NILE-TAC), which in turn led to the Nile Basin Initiative (NBI). The NBI was formally established on February 22, 1999, in Dar es Salaam, Tanzania. At that same meeting policy guidelines were adopted and NBI and NILE-COM instructed the delegates "to prepare a portfolio of priority SVP [Shared Vision Projects]."[54]

THE NILE BASIN INITIATIVE

In the years since its establishment the NBI has achieved a number of objectives, but its greatest achievement has been establishing the groundwork for dialogue and negotiation for dealing with the region's water problems. If projections about the basin's looming water crises turn out to be true, then the precedents set by the NBI in its first 10 years may effectively guide NILE-COM and the various governments in a spirit of cooperation.

In May 1999, the first SVP planning meeting was held in Sodere, Ethiopia, and the following month the NBI Secretariat (NILE-SEC) was established in Entebbe, Uganda. The formal structure remains in place: NILE-COM is the policy guidance and decision-making body; the Technical Advisory Committee "renders technical advice and assistance to NILE-COM"; and the NBI Secretariat is the administrative branch of the organization and the executive agency for the SVPs.[55] In addition, NILE-SEC directly supports and coordinates the Eastern Nile Subsidiary Action Program (ENSAP), based in Addis Ababa, Ethiopia, and the Nile Equatorial Lakes Subsidiary Action Program (NELSAP), based in Kigali, Rwanda. The chair of NILE-COM is on an annual rotational basis, while the executive directorship of NILE-SEC is on a two-year rotational basis.

Shared Vision Projects

As of 2009, NILE-SEC is administering eight SVPs. These include:

- Applied Training Project, whose goal is "to ensure that water professionals manage water in a sustainable and integrated manner."[56]
- Confidence-Building and Stakeholder Involvement Project, which is comprised of four components: setting up management and coordination structures at regional, subregional, and national levels for the implementation and facilitation of the project; public information; stakeholder involvement; and confidence building among the basin countries and in the entire NBI.[57]
- Regional Power Trade Project, which has two important goals. The first is to establish "institutional means to coordinate the development of regional power markets among the Nile Basin countries" by providing technical assistance where needed. The second is to reduce the region's poverty by providing low-cost power.[58]
- Shared Vision Coordination Project whose goals are "to strengthen the capacity of the NBI to execute basin-wide programs [and] ensure effective oversight and coordination of the NBI's Shared Vision Program."[59]
- Socioeconomic Development and Benefits Sharing Project utilizes "a network of professionals from economic planning and research institutions, technical experts from both the public and private sectors, academics, sociologists, and representatives from civic groups and NGOs from across the basin to explore alternative Nile development scenarios and benefit-sharing schemes."[60]

- Trans-boundary Environmental Action Project whose multiple goals include increasing regional cooperation on environmental issues and water management, basin-wide community action and basin-wide networks. In addition the project also seeks to "create a greater appreciation of river hydrology . . . expand information, [the] knowledge base and know-how on land and water resources available to professionals and non governmental organizations (NGOs) in the riparian countries . . . [create a] greater awareness and increased capacity on trans-boundary water quality threats."[61]
- Efficient Water Use for Agriculture Project uses experts within each of the Nile Basin countries to determine the best water harvesting and irrigation practices for that particular area.[62]
- Water Resources Management Project whose goal "is to ensure that Nile Basin water resources are developed and managed in an equitable, optimal, integrated, and sustainable manner to support socio-economic development in the region," all planned through a basin-wide perspective.[63]

NBI Investment Programs

Under the auspices of the NBI, and with World Bank assistance, the governments of Egypt, Sudan, and Ethiopia have invested in the Eastern Nile Subsidiary Action Program (ENSAP), with the goal of developing the area's water resources through efficiency and equitable use and the elimination of poverty. This is to be achieved through cooperation by the eastern Nile countries, not only regarding their water resources but in various economic sectors. Among the projects ENSAP has fast-tracked are the initial phase of a flood preparedness and early warning program, irrigation and drainage for Egypt and Ethiopia, a watershed management program, and the Ethiopia-Sudan [power] Transmission Interconnection.[64]

The second NBI investment program is the Nile Equatorial Lakes Subsidiary Action Program (NELSAP), comprised of the six upriver countries—Burundi, the Democratic Republic of Congo, Kenya, Rwanda, Tanzania, Uganda—and two of the downriver countries, Sudan and Egypt. Essentially, these are the White Nile nations, and the goals of this investment program are basically the same as the ones for ENSAP, with the additional goal of reversing environmental degradation. In all, NELSAP oversees 12 projects, equally divided under the general groupings of Natural Resources Management and Hydropower Development and Power Trade.[65]

ISRAEL, JORDAN, AND THE PALESTINIAN TERRITORIES

It would be disingenuous to blame the political and social troubles of this region on the scarcity of water, but that scarcity is without doubt a root of the problems—not to mention one of the causes of some of the past conflicts. The present-day nations of Israel and Jordan came into being in the years just after World War II (1939–45), with the inhabitants of Palestine—the ancient name for the area—remaining in their towns and villages in both countries.[66] Present-day Palestine, is now known as the Palestinian territories. The political and social problems between the Israelis and the Palestinians have largely defined most other countries' foreign policy in that section of the Middle East in the 21st century.[67] Historically—that is, dating back to the post–World War II years—water scarcity played a role in hostilities between Israel and the Palestinians, but it also played a role in earlier clashes with Jordan and Syria, and to a lesser extent Egypt and Iraq.

The primary source of water for Israel, Jordan, and Palestine is the Jordan River, which runs southward for a mere 206 miles from its source in Mount Hermon, on the border of Lebanon and Syria, to the Dead Sea. Complicating matters is the fact that the Jordan is not an abundant river. Its average flow is less than that of the Euphrates, for example, and far less than that of the Nile. The Jordan River Basin is approximately 6,500 square miles, comprising parts of Syria and Lebanon, Israel, Jordan, and, of course, Palestine. Of these five, Lebanon is least dependent on the Jordan's waters followed by Syria, though Syria's water development schemes have had grave consequences for the downriver states. The basin includes numerous tributaries—notably the Yarmuk River system in Syria and Jordan—and several aquifers.

Water development was among the causes of the 1967 Six-Day War. And when the war was over, Israel, the victor, found itself in a more secure position vis-à-vis its neighbors with regard to the region's freshwater. While the war was initiated by an Israeli preemptive strike on Egyptian troops massed on its border, the hostilities can be traced to 1964 when Israel began diverting water from the Jordan River (actually Lake Kinneret) via its National Water Carrier—a vast system of pipes, canals, reservoirs, and pumps constructed by Mekorot, the Israeli national water company—to the Negev Desert in the southern part of the country. The National Water Carrier took 11 years to build, so water diversion was no surprise to the other nations of the Jordan Basin. Nevertheless, it added to a litany of grievances that Arabs and Jews maintained against one another and which, in fact, led to actual firefights during the 1950s and 1960s between Israel and Syria. A Radio Damascus editorial, broadcast on February 3, 1964, condemned the National Water Carrier project, ending with the dire proclamation, "we consider this challenge more serious with more far-reaching aims than the challenge of Israel's establishment."[68] For its part, Syria (with Lebanese assistance and funding from Egypt and Saudi Arabia) began planning its own water diversion canals, appropriating water from two tributaries of the Jordan River—the Hasbani and Wazzani springs.

Since Israel was a downriver state, the primary purpose of the All-Arab Diversion, as the project became known, was not to increase the two nations' water supply but to strike back at Israel and hamper its efforts to "make the desert bloom." However, the situation was more complicated than that because Jordan is both an upriver nation, with regard to the Yarmuk River, and a downriver nation on the Jordan River, and the diversions as they were originally proposed would cause problems with its water supply also, particularly as a good deal of its share of the Jordan water was planned to supply Amman, the capital. A second plan, proposed by Jordan, was agreed upon; it would leave the Kingdom of Jordan its allotment of Jordan River water while diverting as much as 53 percent of the available upper Jordan water.[69] Thus, about the time the National Water Carrier was set to begin operation Jordan was preparing to open the East Ghor Canal to divert Yarmuk River water for its own purposes.[70] The plan was never enacted because Israel destroyed the construction equipment at both diversion sites during air strikes in March, May, and August 1965; the military incursions—in actuality a water war—more or less served as a prelude to the Six-Day War. Ironically, another plan, formulated in the mid-1950s and known as the Johnston Plan after U.S. special envoy Eric Johnston, called for a more equitable sharing of Jordan water: Jordan would receive 45 percent, Israel 40 percent, and Syria and Lebanon would divvy up the remaining 15 percent. (At that time the Palestinian territories were not politically constituted as they are now, but rather were territories within Egypt and Jordan.) The plan was rejected for political reasons.

A number of aquifers lie beneath the Kingdom of Jordan. Some of these Jordan shares with Saudi Arabia and even Kuwait, but others are completely under Jordanian territory. Hydrologists have estimated that Jordan's groundwater totals approximately 530 million cubic meters (nearly 140 billion gallons), but of this amount some 230 million cubic meters (60.7 billion gallons), or nearly half, is fossil water, which is nonrenewable.[71] It is easy to see, then, how Israel's water diversions from Lake Kinneret (also known as the Sea of Galilee and, less familiarly, Lake Tiberias) would upset its downriver neighbor, Jordan, which also took part in the Six-Day War.[72] The war lasted from June 5 to June 10, 1967, and when it was over Israel had taken control of the

Gaza Strip and the Sinai Peninsula from Egypt, the West Bank (of the Jordan River) from Jordan, and the Golan Heights from Syria. The Sinai Peninsula was later returned to Egypt, and Israel has pulled out of the Gaza Strip, now controlled by the Hamas party. Water is certainly one important reason for Israel's retention of the West Bank, not only because it provides access to the Jordan River but because of the aquifer that lies beneath it. The water war roots of the Six-Day War were admitted by former Israeli general and prime minister Ariel Sharon, who once declared, "People regard June 5, 1967, as the day the Six-Day war began. That is the official date, but in reality it started two-and-one-half years earlier on the day Israel decided to act against the diversion of the Jordan."[73] In fact, Israel's territorial conquests as a result of the war gave it control of two-thirds of the headwaters of the Jordan River.

While the Six-Day War increased the territory and population under direct Israeli control and made Israel more water secure, it did nothing to stabilize the region. In 1973, the combatants were again fighting in what has become known as the Yom Kippur War—this time to a stalemate. By the end of the 1970s, Israel and Egypt had signed a historic peace treaty, but the real breakthrough as far as the Jordan River Basin water crisis is concerned came in 1994, when Prime Minister Abdul Salam Mujali of Jordan and Prime Minister Yitzhak Rabin of Israel signed a peace treaty. Integral to the treaty were the water rights of the two countries, which share most of the Jordan River as a boundary. At the time of the signing of the peace treaty, King Hussein echoed Anwar Sadat's words of more than a decade earlier when he declared that the only thing that would drive his country to declare war against Israel would be a lack of water—a realistic assessment, but also a warning.

As noted, Jordan and Israel both have large Palestinian populations within their territories, and indeed in Jordan Palestinians are a majority. In Israel, oppression, strife, and popular uprisings, intifada, have dominated relations between Palestinians and Israelis—including the Israeli military incursions into Hamas-controlled Gaza.[74] Nevertheless, as the on-again, off-again peace talks between Israeli and Palestinian leaders continue, one thing is certain—water allocation will be a major point of negotiation. This is mainly because the West Bank sits atop some major aquifers in the region. These are the Yarkon-Tanninim Aquifer, also known as the Western Aquifer, from which Israel draws approximately 340 million cubic meters (90 billion gallons) of water per year and the Palestinians draw approximately 20 million cubic meters (5.3 billion gallons) per year; the Nablus-Gilboa, or Northern, Aquifer, from which Israel extracts about 115 million cubic meters (30.4 billion gallons) per year while the Palestinians draw about 23 million cubic meters (6 billion gallons); and the Eastern Aquifer, from which Israel's annual draw is 40 million cubic meters (10.56

billion gallons) and Palestinians' is between 103 and 113 million cubic meters (27.2 billion gallons to 29.8 billion gallons).[75]

Most of Israel's portion of the Nablus-Gilboa Aquifer is used for irrigation, while the majority of its portion of the Yarkon-Tanninim Aquifer is used for urban (both municipal and domestic) supply; the water is principally sent to Jerusalem, Tel Aviv, and Jaffa. In fact, the last two cities sit atop the Coastal Aquifer, which extends only three to ten kilometers from the coast (1.86 to 6.2 miles) in the north, but as much as 20 kilometers (12.4 miles) in the south. However, the groundwater depth along the coast is a mere eight meters (slightly less than nine yards). Thus, there is the ever-present danger of the water table endangering the aquifer by dropping below sea level and becoming saline. There is a fourth, smaller groundwater source, the Gaza Aquifer, whose yield is about one-quarter that of the Coastal Aquifer, to which it is connected. As its name implies, it is the main source of water in Gaza.[76]

Because so much groundwater exists beneath the West Bank and because Israel withdraws so much surface water from the Jordan River, should the peace talks resolve the Israeli–Palestinian hostilities by rendering the area a two-state region, then water allocations treaties will have been signed and these will probably be along the lines of Israel's treaty with Jordan. Conversely, future water allocations could be the sticking point for a lasting treaty. The Palestinians are sure to want to increase their share of the region's water supply—especially if they gain recognized statehood—in order to spur agricultural and industrial development. Furthermore, the populations of Israel, Jordan, and Palestine, to say nothing of Syria and Lebanon, are increasing, which means less water per capita in an already water-critical area—and the region has been overdrawing its supply for years. By the end of the last century, "Israel, Syria, Jordan, the West Bank, and Gaza [were] using between 95 percent and more than 100 percent of their annual renewable freshwater supply." In dry years during the last two decades of the 20th century, "water consumption has routinely exceeded annual supply, the difference usually being made up through over-pumping of fragile groundwater systems."[77]

Counterstrategies for Water Scarcity

That water demand in the region has risen over the last 50 years is a fact, and that it will continue to rise is inevitable. It is estimated that by 2020, Israel, Jordan, and the Palestinian territories will have a combined water deficit of 1 to 2 billion cubic meters (between 264 and 528 billion gallons) per year.[78] Major water developments such as Israel's National Water Carrier, as well as numerous dams and canals throughout the region shift water around to "make the desert bloom" or to feed thirsty urban populations, but they cannot add water to the area, nor can they prevent water loss through evaporation.

Since the region, and especially Israel, is heavily agricultural, it is incumbent that steps be taken in this sector to conserve water (agriculture uses more water per capita than any other use). Furthermore, with little rainfall, 43 percent of Israel's farmland is irrigated, while more than 50 percent of the cropland in the Palestinian territory is irrigated. Jordan's percentage of irrigated cropland is 16 percent.[79] Both Israel and Jordan use improved irrigation methods including drip irrigation, developed by Israel. Drip irrigation involves feeding water "through hoses or pipes, under pressure [the water] drips through holes or nozzles. Loss through evaporation is just 5 percent, and because water can be directed to the roots of the plants it usually leads to increased yields."[80] Drip irrigation is augmented by computerized irrigation, which employs sensors to measure the amount of moisture in soil. In Jordan "drip irrigation combined with soil-moisture sensors increased cucumber and tomato crop yields 15 percent to 20 percent—an overall increase in water-use efficiency of 44 percent to 140 percent."[81] The bad news about drip irrigation is that, like desalination, it is expensive. For instance, while this method is used on more than half of Israel's irrigated land, in Lebanon to the north drip irrigation is used on less than a quarter of irrigated cropland.

Aside from drip irrigation, water reuse, or the use of "gray water" for agricultural purposes, is beginning to gain acceptance worldwide. Naturally, the gray water needs to be treated and the financial cost of treatment plants is prohibitive in the Palestinian territories. Desalination plants are another possibility for increasing the future water supply of Israel and the Palestinian territories—and Jordan, too, through water trading. By mid-2008, Israel was recycling 60 percent of its gray water, and desalination plants were "expected to meet 15 percent of the country's needs."[82] But according to the head scientist of Israel's Environment Protection Ministry, Dr. Yeshayahu Bar-Or, the desalinated water is not nearly enough to counter the growing water shortage, even if the current rate was increased to between 500 and 800 million cubic meters (132 billion to 211.2 billion gallons).[83] Additionally, the desalination process puts a high price tag on the water, which in turn makes purchasing it prohibitive for Jordan and the Palestinian territories. This fact was known in the late 1990s.[84] Nevertheless, the series of Israeli desalination plants along the Mediterranean coast make it possible for Israel to withdraw less water than it would otherwise have to from the aquifers and Lake Kinneret, making more water from these sources available to the Jordanians and the Palestinians. This, of course, is not to say that there is actually more water now in those sources than there was 20 or 30 years ago, just more available water in a dwindling supply.

Probably the simplest conservation measure is to refurbish the infrastructure. In Jordan, for example, "The Ministry of Water . . . has set a target of reducing leakage in the domestic water system from the current [2005] level of 30 percent to 18 percent by 2015. The move is backed by new studies showing that each dinar invested in efficiency improvements would yield 1.9 dinar in health, environmental, and water supply benefits."[85]

Two other possibilities for conserving water are closely linked and can apply to any area where freshwater is scarce. The first strategy is to grow food and other crops or raise animals for human consumption that themselves consume less water. It takes a minimum of 132 gallons and 238 gallons of water respectively to produce 2.2 pounds (a kilogram) of potatoes and wheat, while sorghum, soybeans, and rice require 290, 436, and 501 gallons of water, respectively, to produce 2.2 pounds of each. Other high water consumption crops are corn and cotton. Naturally, more water is involved in raising animals than crops because the equation includes the amount of water used in growing the grain to feed animals. Poultry requires a minimum of 924 gallons of water per 2.2 pounds, while it takes a whopping 3,960 gallons of water to produce a mere 2.2 pounds of beef. A second option for Israel, Jordan, and the Palestinian territories would be to import even more foodstuffs than they already do. This would free up the majority of water for industrial, domestic, and municipal purposes. As the World Water Forum and the World Bank strive to spread the gospel of worldwide commodification of freshwater, countries are forced to think in terms of cost benefit. In the case of Israel, Jordan, and Palestine, is the cost of water for agriculture and ranching less than the cost of importing food products? Again, this is an option that will probably be explored by many governments as freshwater becomes scarcer. However, there may also be social ramifications from diminished agriculture. Job loss, cultural disconnection, and increased government entitlements—or, barring them, increased urban poverty as more people move to the cities in search of work—are among the social problems that might result from diminished agriculture in the Middle East and elsewhere.

In Israel, two canal projects, both of which have been on and off the drawing board for decades, are the antithesis of the National Water Carrier. The first canal, to be cut from the Mediterranean Sea to the Dead Sea, was proposed as far back as the late 1940s; the second, which would carry water from the Red Sea to the Dead Sea, gained credence following the signing of the peace treaty between Israel and Jordan.

THE MED-DEAD CANAL

The best-known plan for digging a canal to run from the Mediterranean Sea east to the Dead Sea was proposed by a U.S. citizen, Walter Clay Lowdermilk, as part of a larger plan he developed in the late 1930s at the urging of the U.S. Department of Agriculture. That plan called for diverting the Jordan and Yarmuk Rivers for hydroelectric power; diverting water to the Negev Desert, which became a reality via the National Water Carrier; Palestinian and Jewish use of the Litani River, which is in Lebanon; as well as the Med-Dead Canal.[86] Though Lowdermilk, a soil conservationist by training with extensive field experience in the United States and China, was enthralled by the possibility of building a new society in the desert, he appeared to be oblivious to the political situation on the ground, though in his defense his plan may have been an attempt to regionalize rather than nationalize water.

Lowdermilk's plan was not the first attempt to tackle the region's water problems, nor was Lowdermilk the first to propose a Med-Dead canal. The Zionist leader Theodor Herzl made a similar proposal in 1902, and in 1855 British admiral William Allen proposed a more ambitious system of canals that would connect the Mediterranean Sea, the Dead Sea, and the Red Sea.[87] It was thought that this would have served as a less expensive alternative to the Suez Canal until it was discovered that the Dead Sea was more than 1,150 feet below sea level. Nevertheless, every few decades the plan has been revised—usually in conjunction with agricultural and/or emerging technology. The incarnation of the plan debated during the 1990s and early 2000s, on the heels of the Israel-Jordanian peace treaty, involved a combination of desalination and raising the level of the Dead Sea. In the new scheme, the Med-Dead Canal, renamed the Qatif (or Katif) Alignment, is the middle of three canal and tunnel projects. The others are the Northern Alignment, which connects the Mediterranean Sea with Lake Kinneret, and the most ambitious of the three, the Red Sea–Dead Sea Canal.[88]

The idea behind this ambitious project is that some of the water pumped from the Mediterranean Sea would be desalinated, with the majority going to Amman, Jordan. The rest would go to Israel and the Palestinian territories. A good amount of the water pumped from the Mediterranean would be sent to the Dead Sea to raise its level. Since the mid-19th century, when a Mediterranean Sea–Dead Sea–Red Sea canal system was first proposed, the level of the Dead Sea has dropped by some 66 feet—and it was already the lowest point on the earth's surface. The drop is caused partly by evaporation and partly by the fact that the sea receives less flow from the Jordan River because of diversions. Evaporation and lower flow have increased the salinity of the Dead Sea to approximately 33 percent, making it useless for irrigating even salt-resistant crops. By contrast, the Mediterranean Sea is about 3 percent saline. The lower water level and continuing evaporation constitute an ecological and health threat à la the Aral Sea and Owens Lake, two notable ecological disasters of the 20th century. Nevertheless, potable water and raising the level of the Dead Sea were ancillary to the canal's true purpose, which was "power generation, rather than water," and because it was "politically unilateral, bringing benefits only to Israel" (though raising the level of the Dead Sea would benefit Jordan as well) its detractors—Israeli, Jordanian, and Palestinian—were many.[89]

By the end of the first decade of the 21st century, the idea of the Med-Dead Canal had been debated nearly to death, but still not acted upon. In fact its sister project, the Red–Dead Canal, had gained more popularity.

THE RED–DEAD CANAL

As the Mediterranean Sea–Dead Sea Canal began to lose impetus, the Red Sea–Dead Sea Canal gained backers, notably the World Bank. Its purpose would essentially be the same as that of the Med-Dead Canal—provide as much as 850 million cubic meters (224.4 billion gallons) of potable water, primarily for Jordan and the Palestinian territories, hydroelectricity, and raise the level of the Dead Sea. While this route is longer than the Med-Dead Canal, it is estimated to be less expensive than the earlier project because of the tunnel that would have had to be cut through the Judean Hills.

The Red–Dead Canal is not without its detractors, both in Israel and abroad. Over the years, critics in Egypt have opposed it for a number of reasons including fears that it would lessen the importance of the Suez Canal, provide coolant water for Israel's nuclear reactor at Dimona, and contribute to Israeli growth in the Negev. A possible increase in seismic activity as a result of the construction is also a concern. Meanwhile, environmentalists and scientists who were dissatisfied by a feasibility study done in the 1990s voiced their concerns about the project. And as with nearly every other aspect of life in the region, politics invaded the discourse. This latter stumbling block was eased to a great extent when the Palestinians were given representation on the committee responsible not only for the Red–Dead Canal project, but for water development in the entire area. With all of the riparian stakeholders in place, the Palestinian Authority, Jordan, and Israel signed an agreement for a new feasibility study of the Red–Dead Canal on May 9, 2005. A second study has not dissuaded some of the canal's critics, who claim the Dead Sea will not evaporate into extinction and point out that less extraction from the Jordan River would again raise the Dead Sea's level. They also worry about the environmental hazards and threats to cultural sites. Among the environmental dangers are the possible destruction of the coral reefs in the Gulf of Aqaba, from which the Red Sea water will be pumped, damage to the landscape of Arabah (the section of the Jordan Rift that separates the Dead and Red Seas) and salination of its aquifer, and possible deleterious change to the Dead Sea itself by mixing Red Sea water into it.[90] It has also been noted that like Egypt's Aswan Dam project, an important archaeological site stands in the way of Red–Dead canal construction—namely the Wadi Finan, the world's earliest known copper mine.

The Red–Dead Canal's supporters are numerous, and with the World Bank backing the project, it seems closer to becoming a reality than ever before despite its cost, which may run as high as $10 billion. (On the low end, the cost has been estimated at $1 billion.) The canal's distance would be some 120 miles, with uphill pumping required for the first 750 feet (230 meters). The hydroelectricity produced by the project will be used for the pumping station, while the desalination plants will rely on existing sources of power. This, no doubt, is another sore point for the canal's domestic detractors, as there would actually be a net loss of electrical power among the present users.

Whether or not the project is finally realized, it has shown that where an important issue of regional survival is concerned the factions can set aside political and sectarian differences (and in some cases, animosity) and come together for the greater good. Indeed, in the Jordan River basin, regional cooperation may be the only chance of overcoming freshwater scarcity. Fortunately, the precedents for diplomatic solutions already exist, in the forms of Israel's treaty with Jordan and the Declaration of Principles (also known as the Oslo Accords) between Israel and the Palestinian Liberation Organization.[91] Although water was the final sticking point between Israel and Jordan, the two countries did eventually reach agreement, and one of the abiding principles of the declaration was cooperation between the signees in the area of water.

INDIA, PAKISTAN, AND BANGLADESH

The three most populous countries of the Asian subcontinent, India, Pakistan and Bangladesh, suffer from every major problem related to water: overpopulation, pollution, overdrawing, overdevelopment, inequality, social issues, climate change, and globalization. Furthermore, these problems have been ignored for decades. To make matters worse, laws have been manipulated to make it appear as though positive change was being affected and new technology has, in some cases, been used for short-term agricultural gains rather than, for example, long-term groundwater storage. It is estimated that by 2050 India, Pakistan, and Bangladesh will all be chronically short of water.[92]

Southern Asia is the home of two of the most revered rivers in the world—the Indus and the Ganges. Like the Nile, the Tigris and Euphrates, and the Yellow Rivers, they were the wellsprings of ancient civilizations and are the hopes of millions of people on the subcontinent today. In fact, within India there are more than 600 "holy water places," or *tirthas.*[93] The Ganges in particular, with its 108 names, is one of the most revered bodies of water on Earth; unfortunately, it is also one of the most polluted. In 1999, it registered a 100 percent violation of pollution standards as recorded by 19 monitoring stations along the river. The Godavari, Sabarmati, and Tapti Rivers in India also recorded 100 percent violations. While not nearly as polluted as the Ganges, the Indus River, which mostly flows through Pakistan, also has pollution issues, most having to do with water salinity caused by neighboring India.

When it comes to the South Asian subcontinent, Western news reports have historically focused on political and social issues, violence, and religious strife, yet freshwater has been and remains a flashpoint between India and the other two lower subcontinent countries—the small Himalayan nations of Nepal and Bhutan, which both have generous freshwater supplies and recharge, are by-and-large unaffected by the water problems of their neighbors. Conversely, water concerns have been the starting point of cooperation between the riparian states of the subcontinent.

In 1923, Nepal, which lies between Tibet and India, was the first of the five subcontinent nations to gain its independence from the United Kingdom, almost two-and-a-half decades before the United Kingdom recognized India's independence, in 1947. From then on the subcontinent countries followed different paths. Of particular importance was the division on August 14, 1947, of the subcontinent between primarily Hindu India and primarily Muslim Pakistan. At this time, the Pakistani state consisted of Muslim-majority regions in the east and northwest of what had formerly been referred to simply as "India." The eastern state, originally known as East Bengal and renamed East Pakistan in 1956, broke away from Pakistan altogether in 1971 and became the independent nation of Bangladesh.

Thus, the current political borders were in place by 1971, but the 1947 partition would have international ramifications when it came to freshwater. Two of the pre-partition Indian states, Punjab and Bengal, were divided between Pakistan and India (and as noted, Pakistani Bengal eventually became Bangladesh). This partitioning made for an anxious period as millions of Muslims, Hindus, and Sikhs uprooted themselves in order to cross the new borders to be with their coreligionists, and it led to a sort of ethnic cleansing on both sides: "A quarter of a million men, women, and children are thought to have been slaughtered along the way."[94] That tragedy, as well as the preexisting strife between Hindus and Muslims in the colonial era, resulted in ethnic tensions between India and Pakistan that have never dissipated, but the partition also created new tensions over water—India being the upriver country for both the Indus and Ganges.

The Indus River Basin

Tensions in the Indus Basin were especially high early on. The Indus River, considered Pakistan's lifeblood, is the country's largest waterway at approximately 2,000 miles and with an annual flow twice that of the Nile.[95] It arises in Tibet (thus its source is under Chinese control) at the confluence of the Sengge and Gar Rivers, travels northward crossing the border into India before reaching its most northern point in Pakistan, then turns southwestward toward the Arabian Sea south of the city of Karāchi, where it discharges at an average rate of 1.7 million gallons per second.[96] The Indus has 20 major tributaries, the most important of which are the Beas, Chenab, Jhelum, Ravi, and Sutlej Rivers, but some, such as the Beas and Sutlej, also flow in India. The 1947 partition placed Pakistan at a disadvantage regarding the Indus and those tributaries that also flowed through India—if the latter were to develop water projects the flow of the Indus would be negatively affected. Complicating the matter from the Indian point of view was that the state of Rajasthan, technically within the Indus River Basin, also includes most of the Thar Desert (Great Indian Desert).[97] India's goal of "making the desert bloom" has led to tensions with Pakistan, since the only way to do so would be to divert water from the Indus or its tributaries.

The first water conflict between India and Pakistan occurred on April 1, 1948, when "India shut off water supplies from Ferozepore [also spelled Ferozepur, it is on the Sutlej River] headworks to the Dipalpur Canal and to the Pakistani portions of the Lahore and Main branches of the UBDC [Upper Bari Doab Canal].... By India's action, about 5.5 percent of the sown area (and almost 8 percent of the culturable commanded area) in West Pakistan [present-day Pakistan] at that time found itself without water.... The city of Lahore was simultaneously deprived of the main source of municipal water, and, incidentally, distribution to West Pakistan of power from the Mandi Hydroelectric scheme was also cut off."[98] In essence, India was asserting "upstream riparian, proprietary rights" for which there was neither a precedent in the subcontinent nor a treaty to define and limit such rights.[99] Pakistan claimed the water by virtue of prior allocation, since it had been receiving the water for years and the payments it made to India were for infrastructure only. The crisis lasted little more than a month and was resolved with the signing of the Inter-Dominion Agreement on May 4, 1948, that allowed for annual payment in return for water, with the figure to be set by the Indian prime minister. Part of the payment would then be sent through the Indian federal government to India's Punjab State. It has been speculated that the water payments were compensation to the Sikhs (who dominated the Punjab) for having to abandon their homes and land in Pakistani Punjab. The Inter-Dominion Agreement remained the working instrument for the next 12 years, until the Indus Waters Treaty superseded it in 1960. However, during that time other water crises arose as Indian water development projects, such as the massive Rajasthani project with its numerous canals and barrages, were undertaken alongside Pakistani irrigation projects.

The Indus Waters Treaty

Long before most of the Indian and Pakistani public works were completed, the two countries were meeting for negotiations related to the waters of the Indus basin. While both countries agreed that the Inter-Dominion Agreement was a temporary solution, both were also intractable when it came to Indus water allocation. By the early 1950s, they could not even agree on how to best settle the problem: "Pakistan wanted to take the matter to the International Court of Justice but India refused, arguing that the conflict required a bilateral resolution."[100] In 1951, the former chairman of the Tennessee Valley Authority and the Atomic Energy Commission, David Lilienthal, visited the subcontinent at a time when Pakistan and India were on the verge of warfare. While the hot topic then as now was the disputed Kashmir region, Lilienthal quickly understood the importance of the Indus basin waters in both exacerbating hostility between the two countries and alleviating it. It was Lilienthal's suggestion that the World Bank become involved in a basinwide development program, which it agreed to do. More important, the two countries agreed to reopen negotiations, and these would be between India and Pakistan alone—the World Bank simply offered assistance and financial support.

It took almost nine years for India and Pakistan to reach an agreement, and even then the World Bank had to play a larger role in brokering the treaty than its president, Eugene Black, had originally planned. In effect, the World Bank became the third signatory to the agreement. What finally transpired was the historic Indus Waters Treaty of 1960, signed by President Mohammed Ayub Khan of Pakistan and Prime Minister Jawaharlal Nehru of India (and W. A. B. Iliff of the World Bank), in which Pakistan controlled the basin's western tributaries and India the eastern tributaries. The treaty includes payments to Pakistan to make up for water it would otherwise lose;[101] it also calls for India to guarantee a certain amount of water to Pakistan, and this precludes water development and over-diversion. There are two remarkable aspects of the Indus Waters Treaty: The first is that it was successfully negotiated by two belligerent countries for whom other political issues have lingered for decades, and the second is that it remains the single most important treaty on the subcontinent, one that has not been abrogated by either side.

Current Water Problems in Pakistan

The crux of Pakistan's water problems has to do with the overdrawing of its water supply for agriculture. Ninety-six percent of Pakistan's water use at the beginning of the 21st century was for agricultural purposes, while domestic and industrial uses each garnered 2 percent. Two things aggravate the problem, and the first, Pakistan's rainfall cycle, cannot be altered. Like many other countries of the world and its subcontinent neighbors India and Bangladesh, Pakistan is subject to a monsoon season. As much as 90 percent of the country's annual rainfall occurs in a matter of months. While this limits freshwater use, it does have its natural benefits such as groundwater renewal and riverine cleansing. The second is the reliance on irrigation and on crops such as sugarcane, which require a lot of water. This has caused problems for farmers, as well as for those living downriver on the Indus. The purpose of all the river damming and diversion was to enable Pakistan to take part in the Green Revolution—the increase in crop output whose original purpose was to alleviate hunger—and to that effect "80 percent of the county's cropland is now irrigated. But people downstream in the delta have paid the price. River flows reaching the Indus delta have declined by 90 percent over the last 60 years. Recent droughts have exacerbated the water shortage, leaving the delta and its dependents bereft of freshwater. With so little freshwater discharge to keep the Arabian Sea at bay, the sea has now inundated some 43 million acres of delta farmland."[102]

Drought is not the only adverse consequence of climate change affecting the Indus Basin. The melting of Himalayan glaciers will also produce changes. According to various studies, glacier melting will have a beneficial effect in the short term as the rivers flow stronger; after the melting has reached its critical point a water crisis of unimaginable scope will most likely occur, especially when coupled with Pakistan's current problem of groundwater depletion. The fact that three-quarters of Pakistan's renewable water supply originates outside of the country will make little difference in the event of glacier melting.

Since the vast majority of Pakistan's water goes to the agricultural sector, extreme water scarcity would affect both method and output, which in turn would adversely affect the country's economy. But water scarcity will take its toll throughout society, not just in the agricultural sector. The supply of hydropower was affected when India blocked the water supply in 1948. This could happen again, in violation of the Indus Waters Treaty, the only difference being that water rationing would be imposed by the government. At the turn of this century, hydropower in Pakistan accounted for 25 percent of the total power generated, a figure that is bound to increase if proposed dam construction is followed through.[103] Otherwise a decrease in Pakistan's hydropower percentage seems inevitable as per India's situation in making use of other sources of power. Furthermore, water scarcity in the Indus Basin may strain the Indus Waters Treaty.

Pakistan also faces a critical soil salination problem because of its outmoded irrigation methods—it has yet to implement drip irrigation on a large scale, probably due to the expense of such a changeover. Soil salination is a problem it shares with, among other nations, India, Egypt, and the United States. By the end of the 20th century, more than one-quarter of Pakistan's total irrigated land had been damaged by salt.[104] Finally, fertilizers have contributed the modern problem of high concentrates of phosphates and nitrates as runoff from the fields reenters the rivers.

The Narmada Dam Project

India's most pressing water problems fall into two broad categories: pollution of the country's freshwater supply and water development schemes that have often done more harm than good at the local level, yet have enriched developers and other insiders. As far as water development projects go, India is in the same place the United States was throughout most of the 20th century—dam and canal construction rule the day. During the final three decades of the last century, India built more than 1,500 large dams.[105] Though the world in general seems to be moving away from this paradigm, India continues to embrace it, as does China to the north. One reason for this is that because India was late in terms of dam-building, it still has good dam-construction sites. A second reason is that the World Bank and the International Monetary Fund (IMF) are willing to provide funding for such projects, though occasionally these institutions discontinue funding when confronted with the findings of second or third feasibility studies. However, damming the rivers (including the Ganges) has became a nationwide issue, exemplified by the ongoing struggle to decertify and halt construction of the Narmada Dam Project, which would see "30 large, 135 medium and 3,000 minor dams on the Narmada River and its tributaries."[106] The large dams will be used to generate hydroelectric power, and the largest, best-known, and most controversial of these is the Sardar Sarovar Dam.

The Narmada River, one of India's five holy rivers,[107] begins in eastern Madhya Pradesh and flows westward across the state and into the western Indian state of Gujarat, from which it eventually empties into the Arabian Sea. Its length is approximately 815 miles.[108] In addition to Madhya Pradesh and Gujarat, the Narmada Basin includes Maharashtra; however when it comes to water allocation, drought-ridden Rajasthan, a nonriparian state, receives twice as much as Maharashtra, as per a 1979 decision of the Narmada Waters Disputes Tribunal.[109]

The Narmada Dam Project is a massive hydroelectric and irrigation plan that began in 1979 (although the idea, first mentioned in the 1860s by British engineers, has been seriously discussed since the 1940s). It is so tied into the development of the state of Gujarat that the project has often been referred to as "Gujarat's lifeline." The Sardar Sarovar phase, when completed, will irrigate some 4,662,500 acres and generate 1,450 megawatts of electricity—57 percent of which will go to Madhya Pradesh. Maharashtra will receive 27 percent of the hydropower, and the remaining 16 percent will go to Gujarat, where the dam is located. A companion project and second-largest dam on the river, the Narmada Sagar Dam, is under construction in Madhya Pradesh. Water stored in the reservoir created by the Narmada Sagar will be released periodically and flow downriver to Sardar Sarovar.

Back in their planning stages, both dams, though Sardar Sarovar in particular, underwent important changes and received even more important waivers. Early on, in 1983, permission for construction of both was denied because the dams would not meet Ministry of Environment and Forests regulations. Evidently these regulations were waived later in the decade during a period of drought, in which it was easier to have water development schemes approved. That Prime Minister Jawaharlal Nehru had favored the project right up to his death in 1964 and that the World Bank was prepared to invest $450 million in it also boded well for its future. Nevertheless, on the heels of anti-dam protests in the early 1990s, the bank undertook an independent study of the project and decided to withdraw from investment. The financial slack was then taken up by the government of the state of Gujarat. Meanwhile the Indian government was not dissuaded by the World Bank's findings and, indeed, not only pushed on, but pushed to make the Sardar Sarovar dam higher. The dam was originally planned to be approximately 260 feet tall, but in 1999 the Indian Supreme Court handed down a decision that allowed for the height to be increased to 289 feet. The following year the Supreme Court again became involved, allowing for an increase to 295 feet. But this would not be the end of it. The Narmada Control Authority approved a new height for the dam in 2002, of 312 feet, and increased this to 361 feet in 2004. Then in 2006 the Narmada Control Authority approved another height change to 400 feet—more than 50 percent higher than the original figure. The actual proposed height of Sardar Sarovar is 453 feet. With each increase came an increase in the size of the reservoir behind it.

While the main benefits of the Sardar Sarovar, and indeed the entire Narmada Project, are flood control, irrigation, hydroelectricity, and increased drinking water supply, there are also problems that over the years have aroused activists to oppose the project. The most obvious is what occurs practically every time a large dam is constructed: People are displaced from their homes. Proponents of major dams and of the Sardar Sarovar take a utilitarian attitude toward the outcome—the greatest good for the greatest number of people. However, opponents counter that that so-called greatest good effected by dam construction is often short term and creates more problems than it solves. Regarding the Sardar Sarovar Dam, the main opposition is an organization called Narmada Bachao Andolan—Save the Narmada Movement.

NARMADA BACHAO ANDOLAN

Narmada Bachao Andolan (NBA) is probably the best-known anti-dam group in the world, having been fighting against the Sardar Sarovar Dam since 1986. Led by Medha Patkar (b. 1954) and Baba Amte (1914–2008), the NBA was instrumental in the early 1990s in the Indian Supreme Court's decision to temporarily halt construction on the dam and in bringing the dam's potential problems to the World Bank, which ultimately led to the bank's withdrawal from the project in 1993 on the heels of its own investigation, the Morse Report.[110] Among the report's conclusions were that "[t]he Bank and India both failed to carry out adequate assessments of human impacts of the Sardar Sarovar Projects. . . . [which] led to an inadequate understanding of the nature and scale of resettlement. . . . This inadequate understanding was compounded by a failure to consult the people potentially to be affected." The report further acknowledged that "relocation and resettlement of the people of the rock-filled dyke villages was implemented in a way that meant that the Bank's overarching principle of resettlement and rehabilitation, i.e., that no one should suffer a fall in standard of living, was not likely to be achieved."[111]

Despite that success, 1993 was a bittersweet year for the protestors. Not long after the World Bank announced its withdrawal from the project, the Gujarat village of Manibeli was submerged by a combination of the monsoon and the dam. NBA therefore applied more pressure, garnering international attention, and causing the Ministry of Water Resources to appoint what was known as the Five Member Group (FMG)—five experts whose task was "to determine how far local experts would concur with the conclusions of the Morse Report."[112] Release of the FMG report was challenged in the Indian Supreme Court but when eventually it did become public; it "substantially concurred with the Morse Report though in less vigorous terms."[113]

In the beginning NBA's opposition to Sardar Sarovar focused on the displacement and relocation of hundreds of thousands of people, but as the dam was granted waivers and height extensions, its environmental impact was increasingly questioned. Nevertheless, the NBA did not abandon the displaced. In fact, the topic came up again in 2006 when it became known that at least another 35,000 families would lose their homes as a result of the latest increase in the dam's height. As for the environment, the Morse Report essentially confirmed protestors' reservations: "There has been no comprehensive environmental assessment of the canal and water delivery system in the command area. Information we have gathered leads us to believe that there will be serious problems with waterlogging and salinity. We also found that many of the assumptions used in project design and for the development of mitigative measures are suspect."[114] Despite these and other reservations on the part of the World Bank, the government-appointed Five Member Group, Narmada Bachao Andolan, and others, the work on the Sardar Sarovar Dam and the Narmada project continues.

The Narmada project continues because for many, particularly in the state of Gujarat, the project's construction reflects India's need to solve its water and development problems using its untapped resources. Another interesting twist has NBA's detractors all but openly accusing the organization of practicing a kind of inverted globalization in which NBA and like-minded nongovernmental organizations from beyond India's borders take actions (in this case attempting to halt the Narmada project) that weaken India's as well as Gujarat's sovereignty.

The Ganges River Basin

The Ganges, or Ganga River, has its source at the foot of the Gangotri Glacier, from which arises the Bhagirathi River. It then flows southeasterly for about 1,557 miles through the Gangetic Plain and into Bangladesh, where approximately two-thirds of its vast delta forms Bangladesh's entire coastline. The rest of the Ganges delta is in India, below Calcutta. The river flows into the Bay of Bengal. Since the Ganges bisects Bangladesh and one of its main tributaries, the Brahmaputra River, also flows through Bangladesh, the country is entirely within the Ganges basin, and downriver from India, which makes Bangladesh even more vulnerable to India with respect to Ganges water than is Pakistan with respect to the Indus River.

The Ganges has numerous tributaries and two major dams. The Haridwar Dam on the Upper Ganges was built by the British in 1854. Its purpose was to divert Himalayan snowmelt for irrigation, but it has had the unintended consequence of reducing the flow of the Upper Ganges and making this part of the river of less use as a means of transportation. The second major dam is the Farakka Barrage in the state of West Bengal near the border with Bangladesh; it was completed in 1975. Ostensibly, its purpose is to divert freshwater via the canal to alleviate silting at Calcutta, located on the Bhagirathi River, part of the Ganges system. How much this has worked is debatable, but it did cause friction between India and Bangladesh as it decreased the flow of water entering the latter country and therefore affected agriculture and caused saltwater incursion in the Bangladeshi delta system. The problem has accentuated a long-standing complaint from Bangladesh—that the Ganges is an international river and should be jointly managed.[115] This contention, of course, would have wide ramifications throughout the basin.

Development in the Ganges Basin, while not on the level of either the Indus or the Narmada River Basins, is not lacking. Plans are under way to build 220 dams of varying heights in the upper basin area alone, including six large hydroelectric dams within the river's first 78 miles. Supporters of these various dam projects tout the benefits of electricity and water for irrigation, while detractors point to population displacement and lower water flow that will affect the health of a river that already seems to be dying.

POLLUTION OF THE GANGES

Politics aside, the present overall situation of the Ganges is a paradox. It is the most revered river in the world, yet it may also be the world's most polluted. In Hindu cosmology, the river is a manifestation of the goddess Ganga, descended to earth through the hair of the god Shiva. The goddess/river is a purifier, hence the Hindu purification festivals held on the banks of the Ganges.[116] This purification (or healing) is corporal as well as spiritual, and it is Hindu belief that mixing Ganges water with other water will make the other sacred. The sacredness of the Ganges to Hindus explains why so many have settled near it, and because there are so many people agriculture and industry have prospered along or near its banks. What it does not explain is the astonishing amount of pollution, which threatens the river's ecology as well as the health of those millions of people in India and Bangladesh. The river has become an open sewer as a result of a lack of industrial, agricultural, and municipal responsibility. And to all of this the Indian municipal, state, and federal governments turn a blind eye.

Regulations exist but enforcement is lax. For example, despite a 1994 law requiring tanneries in the city of Kanpur in the state of Uttar Pradesh "to do preliminary cleanup before channeling wastewater into a government-run treatment plant, many ignore the costly regulation."[117] And breakdowns in the treatment infrastructure cause further pollution from the law-abiding tanneries. Nevertheless, if (primarily) arsenic and chromium pollution from the tanneries of one city along the Ganges were the sole problem, the river would be almost pristine. But every city and village pollutes the river. It is estimated that because of population growth "the amount of domestic sewage being dumped into the Ganges has doubled since the 1990s; it could double again in a generation."[118] No one, of course, knows the daily amount of untreated sewage and industrial waste that is discharged into the Ganges, but estimates run as high as 1 billion liters (264 million gallons) per day.[119] Waterborne diseases, which affect children and the elderly at much higher rates, are therefore endemic. Aquatic life in the basin also is dying and some areas are completely dead—there is special concern about the plight of the river dolphin. Other areas register rates of deadly bacteria that are thousands of times higher than the standard set by the World Health Organization (WHO). This is particularly true where the river flows past the holiest of Hindu cities, Varanasi (formerly Benares).[120] At festivals and other times, pilgrims come not only to bathe in the waters of "Mother Ganga," but to sip it for purification. It is also the sacred custom to send the ashes of loved ones on their journey in the river. Improper cremation methods have added to the Ganges' troubles—corpses are often seen floating down the river.

The river's pollution does not end at its banks, since a good deal of Ganges water is diverted for agricultural purposes. The soil and crops are therefore affected, and runoff from the farms not only returns some of the pollutants but adds salts, phosphates, and nitrates to the water as well. Furthermore, because of the increased diversions, with each passing year the Ganges is less and less able to repair itself—it seems to have passed the critical point in the early 1990s. There is at least one other serious problem for the Ganges: Climate change has caused the Gangotri glacier, the source of the Ganges, to recede at a rate of about 16 feet per year.

Current Water Problems in Bangladesh

Bangladesh has protested the various water development schemes along the Ganges because they threaten the river's flow, hence Bangladesh's freshwater supply. But the freshwater crisis in Bangladesh is paradoxical, and some of the solutions have contributed to this paradox. For example, glancing at a map of the country it would appear that Bangladesh is not in danger of a water crisis. It is a relatively small country in terms of area and is trisected by two of the three great Himalayan rivers: the Ganges (in Bangladesh the Ganges is called the Padma) and the Brahmaputra. Nevertheless, it is because these rivers are Himalayan—that is, they originate outside of Bangladesh—that the country is usually classified as being water scarce. As of the year 2000, Bangladesh had less than 1,000 cubic meters (264,000 gallons) per person per year of internal renewable freshwater resources—some 91 percent of its renewable freshwater resources comes from outside its borders.

To further complicate matters Bangladesh has experienced some of the worst flooding on Earth. Its floodplain comprises about half of the country, and, like the Nile in Egypt, the Ganges and Brahmaputra Rivers annually overflow their banks, depositing rich soil for farmers. Unlike the Nile they tend to overflow to dangerous levels, and with two rivers overflowing at about the same time those caught in between are at risk. The most devastating recent flood in Bangladesh occurred from July to September 1998.[121] More than 1,300 people were killed and about 31 million were left homeless. The flooding directly affected about half the country, with about one-quarter severely affected. Some 10,000 miles of roads were destroyed. All this was exacerbated by "riverbank defenses, built to prevent flooding [which] lengthened the period of the disaster. Floodwater that had spilled over them and inundated fields had no way of draining back into the river as the water level dropped."[122] The effects of flooding outlast the receding of waters. Besides causing waterborne and other diseases (most of the deaths were caused by disease), crops and fields are damaged, as are drinking water and sewage systems. Since the Ganges is already polluted, its flooding, even under somewhat controlled circumstances, poses a danger. When the flooding is out of control, as in 1998, the toxins cover a wider area and thus affect more people directly. Bangladesh is also a country where arsenic occurs naturally in freshwater, as does fluoride in excessive amounts; it shares both of those problems with India and Pakistan.[123] Another disastrous flood occurred in 2004; it resulted in nearly 1,000 deaths and $7 billion in property damage.[124]

Bangladesh's arsenic problem went undetected until the mid-1990s when the toxin was discovered to be leeching into deep wells dug two decades earlier as a means of using groundwater for domestic purposes. Most of the millions of wells dug during the 1970s were in states that bordered the Ganges River, both above and below the point where the Brahmaputra River flows into it. Test boreholes in 1999 proved that what seemed like a good idea turned into a disaster: From the areas where data was available, in seven states, mostly in the delta region, 75 percent or more of the boreholes had higher concentrations of arsenic than the standard set by WHO of .05 milligrams per liter of water. In another six states, located mostly in the delta or along the river, from 50 percent to 74 percent of the tested boreholes revealed higher concentrations of arsenic than the WHO standard.

Bangladesh has practically the same breakdown for water use as Pakistan: 96 percent is used for agricultural purposes, while domestic use takes 3 percent and industrial use 1 percent. And because Bangladesh is also subject to monsoon and periods of drought, 44 percent of its arable land is irrigated.[125] This leads to the problem of soil salination and salinated water from runoff. Depletion is another problem connected with Bangladesh's groundwater supply. Improved groundwater-drawing equipment such as the treadle pump has enabled farmers to increase output, but the cost is a faster depletion of groundwater, causing some farming land to be retired.

Counterstrategies

As noted, the highest percentage of freshwater use in both Pakistan and Bangladesh goes to the agricultural sector. Two important means of reducing these percentages would be to switch to less water-intensive crops and to improve methods of irrigation. In Pakistan, in particular, switching to drip irrigation would have the double effect of using less water and reversing the trend of soil and water salination. To accomplish this changeover, both countries will require international investment (from the World Bank or otherwise). Within a decade of its independence, Bangladesh had begun taking steps to conserve water and maintain a regular flow on the Ganges. Although the Ganges Barrage Project was planned in 1980, however, it was not until 2009 that construction began. The project's full purpose is to "increase the navigability of the rivers in that region, reduce salinity in the rivers, increase production in agriculture and fisheries, increase the flow of sweet water and ultimately save the Sundarbans [a tidal forest located in the Ganges delta in both Bangladesh and India], a World Heritage Site."[126]

On December 12, 1996, Bangladesh and India signed an important 30-year treaty dealing with water-sharing at the Farakka Barrage. The fact that the treaty expires in 2026 is probably one of the main reasons Bangladesh has instituted its own barrage system. Certainly, a permanent sharing plan is needed, perhaps modeled after the Indus Waters Treaty. This treaty, if it comes about, would undoubtedly create a bilateral Ganges basin authority that would recognize the rights and decision-making responsibilities of both countries regarding the health of the river, the downstream ecology, water use, and many other issues. As Bangladesh becomes more reliant on surface water (because of arsenic in its groundwater), tensions with India will undoubtedly rise (making the need for a treaty creating a bilateral basinwide authority more acute), but again foreign investment to modernize irrigation and construction of desalination plants could help alleviate the country's water crisis. Indeed, Pakistan and Bangladesh would both benefit by increasing their desalination capacities. In this respect, India is the regional leader, with numerous plants, and is one of the world's innovators in the technology. It not only possesses a nuclear-powered plant but constructed the world's first floating desalination plant, off the coast of Chennai (formerly Madras), and is a pioneer in low-temperature thermal desalination (LTTD). In fact, the world's first LTTD plant was opened in 2005 on one of the Lakshadweep Islands off the country's southwest coast.

Modernizing irrigation methods would also help India conserve water, though at 86 percent of total national water use India's agricultural sector uses less water proportionally than its neighbors. Another important measure the subcontinent countries (as well as the rest of the countries under discussion in this book) take to help alleviate their freshwater crises is water harvesting. This involves capturing rainwater runoff from rooftops and local catchments, as well as capturing and conserving runoff from seasonal flooding. Water harvesting also includes watershed and basin management to increase conservation methods.[127] In fact, in 2002 India, Pakistan, and Bangladesh entered into an 18-nation international program involving many East African nations (including some of the Nile basin countries) to help develop water harvesting at the local level.[128] At present, water harvesting remains limited due to factors such as prolonged drought (in East Africa) and, among the subcontinent nations, a feeble infrastructure. But the most important step India can take is to restore the health of the Ganges River, upon which an estimated 500 million people depend. In 1985, the Indian government launched phase one of the Ganga Action Plan, but by the early 1990s it was clear that the plan's initial phase had failed. This was due to the municipal, state, and central governments' failure to enforce antipollution laws and regulations. Finally, in the late 1990s, the state government of Uttar Pradesh "ordered the shutdown of 250 factories, including 127 tanneries in the city of Kanpur," largely as a result of pressure by environmental activist Rakesh Jaiswal (b. 1959), who afterward received anonymous death threats.[129] In the years since then, however, not much has actually been done to reverse the Ganges' spiral toward death. In 2008, Indian prime minister Manmohan Singh declared the Ganges a "national river," which should pave the way for stricter enforcement of antipollution laws, as well as for the construction of water treatment plants.

Presently, India, Pakistan, and their Himalayan neighbors Nepal and Bhutan have plans to dam the Himalayan rivers to offset the future effects of the melting glaciers and to generate hydroelectricity. Hundreds of very large dams have been proposed for construction over the next two decades, and it is estimated that when completed they will generate 150,000 megawatts (MW) of electricity. They include a 3,000 MW project in India, and a 1,000 MW project in Bhutan. Overshadowing these could be Pakistan's proposed $12.6 billion Diamer-Bhasha Dam.[130]

Finally, Nepal and Bhutan are two water-rich countries—only 6 percent of Nepal's renewable freshwater resources originates outside the country, while Bhutan's percentage is zero.[131] In other words, they can afford to build dams and reservoirs in hopes of staving off glacier melting, and in fact this could potentially provide more water to India and Pakistan. Reservoirs in the Himalayas would be subject to less evaporation than those farther south in the subcontinent.

---

[1] Egypt is a part of the Maghreb climatically but has never been grouped with the other countries politically. The reason for this is, obviously, the Nile River.

[2] Actually, climate change has affected Antarctica. Ice shelves have been breaking off and melting at a rate not experienced in recorded human history.

[3] Jeffrey Rothfeder. *Every Drop for Sale: Our Desperate Battle over Water in a World about to Run Out.* New York: Tarcher/Putnam, 2001, p. 162.

[4] Green Cross was founded in 1993 by former Soviet president Mikhail Gorbachev.

[5] Rothfeder, p. 163.

[6] Oscar Olivera with Tom Lewis. *¡Cochabamba!: Water War in Bolivia.* Cambridge, Mass.: South End Press, 2004, p. 12.

[7] See Elizabeth Martinez and Arnoldo Garcia. "What Is Neoliberalism?: A Brief Definition for Activists." CorpWatch. Available online. URL: http://www.corpwatch.org/article.php?id=376. Accessed February 3, 2009.

[8] Olivera, p. 9.

[9] This is the same Bechtel that had a hand in constructing the Hoover Dam. Sixty years later it had grown into a transnational corporation.

[10] The percentage of rate increases has been disputed by observers of and participants in the events and Bechtel.

[11] Olivera, p. 30.

[12] Olivera, pp. 34–37.

[13] The decision to occupy the offices of Aguas del Tunari was more or less an ad hoc one by the crowd.

[14] The army captain's name was Robinson Iriarte de la Fuente. In 2002, a military tribunal acquitted Iriarte of the shooting. According to Oscar Olivera, Daza was coming home from work and had stopped to watch the protest. Olivera, p. 43.

[15] The agreement also called for the release of arrested protesters.

[16] "Timeline: Cochabamba Water Revolt." Frontline World. Available online. URL: http://www.pbs.org/frontlineworld/stories/bolivia/timeline.html. Accessed February 13, 2009.

[17] This was not the only water lawsuit involving a transnational to be heard by ICSID. On December 8, 2003, an ICSID tribunal handed down a decision in *Azurix Corp. v. the Argentine Republic.*

[18] "Secretive World Bank Tribunal Bans Public and Media Participation in Bechtel Lawsuit over Access to Water" (2/12/03). The Center for International Environmental Law (CIEL). Available online. URL: http://www.ciel.org/Ifi/Bechtel_Lawsuit_12Feb03.html. Accessed February 14, 2009.

[19] "Cochabamba Water Dispute Settle" (1/19/06). Bechtel. Available online. URL: http://www.bechtel.com/2006-01-19.html. Accessed February 14, 2009.

[20] Juan Forero. "As Andean Glaciers Shrink, Water Worries Grow." *New York Times* (11/24/02), p. 1

[21] Forero, p. 1.

[22] Forero, p. 1

[23] José Luis Alcázar. "Pilcomayo River to Be Saved from Ruin." Tierramérica. Available online. URL: http://www.tierramerica.net/2005/0521/iacentos.shtml. Accessed June 2, 2009.

[24] Alcázar.

[25] "Law 2878." Global Legal Information Network. Available online. URL: http://www.glin.gov/view.action?glinID=123807. Accessed October 13, 2009.

[26] "Bolivia: Ministry Announces 'Water Belonging to All and for All' Declaration." IRC. Available online. URL: http://www.irc.nl/page/38528. Accessed June 1, 2009.

[27] Those nations are Egypt, Ethiopia, Sudan, Rwanda, Tanzania, Uganda, Burundi, Democratic Republic of the Congo, Kenya, and Eritrea, which is part of the basin but not a riparian country.

[28] The first figure is from Arnon Soffer, *Rivers of Fire: The Conflict over Water in the Middle East,* translated by Murray Rosovsky and Nina Copaken. Lanham, Md.: Rowman & Littlefield, 1999, p. 56. The second figure is the sum of the national estimates of the countries of the basin gleaned from the CIA's World Factbook. Available online. URL: https://www.cia.gov/library/publications/the-world-factbook/geos/xx.html. Accessed February 20, 2009.

[29] Soffer, p. 62.

[30] Evelyn Lirri. "East Africa: Regional Countries to Jointly Manage River Kagera." *The Monitor* (Kampala) (9/9/08). Available online. URL: http://allafrica.com/stories/200809090249.html. Accessed February 22, 2009.

[31] Both quotes in Robert O. Collins. *The Nile.* New Haven, Conn.: Yale University Press, 2002, p. 213.

[32] Soffer, p. 51.

[33] Soffer, p. 51.

[34] "Dams in Sudan." Miraya FM (8/21/08). Available online. URL: http://www.mirayafm.org/reports/reports/_200808214531. Accessed February 23, 2009.

[35] Ali Askouri. "A Culture Drowned: Sudan Dam Will Submerge Historically Rich Area, Destroy Nile Communities." *World Rivers Review* 19, no. 2 (April 2004). Available online. URL: http://internationalrivers.org/files/WRR.V19.N2.pdf. Accessed February 23, 2009.

[36] The information on the Hamdab people comes from Askouri.

[37] Soffer, p. 32.

[38] Soffer, pp. 32–33.

[39] Technically the agreement was between Sudan and the United Arab Republic (UAR), which at that time consisted of Egypt and Syria. The UAR, Nasser's dream of an Egyptian-led pan-Arabism, dissolved in 1961 though Egypt continued to use the name until 1971.

[40] Another reason for the slower pace is that from an engineering standpoint the best sites on many rivers have already been dammed.

[41] All water flow and reservoir level figures are from Soffer, pp. 38–39.

[42] Soffer, p. 41.

[43] The project has also gone by the name of South Valley Agricultural Development Project.

[44] "Toshka Project—Mubarak Pumping Station/Sheikh Zayed Canal, Egypt." Water-Technology.net. Available online. URL: http://www.water-technology.net/projects/mubarak. Accessed March 4, 2009.

[45] "Sinai Development Projects." Sinai Liberation . . . 26 Years. Available online. URL: http://www.sis.gov.eg/VR/sinia/html/esinia10.htm. Accessed March 5, 2009.

[46] Kermit Zarley. "Extending Egypt's Al Salam Canal to the Palestinian State." Available online. URL: http://www.kermitzarley.com/pdf/alsalamcanal.pdf. Accessed March 4, 2009.

[47] Zarley.

[48] The information in this paragraph comes from "Sinai Development Projects."

[49] All figures from Robin Clarke and Jannet King, *The Water Atlas: A Unique Visual Analysis of the World's Most Critical Resource.* New York and London: The New Press, 2004, pp. 22–23.

[50] Clarke and King, p. 54.

[51] Regarding Tanzanian, Burundian, and Kenyan pollution, the Sudd Swamps in Sudan probably filtered some of the industrial and agricultural pollutants that traveled downriver along the White Nile.

[52] Clarke and King, pp. 36–37.

[53] "Sequence of Major Events of the Nile Basin Initiative Process." Available online. URL: http://www.africanwater.org/nbihistory.htm. Accessed March 6, 2009.

[54] "Key Milestones, 1998–2006." Nile Basin Initiative. Available online. URL: http://www.nile-basin.org/index.php?option=com_content&task=view&id=13&Itemid=42. Accessed March 6, 2009.

[55] "Operational Structure." Nile Basin Initiative. Available online. URL: http://www.nile basin.org/index.php?option=com_content&task=view&id=30&Itemid=77. Accessed March 6, 2009.

[56] "Applied Training Project." Nile Basin Initiative. Available online. URL: http://atp.nileba-sin.org/index.php?option=com_content&task=view&id=14&Itemid=27. Accessed March 6, 2009.

[57] "Confidence Building and Stakeholders Involvement." Nile Basin Initiative. Available online. URL: http://cbsi.nilebasin.org/index.php?option=com_content&task=view&id=13& Itemid=27. Accessed March 6, 2009.

[58] "Regional Power Trade Project." Nile Basin Initiative. Available online. URL: http://rpt.nile basin.org/index.php?option=com_content&task=view&id=18&Itemid=99. Accessed March 6, 2009.

[59] "Shared Vision Coordination Project." Nile Basin Initiative. Available online. URL: http://svpcp.nilebasin.org/index.php?option=com_content&task=view&id=12&Itemid=29. Accessed March 6, 2009.

[60] "Socioeconomic Development and Benefits Sharing Project." Nile Basin Initiative. Available online. URL: http://sdbs.nilebasin.org/index.php?option=com_content&task=view&id =16&Itemid=39. Accessed March 6, 2009.

[61] "Trans-boundary Environmental Action Project. Nile Basin Initiative. Available online. URL: http://nteap.nilebasin.org/index.php?option=com_content&task=view&id=50&Itemi d=72. Accessed March 6, 2009.

[62] "Efficient Water Use for Agricultural Production." Nile Basin Initiative. Available online. URL: http://ewuap.nilebasin.org. Accessed March 6, 2009.

[63] "Water Resources Planning and Management Project." Nile Basin Initiative. Available online. URL: http://wrpmp.nilebasin.org/index.php?option=com_content&task=view&id=12 &Itemid=50. Accessed March 6, 2009.

[64] "Eastern Nile Subsidiary Action Program." Nile Basin Initiative. Available online. URL: http://ensap.nilebasin.org. Accessed March 7, 2009.

[65] "Nile Equatorial Lakes Subsidiary Action Program." Nile Basin Initiative. Available online. URL: http://nelsap.nilebasin.org/index.php?option=com_content&task=view&id=20 &Itemid=85. Accessed March 7, 2009.

[66] The British held a ruling mandate over the territory, called Palestine, under discussion until the late 1940s. The mandate covered land on both sides of the Jordan River. The territory east

of the river became, officially, the Hashemite Kingdom of Jordan, while the State of Israel was carved from land west of the river. Palestinians, meanwhile, were left out of the negotiations and therefore received no territory of their own.

[67] The Palestinian territories are the West Bank and the Gaza Strip.

[68] Quoted in Soffer, p. 167.

[69] Soffer, p. 169.

[70] The canal, which also now takes water from Lake Kinneret, is now known as the King Abdullah Canal.

[71] Soffer, p. 131.

[72] Egypt, Israel's main opponent in the Six-Day War, was involved in a rivalry with Syria, its former partner in the United Arab Republic, as to which nation would lead the Arab world. Its own actions of massing troops on its border with Israel were driven by this consideration and by the belief that Israel was concentrating its military against Syria.

[73] Quoted in Paul Simon. *Tapped Out: The Coming World Crisis in Water and What We Can Do about It.* New York: Welcome Rain Publishers, 1998, p. 50.

[74] Hamas, one of two major Palestinian political parties, controls Gaza, while the other, Fatah, controls the West Bank.

[75] "Mountain and Coastal Aquifers." Israel-Palestine Water Issues. Available online. URL: http://mapsomething.com/demo/waterusage/hydrology.php. Accessed March 25, 2009. See also Soffer, p. 149.

[76] Aaron T. Wolf. *Hydropolitics along the Jordan River: Scarce Water and Its Impact on the Arab-Israeli Conflict.* Tokyo: United Nations University Press, 1995, p. 10.

[77] Wolf, p. 1.

[78] Sandra Postel. *Pillar of Sand: Can the Irrigation Miracle Last?* New York and London: W. W. Norton & Company, 1992, pp. 139–141.

[79] Clarke and King, pp. 34–35, 104.

[80] Clarke and King, p. 87.

[81] Peter H. Gleick. "A Soft Path: Conservation, Efficiency, and Easing Conflicts over Water," in Bernadette McDonald and Douglas Jehl, eds. *Whose Water Is It?: The Unquenchable Thirst of a Water-Hungry World.* Washington, D.C.: National Geographic Society, 2003, p. 194.

[82] Amir Mizroch. "Ministry: Desalination Can't Meet Water Needs." *Jerusalem Post* (5/14/08).

[83] Mizroch.

[84] "Seawater Desalination Projects: The Challenge and the Options to Meet the Water Shortage" (March 1999). Jewish Virtual Library. Available online. URL: http://www.jewishvirtuallibrary.org/jsource/History/desal.html. Accessed March 28, 2009.

[85] Sandra Postel. *Liquid Assets: The Critical Need to Safeguard Freshwater Ecosystems.* Washington, D.C.: Worldwatch Institute, 2005, pp. 36–37. As of January 14, 2010, one Jordanian dinar equaled US $1.41.

[86] Marq de Villiers. *Water: The Fate of Our Most Precious Resource.* Boston and New York: Houghton Mifflin Company, 2000, p. 192.

[87] Michael Beyth. "The Red Sea and the Mediterranean-Dead Sea Canals Project" (8/10/02). Israel Ministry of Foreign Affairs. Available online. URL: http://www.mfa.gov.il/MFA/MFAArchive/2000_2009/2002/8/The%20Red%20Sea%20and%20the%20Mediterranean%20Dead%20Sea%20canals. Accessed March 30, 2009.

[88] Beyth.

[89] Wolf, p. 163.

[90] See Reuters. "Dead Sea-Red Sea Canal Could Cause Quakes—Official" (7/27/05). Available online. URL: http://www.planetark.com/dailynewsstory.cfm/newsid/31801/newsDate/27-Jul-2005/story.htm. Accessed June 3, 2009; and Rory Kress. "World Bank Promotes Dead Sea Canal." *Jerusalem Post* (7/25/07). Available online. URL: http://www.jpost.com/servlet/Satellite?cid=1185379003420&pagename=JPost%2FJP Article%2FShowFull. Accessed June 3, 2009.

[91] Formed after Israel annexed Gaza and the West Bank in the wake of the Six-Day War, the Palestinian Liberation Organization (PLO), led by Yasser Arafat, became the military and political authority for the occupied territories. The PLO relinquished control to the Palestinian Authority on signing the Declaration of Principles on September 13, 1993.

[92] Clarke and King, pp. 22–23.

[93] Binayak Ray. *Water: The Looming Crisis in India.* Lanham, Md.: Lexington Books, 2008, p. 23.

[94] Diane Raines Ward. *Water Wars: Drought, Flood, Folly and the Politics of Thirst.* New York: Riverhead Books, 2002, p. 80.

[95] Other sources give the length of the Indus River at approximately 1,800 miles.

[96] "Britannica Concise Encyclopedia: Indus River." Available online. URL: http://www.answers.com/topic/indus-river-1. Accessed June 3, 2009.

[97] The Thar extends into the Indian states of Gujarat, Hayana, and Punjab and the Pakistani provinces of Sindh and Punjab.

[98] Aloys Arthur Michel. *The Indus Rivers: A Study of the Effects of Partition.* New Haven, Conn.: Yale University Press, 1967, p. 196.

[99] Michel, p. 200.

[100] "The Indus Waters Treaty: A History." Available online. URL: http://www.stimson.org/?SN=SA20020116301. Accessed April 6, 2009.

[101] By 1960 India had completed a number of water development projects in the Indus basin, including what later became known as the Indira Gandhi Canal.

[102] Postel, *Liquid Assets,* p. 19.

[103] Clarke and King, p. 107.

[104] The percentages for the United States and Egypt were 23 and 33 percent, respectively.

[105] Vandana Shiva. *Water Wars: Privatization, Pollution, and Profit.* Cambridge, Mass.: South End Press, 2002, p. 62.

[106] Shiva, p. 64.

[107] The other four holy rivers are Ganges, Godavari, Kaveri, and Yamuna.

[108] "Narmada River." Available online. URL: http://www.india9.com/i9show/Narmada-River-17173.htm. Accessed June 3, 2009.

[109] "Final Order and Decision of the Tribunal." Available online. URL: http://www.sscac.gov.in/NWDT.pdf. Accessed June 3, 2009.

[110] The review committee was headed by Bradford Morse, former head of the United Nations Development Program.

[111] "Extracts from 'Sardar Sarovar' the Report of the Independent Review (Morse Committee) Chapter 17: The Findings: Resettlement and Rehabilitation." Available online. URL: http://narmada.aidindia.org/content/view/52/1. Accessed April 13, 2009.

[112] Phillipe Cullet. "The Sardar Sarovar Dam Project: An Overview." Available online. URL: https://eprints.soas.ac.uk/2985/1/a0704.pdf. Accessed April 13, 2009.

[113] Cullet.

[114] "Extracts from 'Sardar Sarovar' the Report of the Independent Review (Morse Committee) Chapter 17: The Findings: Resettlement and Rehabilitation."

[115] "River Ganges." The Water Page. Available online. URL: http://www.africanwater.org/ganges.htm#Dams and the Farakka Barrage. Accessed April 7, 2009.

[116] Alain Daniélou. *The Gods of India.* New York: Inner Traditions International, 1985, p. 215.

[117] Joshua Hammer. "A Prayer for the Ganges: Across India, Environmentalists Battle a Tide of Troubles to Clean Up a River Revered as the Source of Life." *Smithsonian* (November 2007).

[118] Hammer.

[119] "About the Ganges River." Available online. URL: http://web.bryant.edu/~langlois/ecology/gangesmap.htm. Accessed April 7, 2009.

[120] Hammer.

[121] That same year a devastating flood occurred along the Yangtze River in China.

[122] Clarke and King, p. 71.

[123] Clarke and King, pp. 57–58. Other countries that have excessive amounts of arsenic are China, Thailand, Cambodia, Vietnam, Myanmar, Hungary, Brazil, Argentina, Chile, Mexico, and the United States (in areas of Oregon, California, Nevada, Texas, Minnesota, and Florida). Too much fluoride in water causes fluorosis, wherein ligaments calcify and the bone structure is altered. Countries with excessive fluoride are more numerous. Those under consideration in this book include: Egypt, Ethiopia, Kenya, Uganda, Tanzania, Jordan, and the Palestinian Territories.

[124] "Bangladesh." *The World Almanac and Book of Facts, 2007.* New York: World Almanac Books, 2007, p. 751.

[125] Clarke and King, pp. 102–103.

[126] "Govt. Begins Process on Ganges Barrage." Available online. URL: http://www.southasianmedia.net/cnn.cfm?id=571021&category=Development&Country=BANGLADESH. Accessed April 15, 2009.

[127] "What Is Water Harvesting?" Available online. URL: http://www.rainwaterharvesting.org/whatiswh.htm. Accessed October 7, 2009.

[128] "Rainwater Harvesting Network—East and Southern Africa and Asia." The Communication Initiative Network. Available online. URL: http://www.comminit.com/en/node/122165/36. Accessed October 7, 2009.

[129] Hammer.

[130] For more on Himalayan dam projects, see Ann-Kathrin Schneider. "Dam Boom in Himalayas Will Create Mountains of Risk." *World Rivers Review.* Available online. URL: http://internationalrivers.org/en/node/3924. Accessed April 15, 2009.

[131] Clarke and King, pp. 95, 99.

PART II

Primary Sources

4

United States Documents

The primary sources in this chapter are arranged in chronological order. They include laws, compacts, agreements, treaties, and U.S. Supreme Court decisions. Documents that have been excerpted are identified as such; all others are reproduced in full.

The Rivers and Harbors Act of 1899 (excerpts)

Formally known as "The Rivers and Harbors Appropriation Act of 1899," it was signed into law by President William McKinley. While navigation and strengthening ports were the primary purposes of the act, its authors, on the brink of the 20th century, also laid the groundwork for the management of two important water issues. Sections 9 and 10 deal with dam and bridge construction, while Section 13 discusses pollution. The first two sections provided the impetus for the U.S. Army Corps of Engineers and especially the U.S. Bureau of Reclamation to carry out their various water development projects. Section 13 would become the inspiration for the Clean Water Act.

> SEC. 9. It shall not be lawful to construct or commence the construction of any bridge, causeway, dam, or dike over or in any port, roadstead, haven, harbor, canal, navigable river, or other navigable water of the United States until the consent of Congress to the building of such structures shall have been obtained and until the plans for (1) the bridge or causeway shall have been submitted to and approved by the Secretary of Transportation, or (2) the dam or dike shall have been submitted to and approved by the Chief of Engineers and Secretary of the Army. However, such structures may be built under authority of the legislature of a State across rivers and other waterways the navigable portions of which lie wholly within the limits of a single State, provided the location and plans thereof are submitted to and approved by the Secretary of Transportation or by the Chief of Engineers and Secretary of the Army before construction is commenced. . . .

SEC. 10. That the creation of any obstruction not affirmatively authorized by Congress, to the navigable capacity of any of the waters of the United States is hereby prohibited; and it shall not be lawful to build or commence the building of any wharf, pier, dolphin [a mooring for boats], boom, weir, breakwater, bulkhead, jetty, or other structures in any port, roadstead, haven, harbor, canal, navigable river, or other water of the United States, outside established harbor lines, or where no harbor lines have been established, except on plans recommended by the Chief of Engineers and authorized by the Secretary of War; and it shall not be lawful to excavate or fill, or in any manner to alter or modify the course, location, condition, or capacity of, any port, roadstead, haven, harbor, canal, lake, harbor of refuge, or inclosure within the limits of any breakwater, or of the channel of any navigable water of the United States, unless the work has been recommended by the Chief of Engineers and authorized by the Secretary of War prior to beginning the same.

SEC. 13. That it shall not be lawful to throw, discharge, or deposit, or cause, suffer, or procure to be thrown, discharged, or deposited either from or out of any ship, barge, or other floating craft of any kind, or from the shore, wharf, manufacturing establishment, or mill of any kind, and refuse matter of any kind or description whatever other than that flowing from streets and sewers and passing therefrom in a liquid state, into any navigable water of the United States, or into any tributary of any navigable water from which the same shall float or be washed into such navigable water; and it shall not be lawful to deposit, or cause, suffer, or procure to be deposited material of any kind in any place on the bank of any navigable water, or on the bank of any tributary of any navigable water, where the same shall be liable to be washed into such navigable water, either by ordinary or high tides, or by storms or floods, or otherwise, whereby navigation shall or may be impeded or obstructed: *Provided,* That nothing herein contained shall extend to, apply to, or prohibit the operations in connection with the improvement of navigable waters or construction of public works, considered necessary and proper by the United States officers supervising such improvement or public work: *And provided further,* That the Secretary of War, whenever in the judgment of the Chief of Engineers anchorage and navigation will not be injured thereby, may permit the deposit of any material above mentioned in navigable waters, within limits to be defined and under conditions to be prescribed by him, provided application is made to him prior to depositing such material; and whenever any permit is so granted the conditions thereof shall be strictly complied with, and any violation thereof shall be unlawful.

Source: U.S. Senate Committee on Environment & Public Works. "Sections 9 to 20 of the Act of March 3, 1899." Available online. URL: http://epw.senate.gov/rivers.pdf. Accessed May 18, 2009.

The Boundary Waters Treaty, 1909 (excerpts)

On January 11, 1909, the "Treaty between the United States and Great Britain Relating to Boundary Waters, and Questions Arising between the United States and Canada" was signed by Secretary of State Elihu Root and James Brice, O.M. It set up the International Joint Commission to investigate, advise the governments, and arbitrate on matters related to the lakes and rivers. The treaty has 21st-century ramifications because of the precedents it set regarding water diversion.

ARTICLE I

The High Contracting Parties agree that the navigation of all navigable boundary waters shall forever continue free and open for the purposes of commerce to the inhabitants and to the ships, vessels, and boats of both countries equally, subject, however, to any laws and regulations of either country, within its own territory, not inconsistent with such privilege of free navigation and applying equally and without discrimination to the inhabitants, ships, vessels, and boats of both countries.

It is further agreed that so long as this treaty shall remain in force, this same right of navigation shall extend to the waters of Lake Michigan and to all canals connecting boundary waters, and now existing or which may hereafter be constructed on either side of the line. . . .

ARTICLE II

Each of the High Contracting Parties reserves to itself or to the several State Governments on the one side and the Dominion or Provincial Governments on the other as the case may be, subject to any treaty provisions now existing with respect thereto, the exclusive jurisdiction and control over the use and diversion, whether temporary or permanent, of all waters on its own side of the line which in their natural channels would flow across the boundary or into boundary waters; but it is agreed that any interference with or diversion from their natural channel of such waters on either side of the boundary, resulting in any injury on the other side of the boundary, shall give rise to the same rights and entitle the injured parties to the same legal remedies as if such injury took place in the country where such diversion or interference occurs; but this provision shall not apply to cases already existing or to cases expressly covered by special agreement between the parties hereto. . . .

ARTICLE III

It is agreed that, in addition to the uses, obstructions, and diversions heretofore permitted or hereafter provided for by special agreement between the Parties hereto, no further or other uses or obstructions or diversions, whether temporary or permanent, of boundary waters on either side of the line, affecting the natural level or flow of boundary waters on the other side of the line shall be made except by authority of the United States or the Dominion of Canada within their respective jurisdictions and with the approval, as hereinafter provided, of a joint commission, to be known as the International Joint Commission.

The foregoing provisions are not intended to limit or interfere with the existing rights of the Government of the United States on the one side and the Government of the Dominion of Canada on the other, to undertake and carry on governmental works in boundary waters for the deepening of channels, the construction of breakwaters, the improvement of harbours, and other governmental works for the benefit of commerce and navigation, provided that such works are wholly on its own side of the line and do not materially affect the level or flow of the boundary waters on the other, nor are such provisions intended to interfere with the ordinary use of such waters for domestic and sanitary purposes.

ARTICLE VII

The High Contracting Parties agree to establish and maintain an International Joint Commission of the Untied States and Canada composed of six commissioners, three on the part of the United States appointed by the President thereof, and three on the part of the United Kingdom appointed by His Majesty on the recommendation of the Governor in Council of the Dominion of Canada.

ARTICLE VIII

This International Joint Commission shall have jurisdiction over and shall pass upon all cases involving the use or obstruction or diversion of the waters with respect to which under Article III or IV of this Treaty the approval shall be governed by the following rules of principles which are adopted by the High Contracting Parties for this purpose:

The High Contracting Parties shall have, each on its own side of the boundary, equal and similar rights in the use of the waters hereinbefore defined as boundary waters.

The following order of precedence shall be observed among the various uses enumerated hereinafter for these waters, and no use shall be permitted which tends materially to conflict with or restrain any other use which is given preference over it in this order of precedence:

1. Uses for domestic and sanitary purposes;
2. Uses for navigation, including the service of canals for the purposes of navigation;
3. Uses for power and for irrigation purposes.

The foregoing provisions shall not apply to or disturb any existing uses of boundary waters on either side of the boundary. The requirement for an equal division may in the discretion of the Commission be suspended in cases of temporary diversions along boundary waters at points where such equal division can not be made advantageously on account of local conditions, and where such diversion does not diminish elsewhere the amount available for use on the other side.

The Commission in its discretion may make its approval in any case conditional upon the construction of remedial or protective works to compensate so far as possible for the particular use or diversion proposed, and in such cases may require that suitable and adequate provision, approved by the Commission, be made for the protection and indemnity against injury of all interests on the other side of the line which may be injured thereby.

In cases involving the elevation of the natural level of waters on either side of the line as a result of the construction or maintenance on the other side of remedial or protective works or dams or other obstructions in boundary waters flowing there from or in waters below the boundary in rivers flowing across the boundary, the Commission shall require, as a condition of its approval thereof, that suitable and adequate provision, approved by it, be made for the protection and indemnity of all interests on the other side of the line which may be injured thereby.

The majority of the Commissioners shall have power to render a decision. In case the Commission is evenly divided upon any question or matter presented to it for decision, separate reports shall be made by the Commissioners on each side to their own Government. The High Contracting Parties shall thereupon endeavour to agree upon an adjustment of the question or matter of difference, and if an agreement is reached between them, it shall be reduced to writing in the form of a protocol, and shall be communicated to the Commissioners, who shall take such further proceedings as may be necessary to carry out such agreement.

Source: Great Lakes Directory. "What Is the Boundary Waters Treaty?" Available online. URL: http://www.greatlakesdirectory.org/zarticles/102602_great_lakes.htm. Accessed May 18, 2009.

The Colorado River Compact, 1922 (excerpts)

In 1922, representatives from the states of Colorado, Wyoming, Utah, New Mexico, Arizona, Nevada, and California met in Santa Fe, New Mexico, to apportion the water of the Colorado River among themselves, under the auspices of the federal government as represented by Secretary of Commerce Herbert Hoover. After the provisions had been hammered out, six of the states ratified the compact, thus putting it into effect. Arizona finally ratified the compact in 1944.

The following excerpts define the purpose of the Colorado River Compact (Article I); set down and define the terms, including which states are upper basin and which lower basin (Article II); set out the apportionment formula (Article III); and lay the legal groundwork for future dam construction (Article IV). Of particular interest is the brief mention of Mexico in Article III, section c. In later years, this allowed Mexico to assert a rightful claim to Colorado River water. Article VII unwittingly asserts Native American claims to Colorado River water that were upheld by the Supreme Court decades later.

ARTICLE I

The major purposes of this compact are to provide for the equitable division and apportionment of the use of the waters of the Colorado River System; to establish the relative importance of different beneficial uses of water, to promote interstate comity; to remove causes of present and future controversies; and to secure the expeditious agricultural and industrial development of the Colorado River Basin, the storage of its waters, and the protection of life and property from floods. To these ends the Colorado River Basin is divided into two Basins, and an apportionment of the use of part of the water of the Colorado River System is made to each of them with the provision that further equitable apportionments may be made.

ARTICLE II

As used in this compact—

(a) The term "Colorado River System" means that portion of the Colorado River and its tributaries within the United States of America.

(b) the term "Colorado River Basin" means all of the drainage area of the Colorado River System and all other territory within the United States of America to which the waters of the Colorado River System shall be beneficially applied.

(c) The term "States of the Upper Division" means the States of Colorado, New Mexico, Utah, and Wyoming.

(d) The term "States of the Lower Division" means the States of Arizona, California, and Nevada.
(e) The term "Lee Ferry" means a point in the main stream of the Colorado River one mile below the mouth of the Paria River.
(f) The term "Upper Basin" means those parts of the States of Arizona, Colorado, New Mexico, Utah, and Wyoming within and from which waters naturally drain into the Colorado River System above Lee Ferry, and also all parts of said States located without the drainage area of the Colorado River System which are now or shall hereafter be beneficially served by waters diverted from the System above Lee Ferry.
(g) The term "Lower Basin" means those parts of the States of Arizona, California, Nevada, New Mexico, and Utah within and from which waters naturally drain into the Colorado River System below Lee Ferry, and also all parts of said States located without the drainage area of the Colorado River System which are now or shall hereafter be beneficially served by waters diverted from the System below Lee Ferry.
(h) The term "domestic use" shall include the use of water for household, stock, municipal, mining, milling, industrial, and other like purposes, but shall exclude the generation of electrical power.

ARTICLE III

(a) There is hereby apportioned from the Colorado River System in perpetuity to the Upper Basin and to the Lower Basin, respectively, the exclusive beneficial consumptive use of 7,500,000 acre-feet of water per annum, which shall include all water necessary for the supply of any rights which may now exist.
(b) In addition to the apportionment in paragraph (a), the Lower Basin is hereby given the right to increase its beneficial consumptive use of such waters by one million acre-feet per annum.
(c) If, as a matter of international comity, the United States of America shall hereafter recognize in the United States of Mexico any right to the use of any waters of the Colorado River System, such waters shall be supplied first from the waters which are surplus over and above the aggregate of the quantities specified in paragraphs (a) and (b); and if such surplus shall prove insufficient for this purpose, then, the burden of such deficiency shall be equally borne by the Upper Basin and the Lower Basin, and whenever necessary

the States of the Upper Division shall deliver at Lee Ferry water to supply one-half of the deficiency so recognized in addition to that provided in paragraph (d).

(d) The States of the Upper Division will not cause the flow of the river at Lee Ferry to be depleted below an aggregate of 75,000,000 acre-feet for any period of ten consecutive years reckoned in continuing progressive series beginning with the first day of October next succeeding the ratification of this compact.

(e) The States of the Upper Division shall not withhold water, and the States of the Lower Division shall not require the delivery of water, which cannot reasonably be applied to domestic and agricultural uses.

(f) Further equitable apportionment of the beneficial uses of the waters of the Colorado River System unapportioned by paragraphs (a), (b), and (c) may be made in the manner provided in paragraph (g) at any time after October first, 1963, if and when either Basin shall have reached its total beneficial consumptive use as set out in paragraphs (a) and (b).

(g) In the event of a desire for a further apportionment as provided in paragraph (f) any two signatory States, acting through their Governors, may give joint notice of such desire to the Governors of the other signatory States and to The President of the United States of America, and it shall be the duty of the Governors of the signatory States and of The President of the United States of America forthwith to appoint representatives, whose duty it shall be to divide and apportion equitably between the Upper Basin and Lower Basin the beneficial use of the unapportioned water of the Colorado River System as mentioned in paragraph (f), subject to the legislative ratification of the signatory States and the Congress of the United States of America.

ARTICLE IV

(b) Subject to the provisions of this compact, water of the Colorado River System may be impounded and used for the generation of electrical power, but such impounding and use shall be subservient to the use and consumption of such water for agricultural and domestic purposes and shall not interfere with or prevent use for such dominant purposes.

(c) The provisions of this article shall not apply to or interfere with the regulation and control by any State within its boundaries of the appropriation, use, and distribution of water.

ARTICLE VII

Nothing in this compact shall be construed as affecting the obligations of the United States of America to Indian tribes.

Source: "Colorado River Compact, 1922." Available online. URL: http://www.usbr.gov/lc/region/g1000/pdffiles/crcompct.pdf. Accessed May 18, 2009.

The Boulder Canyon Project Act, 1928 (excerpts)

By this act, the U.S. federal government, in 1928, recognized and ratified the provisions of the Colorado River Compact and devised a formula for putting the compact into effect (Section 13). It also effectively ushered in the era of high dam building. The two great construction projects for which the act is directly responsible are the Boulder Dam (later Hoover Dam) along with its reservoir, Lake Mead, and the All-American Canal. Boulder Dam would be dedicated in 1935, while the All-American Canal was completed in 1942.

AN ACT To provide for the construction of works for the protection and development of the Colorado River Basin, for the approval of the Colorado River compact, and for other purposes.

Be it enacted by the Senate and House of Representatives of the United States of America in Congress assembled, That for the purpose of controlling the floods, improving navigation and regulating the flow of the Colorado River, providing for storage and for the delivery of the stored waters thereof for reclamation of public lands and other beneficial uses exclusively within the United States, and for the generation of electrical energy as a means of making the project herein authorized a self-supporting and financially solvent undertaking, the Secretary of the Interior, subject to the terms of the Colorado River compact hereinafter mentioned, is hereby authorized to construct, operate, and maintain a dam and incidental works in the main stream of the Colorado River at Black Canyon or Boulder Canyon adequate to create a storage reservoir of a capacity of not less than twenty million acre-feet of water and a main canal and appurtenant structures located entirely within the United States connecting the Laguna Dam, or other suitable diversion dam, which the Secretary of the Interior is hereby authorized to construct if deemed necessary or advisable by him upon engineering or economic considerations, with the Imperial and Coachella Valleys in California, the expenditures for said main canal and appurtenant structures to be reimbursable, as provided in the reclamation law, and shall not be paid out of revenues derived from the sale or disposal of water power or electric energy at the dam authorized to be constructed at said Black Canyon or Boulder Canyon, or for water for potable purposes outside of the Imperial and Coachella Valleys: *Provided, however,* That no charge shall be made for water or for the use, storage, or delivery of water for irrigation or water for potable purposes in the Imperial or Coachella Valleys; also to construct and equip, operate, and maintain at or near said dam, or cause to be constructed, a complete plant and incidental structures suitable for the fullest economic development of electrical energy from the water discharged from said reservoir; and to acquire by proceedings in eminent domain, or otherwise all lands, rights-of-way, and other property necessary for said purposes.

SEC. 8. (a) The United States, its permittees, licensees, and contractees, and all users and appropriators of water stored, diverted, carried, and/or distributed by the reservoir, canals, and other works herein authorized, shall observe and be subject to and controlled by said Colorado River compact in the construction, management, and operation of said reservoir, canals, and other works and the storage, diversion, delivery, and use of water for the generation of power, irrigation, and other purposes, anything in this Act to the contrary notwithstanding, and all permits, licenses, and contracts shall so provide.

SEC. 13. (a) The Colorado River compact signed at Santa Fe, New Mexico, November 24, 1922, pursuant to Act of Congress approved August 19, 1921, entitled "An Act to permit a compact or agreement between the States of Arizona, California, Colorado, Nevada, New Mexico, Utah, and Wyoming respecting the disposition and apportionment of the waters of the Colorado River, and for other purposes," is hereby approved by the Congress of the United States, and the provisions of the first paragraph of article II of the said Colorado River compact, making said compact binding and obligatory when it shall have been approved by the legislature of each of the signatory States, are hereby waived, and this approval shall become effective when the State of California and at least five of the other States mentioned, shall have approved or may hereafter approve said compact as aforesaid and shall consent to such waiver, as herein provided.

Source: "Boulder Canyon Project Act." Available online. URL: http://www.usbr.gov/lc/region/g1000/pdfiles/bcpact.pdf. Accessed May 18, 2009.

The Mexico-United States Treaty of 1944 (excerpts)

This treaty served numerous purposes. It strengthened the U.S. southern flank during World War II, but it also dealt with the Rio Grande and Tijuana Rivers and with Mexico's share of Colorado River water. With regard to the last, it recognized Mexico's claim as a riparian state, the precedent for this being its mention in the Colorado River Compact. Philosophically, the treaty was a first step in repudiating the Harmon Doctrine.

III—COLORADO RIVER

ARTICLE 10

Of the waters of the Colorado River, from any and all sources, there are allotted to Mexico:

(a) A guaranteed annual quantity of 1,500,000 acre-feet (1,850,234,000 cubic meters) to be delivered in accordance with the provisions of Article 15 of this Treaty.

(b) Any other quantities arriving at the Mexican points of diversion, with the understanding that in any year in which, as determined by the United States Section, there exists a surplus of waters of the Colorado River in excess of the amount necessary to supply uses in the United States and the guaranteed quantity of 1,500,000 acre-feet (1,850,234,000 cubic meters) annually to Mexico, the United States undertakes to deliver to Mexico, in the manner set out in Article 15 of this Treaty, additional waters of the Colorado River system to provide a total quantity not to exceed 1,700,000 acre-feet (2,096,931,000 cubic meters) a year. Mexico shall acquire no right beyond that provided by this subparagraph by the use of the waters of the Colorado River system, for any purpose whatsoever, in excess of 1,500,000 acre-feet (1,850,234,000 cubic meters) annually. In the event of extraordinary drought or serious accident to the irrigation system in the United States, thereby making it difficult for the United States to deliver the guaranteed quantity of 1,500,000 acre-feet (1,850,234,000 cubic meters) a year, the water allotted to Mexico under subparagraph (a) of this Article will be reduced in the same proportion as consumptive uses in the United States are reduced.

ARTICLE 15

A. The water allotted in subparagraph (a) of Article 10 of this Treaty shall be delivered to Mexico at the points of delivery specified in Article 11, in accordance with the following two annual schedules of deliveries by months, which the Mexican Section shall formulate and present to the Commission before the beginning of each calendar year:

SCHEDULE I

Schedule I shall cover the delivery, in the limitrophe section [the boundary area] of the Colorado River, of 1,000,000 acre-feet (1,233,489,000 cubic meters) of water each year from the date Davis dam and reservoir are placed in operation until January 1, 1980 and the delivery of 1,125,000 acre-feet (1,387,675,000 cubic meters) of water each year thereafter....

SCHEDULE II

Schedule II shall cover the delivery at the boundary line by means of the All-American Canal of 500,000 acre-feet (616,745,000 cubic meters) of water each year from the date Davis dam and reservoir are placed in operation until January 1, 1980 and the delivery of 375,000 acre-feet (462,558,000 cubic meters) of water each year thereafter....

Source: "The Mexican Water Treaty of 1944." Available online. URL: http://www.usbr.gov/lc/region/g1000/pdfiles/mextrety.pdf. Accessed May 19, 2009.

Upper Colorado River Basin Compact, 1948 (excerpts)

On October 11, 1948, representatives from the upper Colorado River basin states of Colorado, New Mexico, Utah, and Wyoming as well the lower basin state of Arizona and a federal representative appointed by President Harry S. Truman negotiated a water agreement "to determine the rights and obligations of each signatory state respecting the uses and deliveries of the water of the upper basin of the Colorado River...."

ARTICLE I

(a) The major purposes of this compact [this section] are to provide for the equitable division and apportionment of the use of the waters of the Colorado river system, the use of which was apportioned in perpetuity to the upper basin by the Colorado River Compact; to establish the obligations of each state of the upper division with respect to the deliveries of water required to be made at Lee Ferry by the Colorado River Compact; to promote interstate comity; to remove causes of present and future controversies; to secure the expeditious agricultural and industrial development of the upper basin, the storage of water and to protect life and property from floods.

(b) It is recognized that the Colorado River Compact is in full force and effect and all of the provisions hereof are subject thereto.

ARTICLE III

(a) Subject to the provisions and limitations contained in the Colorado River Compact and in this compact, there is hereby apportioned from the upper Colorado river system in perpetuity to the states of Arizona, Colorado, New Mexico, Utah and Wyoming, respectively, the consumptive use of water as follows:

(1) to the state of Arizona the consumptive use of 50,000 acre-feet of water per annum;

(2) to the states of Colorado, New Mexico, Utah and Wyoming, respectively, the consumptive use per annum of the quantities resulting from the application of the following percentages to the total quantity of consumptive use per annum apportioned in perpetuity to and available for use each year by upper basin under the Colorado River Compact and remaining after the deduction of the use, not to exceed 50,000 acre-feet per annum, made in the state of Arizona.

state of Colorado	51.75 percent.
state of New Mexico	11.25 percent.
state of Utah	23.00 percent.
state of Wyoming	14.00 percent.

(b) The apportionment made to the respective states by Paragraph (a) of this article is based upon, and shall be applied in conformity with, the following principles and each of them:

(1) the apportionment is of any and all man-made depletions;

(2) beneficial use is the basis, the measure and the limit of the right to use;

(3) no state shall exceed its apportioned use in any water year when the effect of such excess use, as determined by the commission, is to deprive another signatory state of its apportioned use during that water year; provided, that this Subparagraph (b) (3) shall not be construed as:

(i) altering the apportionment of use, or obligations to make deliveries as provided in Article XI, XII, XIII or XIV of this compact;

(ii) purporting to apportion among the signatory states such uses of water as the upper basin may be entitled to under Paragraphs (f) and (g) of Article III of the Colorado River Compact; or

(iii) countenancing average uses by any signatory state in excess of its apportionment;

(4) the apportionment to each state includes all water necessary for the supply of any rights which now exist.

(c) No apportionment is hereby made, or intended to be made, of such uses of water as the upper basin may be entitled to under Paragraphs (f) and (g) of Article III of the Colorado River Compact.

(d) The apportionment made by this article shall not be taken as any basis for the allocation among the signatory states of any benefits resulting from the generation of

ARTICLE VII

The consumptive use of water by the United States of America or any of its agencies, instrumentalities or wards shall be charged as a use by the state in which the use is made; provided, that such consumptive use incident to the diversion, impounding or conveyance of water in one state for use in another shall be charged to such latter state.

ARTICLE IX

(a) No state shall deny the right of the United States of America and, subject to the conditions hereinafter contained, no state shall deny the right of another signatory state, any person or entity of any signatory state to acquire rights to the use of water, or to construct or participate in the construction and use of diversion works and storage reservoirs with appurtenant works, canals and conduits in one state for the purpose of diverting, conveying, storing, regulating and releasing water to satisfy the provisions of the Colorado River Compact relating to the obligation of the states of the upper division to make deliveries of water at Lee Ferry, or for the purpose of diverting, conveying, storing or regulating water in an upper signatory state for consumptive use in a lower signatory state, when such use is within the apportionment to such lower state made by this compact. Such rights shall be subject to the rights of water users, in a state in which such reservoir or works are located, to receive and use water, the use of which is within the apportionment to such state by this compact.

ARTICLE XI

Subject to the provisions of this compact, the consumptive use of the water of the Little Snake River and its tributaries is hereby apportioned between the states of Colorado and Wyoming . . .

ARTICLE XII

Subject to the provisions of this compact, the consumptive use of the waters of Henry's Fork, a tributary of Green River originating in the state of Utah and flowing into the state of Wyoming and thence into the Green River in the state of Utah; Beaver Creek, originating in the state of Utah and flowing into Henry's Fork in the state of Wyoming; Burnt Fork, a tributary of Henry's Fork, originating in the state of Utah and flowing into Henry's Fork in the state of Wyoming; Birch Creek, a tributary of Henry's Fork, originating in the state of Utah and flowing into Henry's Fork in the state of Wyoming; and Sheep Creek, a tributary of Green River in the state of Utah, and their tributaries, are hereby apportioned between the states of Utah and Wyoming . . .

ARTICLE XIII

Subject to the provisions of this compact, the rights to the consumptive use of the water of the Yampa River, a tributary entering the Green River in the state of Colorado, are hereby apportioned between the states of Colorado and Utah . . .

ARTICLE XIV

Subject to the provisions of this compact, the consumptive use of the waters of the San Juan River and its tributaries is hereby apportioned between the states of Colorado and New Mexico . . .

ARTICLE XV

(a) Subject to the provisions of the Colorado River Compact and of this compact, water of the upper Colorado River system may be impounded and used for the generation of electrical power, but such impounding and use shall be subservient to the use and consumption of such water for agricultural and domestic purposes and shall not interfere with or prevent use for such dominant purposes.

ARTICLE XVI

The failure of any state to use the water, or any part thereof, the use of which is apportioned to it under the terms of this compact, shall not constitute a relinquishment of the right to such use to the lower basin or to any other state, nor shall it constitute a forfeiture or abandonment of the right to such use.

ARTICLE XVIII

(a) The state of Arizona reserves its rights and interests under the Colorado River Compact as a state of the lower division and as a state of the lower basin.

(b) The state of New Mexico and the state of Utah reserve their respective rights and interests under the Colorado River Compact as states of the lower basin.

Source: "Upper Colorado River Basin Compact." Available online. URL: http://wrri.nmsu.edu/wrdis/compacts/Upper-Colorado-River-Basin-Compact.pdf. Accessed May 21, 2009.

Two U.S. Supreme Court Decisions in *Arizona v. California,* 1963 (excerpts)

The Supreme Court, on June 3, 1963 made two separate rulings for the complicated case of Arizona v. California. *The first ruling deals with the court's exceptions (and agreements that were in the majority) to a special master's ruling in the case. The second spells out the court's decision with reference to federal versus state responsibilities. The excerpts here note Native American rights to Colorado River water, the first in general terms and the second by specific tribe and reservation.*

373 U.S. 546
ARIZONA v. CALIFORNIA ET AL.

The Government, on behalf of five Indian Reservations in Arizona, California, and Nevada, asserted rights to water in the mainstream of the Colorado River. The Colorado River Reservation, located partly in Arizona and partly in California, is the largest. It was originally created by an Act of Congress in 1865, but its area was later increased by Executive Order. Other reservations were created by Executive Orders and amendments to them, ranging in dates from 1870 to 1907. The Master found both as a matter of fact and law that when the United States created these reservations or added to them, it reserved not only land but also the use of enough water from the Colorado to irrigate the irrigable portions of the reserved lands. The aggregate quantity of water which the Master held was reserved for all the reservations is about 1,000,000 acre-feet, to be used on around 135,000 irrigable acres of land. Here, as before the Master, Arizona argues that the United States had no power to make a reservation of navigable waters after Arizona became a State; that navigable waters could not be reserved by Executive Orders; that the United States did not intend to reserve water for the Indian Reservations; that the amount of water reserved should be measured by the reasonably foreseeable needs of the Indians living on the reservation rather than by the number of irrigable acres; and, finally, that the judicial doctrine of equitable apportionment should be used to divide the water between the Indians and the other people in the State of Arizona.

The last argument is easily answered. The doctrine of equitable apportionment is a method of resolving water disputes between States. It was created by this Court in the exercise of its original jurisdiction over controversies in which States are parties. An Indian Reservation is not a State. And while Congress has sometimes left Indian Reservations considerable power to manage their own affairs, we are not convinced by Arizona's argument that each reservation is so much like a State that its rights to water should be determined by the doctrine of equitable apportionment. Moreover, even were we to treat an Indian Reservation like a State, equitable apportionment would still not control since, under our view, the Indian claims here are governed by the statutes and Executive Orders creating the reservations. . . .

Arizona also argues that, in any event, water rights cannot be reserved by Executive Order. Some of the reservations of Indian lands here involved were made almost 100 years ago, and all of them were made over 45 years ago. In our view, these reservations, like those created directly by Congress, were not limited to land, but included waters as well. Congress and the Executive have ever since recognized these as Indian Reservations. Numerous appropriations, including appropriations for irrigation projects, have been made by Congress. They have been uniformly and universally treated as reservations by map makers, surveyors, and the public. We can give but short shrift at this late date to the argument that the reservations either of land or water are invalid because they were originally set apart by the Executive.

Arizona also challenges the Master's holding as to the Indian Reservations on two other grounds: first, that there is a lack of evidence showing that the United States in establishing the reservations intended to reserve water for them; second, that even if water was meant to be reserved the Master has awarded too much water. We reject both of these contentions. Most of the land in these reservations is and always has been arid. If the water necessary to sustain life is to be had, it must come from the Colorado River or its tributaries. It can be said without overstatement that when the Indians were put on these reservations they were not considered to be located in the most desirable area of the Nation. It is impossible to believe that when Congress created the great Colorado River Indian Reservation and when the Executive Department of this Nation created the other reservations they were unaware that most of the lands were of the desert kind—hot, scorching sands—and that water from the river would be essential to the life of the

Indian people and to the animals they hunted and the crops they raised. In the debate leading to approval of the first congressional appropriation for irrigation of the Colorado River Indian Reservation, the delegate from the Territory of Arizona made this statement:

> Irrigating canals are essential to the prosperity of these Indians. Without water there can be no production, no life; and all they ask of you is to give them a few agricultural implements to enable them to dig an irrigating canal by which their lands may be watered and their fields irrigated, so that they may enjoy the means of existence. You must provide these Indians with the means of subsistence or they will take by robbery from those who have. During the last year I have seen a number of these Indians starved to death for want of food. Cong. Globe, 38th Cong., 2d Sess. 1321 (1865).

The question of the Government's implied reservation of water rights upon the creation of an Indian Reservation was before this Court in *Winters v. United States,* 207 U.S. 564, decided in 1908. Much the same argument made to us was made in *Winters* to persuade the Court to hold that Congress had created an Indian Reservation without intending to reserve waters necessary to make the reservation livable. The Court rejected all of the arguments. As to whether water was intended to be reserved, the Court said, at p. 576:

> The lands were arid and, without irrigation, were practically valueless. And yet, it is contended, the means of irrigation were deliberately given up by the Indians and deliberately accepted by the Government. The lands ceded were, it is true, also arid; and some argument may be urged, and is urged, that with their cession there was the cession of the waters, without which they would be valueless, and 'civilized communities could not be established thereon.' And this, it is further contended, the Indians knew, and yet made no reservation of the waters. We realize that there is a conflict of implications, but that which makes for the retention of the waters is of greater force than that which makes for their cession.

The Court in *Winters* concluded that the Government, when it created that Indian Reservation, intended to deal fairly with the Indians by reserving for them the waters without which their lands would have been useless. *Winters* has been followed by this Court as recently as 1939 in *United States v. Powers,* 305 U.S. 52. We follow it now and agree that the United States did reserve the water rights for the Indians effective as of the time the Indian Reservations were created. This means, as the Master held, that these water rights, having vested before the Act became effective on June 25, 1929, are "present perfected rights" and as such are entitled to priority under the Act.

We also agree with the Master's conclusion as to the quantity of water intended to be reserved. He found that the water was intended to satisfy the future as well as the present needs of the Indian Reservations and ruled that enough water was reserved to irrigate all the practicably irrigable acreage on the reservations. Arizona, on the other hand, contends that the quantity of water reserved should be measured by the Indians' "reasonably foreseeable needs," which, in fact, means by the number of Indians. How many Indians there will be and what their future needs will be can only be guessed. We have concluded, as did the Master, that the only feasible and fair way by which reserved water for the reservations can be measured is irrigable acreage. The various acreages of irrigable land which the Master found to be on the different reservations we find to be reasonable.

376 U.S. 340

ARIZONA v. CALIFORNIA ET AL.

IT IS ORDERED, ADJUDGED AND DECREED THAT . . .

II. The United States, its officers, attorneys, agents and employees be and they are hereby severally enjoined:

(A) From operating regulatory structures controlled by the United States and from releasing water controlled by the United States other than in accordance with the following order of priority:

(D) From releasing water controlled by the United States for use in the States of Arizona, California, and Nevada for the benefit of any federal establishment named in this subdivision (D) except in accordance with the allocations made herein; provided, however, that such release may be made notwithstanding the provisions of Paragraph (5) of subdivision (B) of this Article; and provided further that nothing herein shall prohibit the United States from making future additional reservations of mainstream water for use in any of such States as may be authorized by law and subject to present perfected rights and rights under contracts theretofore made with water users in such State under Section 5 of the Boulder Canyon Project Act or any other applicable federal statute:

(1) The Chemehuevi Indian Reservation in annual quantities not to exceed (i) 11,340 acre-feet of diversions from the mainstream or (ii) the quantity of mainstream water necessary to supply the consumptive use required for irrigation of 1,900 acres and for the satisfaction of related uses, whichever of (i) or (ii) is less, with a priority date of February 2, 1907;

(2) The Cocopah Indian Reservation in annual quantities not to exceed (i) 2,744 acre-feet of diversions from the mainstream or (ii) the quantity of mainstream water necessary to supply the consumptive use required for irrigation of 431 acres and for the satisfaction of related uses, whichever of (i) or (ii) is less, with a priority date of September 27, 1917;

(3) The Yuma Indian Reservation in annual quantities not to exceed (i) 51,616 acre-feet of diversions from the mainstream or (ii) the quantity of mainstream water necessary to supply the consumptive use required for irrigation of 7,743 acres and for the satisfaction of related uses, whichever of (i) or (ii) is less, with a priority date of January 9, 1884;

(4) The Colorado River Indian Reservation in annual quantities not to exceed (i) 717,148 acre-feet of diversions from the mainstream or (ii) the quantity of mainstream water necessary to supply the consumptive use required for irrigation of 107,588 acres and for the satisfaction of related uses, whichever of (i) or (ii) is less, with priority dates of March 3, 1865, for lands reserved by the Act of March 3, 1865 (13 Stat. 541, 559); November 22, 1873, for lands reserved by the Executive Order of said date; November 16, 1874, for lands reserved by the Executive Order of said date, except as later modified; May 15, 1876, for lands reserved by the Executive Order of said date; November 22, 1915, for lands reserved by the Executive Order of said date;

(5) The Fort Mohave Indian Reservation in annual quantities not to exceed (i) 122,648 acre-feet of diversions from the mainstream or (ii) the quantity of mainstream water necessary to supply the consumptive use required for irrigation of 18,974 acres and for the satisfaction of related uses, whichever of (i) or (ii) is less, and, subject to the next succeeding proviso, with priority dates of Septem-

ber 19, 1890, for lands transferred by the Executive Order of said date; February 2, 1911, for lands reserved by the Executive Order of said date; provided, however, that lands conveyed to the State of California pursuant to the Swamp Land Act [9 Stat. 519 (1850)] as well as any accretions thereto to which the owners of such land may be entitled, and lands patented to the Southern Pacific Railroad pursuant to the Act of July 27, 1866 (14 Stat. 292), shall not be included as irrigable acreage within the Reservation and that the above specified diversion requirement shall be reduced by 6.4 acre-feet per acre of such land that is irrigable; provided that the quantities fixed in this paragraph and paragraph (4) shall be subject to appropriate adjustment by agreement or decree of this Court in the event that the boundaries of the respective reservations are finally determined;

Source: "U.S. Supreme Court. *ARIZONA v. CALIFORNIA,* 373 U.S. 546 (1963)." FindLaw.com. Available online. URL: http://caselaw.lp.findlaw.com/scripts/getcase.pl?court=us&vol=373&invol=546. "U.S. Supreme Court *ARIZONA v. CALIFORNIA,* 376 U.S. 340 (1963)." FindLaw.com. Available online. URL: http://caselaw.lp.findlaw.com/scripts/getcase.pl?court=us&vol=376&invol=340. Accessed May 19, 2009.

The Great Lakes Water Quality Agreement of 1972 (excerpts)

The eutrophication of Lake Erie caused an outcry on both sides of the U.S.–Canadian border and led to an investigation by the International Joint Commission (IJC). The IJC released its report in 1970 on "Pollution of Lake Erie, Lake Ontario, and the International Section of the St. Lawrence River." This report became the basis of the first Great Lakes Water Quality Agreement, signed in Ottawa on April 15, 1972, and excerpted below.

Article II
General Water Quality Objectives

The following general water quality objectives for the boundary waters of the Great Lakes System are adopted. These waters should be:

(a) Free from substances that enter the waters as a result of human activity and that will settle to form putrescent or otherwise objectionable sludge deposits, or that will adversely affect aquatic life or waterfowl;

(b) Free from floating debris, oil, scum and other floating materials entering the waters as a result of human activity in amounts sufficient to be unsightly or deleterious;

(c) Free from materials entering the waters as a result of human activity producing colour, odour or other conditions in such a degree as to create a nuisance;
(d) Free from substances entering the waters as a result of human activity in concentrations that are toxic or harmful to human, animal or aquatic life;
(e) Free from nutrients entering the waters as a result of human activity in concentrations that create nuisance growths of aquatic weeds and algae.

Article III
Specific Water Quality Objectives

1. The specific water quality objectives for the boundary waters of the Great Lakes System set forth in annex 1 are adopted.
2. The specific water quality objectives may be modified and additional specific water quality objectives for the boundary waters of the Great Lakes System or for particular sections thereof may be adopted by the Parties in accordance with the provisions of articles IX and XII of this Agreement.
3. The specific water quality objectives adopted pursuant to this article represent the minimum desired levels of water quality in the boundary waters of the Great Lakes System and are not intended to preclude the establishment of more stringent requirements.
4. Notwithstanding the adoption of specific water quality objectives, all reasonable and practicable measures shall be taken to maintain the levels of water quality existing at the date of entry into force of this Agreement in those areas of the boundary waters of the Great Lakes System where such levels exceed the specific water quality objectives.

ARTICLE IV
Standards and Other Regulatory Requirements

Water quality standards and other regulatory requirements of the Parties shall be consistent with the achievement of the water quality objectives. The parties shall use their best efforts to ensure that water quality standards and other regulatory requirements of the State and Provincial Governments shall similarly be consistent with the achievement of the water quality objectives.

Source: "United States of America and Canada: Agreement on Great Lakes Water Quality (with annexes and attachments)." International Freshwater Treaties Database. Oregon State University. URL: http://ocid.nacse.org/tfdd/tfdddocs/321ENG.pdf. Accessed May 19, 2009.

U.S.–Mexican Treaty Regarding Colorado River Water Salinity, 1973

Throughout the 1950s and 1960s, Mexico's allotment of Colorado River water had become increasingly saline. During the latter decade, official protests were lodged but little action was taken. Finally, in the early 1970s the two countries sat down at the negotiating table and reached an agreement by which the United States promised to deliver less saline water to Mexico (section 1a). The United States built a desalination plant at Yuma, Arizona, to meet the requirements of the agreement. Below are minutes of the meeting at which the treaty was agreed to.

MEXICO-US AGREEMENT ON THE PERMANENT AND DEFINITIVE SOLUTION TO THE SALINITY OF THE COLORADO RIVER (MINUTE NO. 242), MEXICO CITY 1973.

International Boundary and Water Commission
United States and Mexico
Mexico, D.F., August 30, 1973.
Minute No. 242
Permanent and Definitive Solution to the International Problem of the Salinity of the Colorado River.

The Commission met at the Secretariat of Foreign Relations, at Mexico, D.F., at 5:00 p.m. on August 30, 1973, pursuant to the instructions received by the two Commissioner from their respective Governments, in order to incorporate in a Minute of the Commission the joint recommendations which were made to their respective Presidents by the Special Representative of President Richard Nixon, Ambassador Herbert Brownell, and the Secretary of Foreign Relations of Mexico, Lic. Emilio O. Rabasa, and which have been approved by the Presidents, for a permanent and definitive solution of the international problem of the salinity of the Colorado River, resulting from the negotiations which they, and their technical and juridical advisers, held in June, July and August of 1973, in compliance with the references to this matter contained in the Joint Communique of Presidents Richard Nixon and Luis Echeverria of June 17, 1972.

Accordingly, the Commission submits for the approval of the two Governments the following

Resolution:

1. Referring to the annual volume of Colorado River waters guaranteed to Mexico under the Treaty of 1944, of 1,500,000 acre-feet (1,850,234,000 cubic meters):

(a) The United States shall adopt measures to assure that not earlier than January 1, 1974, and no later than July 1, 1974, the approximately 1,360,000 acre-feet (1,677,545,000 cubic meters) delivered to Mexico upstream of Morelos Dam, have an annual average salinity of no more than 115 p.p.m.+30 p.p.m. Mexican count) over the annual average salinity of Colorado River waters which arrive at Imperial Dam, with the understanding that any waters that may be delivered to Mexico under the Treaty of 1944 by means of the All-American Canal shall be considered as having been delivered upstream of Morelos Dam for the purpose of computing this salinity.

(b) The United States will continue to deliver to Mexico on the land boundary at San Luis and in the limitrophe section of the Colorado River downstream from Morelos Dam approximately 140,000 acre-feet (172,689,000 cubic meters) annually with a salinity substantially the same as that of the waters customarily delivered there.

(c) Any decrease in deliveries under point 1(b) will be made up by an equal increase in deliveries under point (1a).

(d) Any other substantial changes in the aforementioned volumes of water at the stated locations must be agreed to by the Commission.

(e) Implementation of the measures referred to in point (1a) above is subject to the requirement in point 10 of the authorization of the necessary works.

2. The life of Minute No. 241 shall be terminated upon approval of the present Minute. From September 1, 1973, until the provisions of point 1(a) become effective, the United States shall discharge to the Colorado River downstream from Morelos Dam volumes of drainage waters from the Wellton-Mohawk District at the annual rate of 118,000 acre-feet (145,551,000 cubic meters) and substitute therefor an equal volume of other waters to be discharged to the Colorado River above Morelos Dam; and, pursuant to the decision of President Echeverria expressed in the Joint Communique of June 17, 1972, the United States shall discharge to the Colorado River downstream from Morelos Dam the drainage waters of the Wellton-Mohawk District that do not form a part of the volumes of drainage waters referred to above, with the understanding that this

remaining volume will not be replaced by substitution waters. The Commission shall continue to account for the drainage waters discharged below Morelos Dam as part of those described in the provisions of Article 10 of the Water Treaty of February 3, 1944.

3. As a part of the measures referred to in point 1(a), the United States shall extend in its territory the concrete-lined Wellton-Mohawk bypass drain from Morelos Dam to the Arizona-Sonora international boundary, and operate and maintain the portions of the Wellton-Mohawk bypass drain located in the United States.

4. To complete the drain referred to in point 3, Mexico, through the Commission and at the expense of the United States, shall construct, operate and maintain an extension of the concrete-lined bypass drain from the Arizona-Sonora international boundary to the Santa Clara Slough of a capacity of 353 cubic feet (10 cubic meters) per second. Mexico shall permit the United States to discharge through this drain to the Santa Clara Slough all or a portion of the Wellton-Mohawk drainage waters, the volumes of brine from such desalting operations in the United States as are carried out to implement the Resolution of this Minute, and any other volumes of brine which Mexico may agree to accept. It is understood that no radioactive material or nuclear wastes shall be discharged through this drain, and that the United States shall acquire no right to navigation, servitude or easement by reason of the existence of the drain, nor other legal rights, except as expressly provided in this point.

5. Pending the conclusion by the Governments of the United States and Mexico of a comprehensive agreement on groundwater in the border areas, each country shall limit pumping of groundwaters in its territory within five miles (eight kilometers) of the Arizona-Sonora boundary near San Luis to 160,000 acre-feet (197,358,000 cubic meters) annually.

6. With the objective of avoiding future problems, the United States and Mexico shall consult with each other prior to undertaking any new development of either the surface or the groundwater resources, or undertaking substantial modifications of present developments, in its own territory in the border area that might adversely affect the other country.

7. The United States will support efforts by Mexico to obtain appropriate financing on favorable terms for the improvement and rehabilitation of

the Mexicali Valley. The United States will also provide non-reimbursable assistance on a basis mutually acceptable to both countries exclusively for those aspects of the Mexican rehabilitation program of the Mexicali Valley relating to the salinity problem, including tile drainage. In order to comply with the above-mentioned purposes, both countries will undertake negotiations as soon as possible.

8. The United States and Mexico shall recognize the undertakings and understandings contained in this Resolution as constituting the permanent and definitive solution of the salinity problem referred to in the Joint Communique of President Richard Nixon and President Luis Echeverria dated June 17, 1972.

9. The measures required to implement this Resolution shall be undertaken and completed at the earliest practical date.

10. This Minute is subject to the express approval of both Governments by exchange of Notes. It shall enter into force upon such approval; provided, however, that the provisions which are dependent for their implementation on the construction of works or on other measures which require expenditure of funds by the United States, shall become effective upon the notification by the United States to Mexico of the authorization by the United States Congress of said funds, which will be sought promptly.

Thereupon, the meeting adjourned.

Source: "Mexico–U.S. Agreement on the Permanent and Definitive Solution to the Salinity of the Colorado River." International Freshwater Treaties Database. Oregon State University. Available online. URL: http://ocid.nacse.org/tfdd/tfdddocs/315ENG.htm. Accessed May 21, 2009.

The Clean Water Act of 1972 as Amended in 1977 (excerpts)

The growing influence of the environmental movement in the United States, coupled with such environmental disasters in the 1960s as the eutrophication of Lake Erie and the fire on the Cuyahoga River, brought about an awareness of the danger of pollution to precious natural resources. The Clean Water Act of 1972 was a major congressional attempt to address the problem and update earlier federal water laws; the first major amendments to the act came in 1977. Below are the first two sections of the amended act.

The Federal Water Pollution Control Act is amended to read as follows:

SEC. 101 [33 U.S.C. 1251] Declaration of Goals and Policy

(a) The objective of this Act is to restore and maintain the chemical, physical, and biological integrity of the Nation's waters. In order to achieve this objective it is hereby declared that, consistent with the provisions of this Act—

(1) it is the national goal that the discharge of pollutants into the navigable waters be eliminated by 1985;
(2) it is the national goal that wherever attainable, an interim goal of water quality which provides for the protection and propagation of fish, shellfish, and wildlife and provides for recreation in and on the water be achieved by July 1, 1983;
(3) it is the national policy that the discharge of toxic pollutants in toxic amounts be prohibited;
(4) it is the national policy that Federal financial assistance be provided to construct publicly owned waste treatment works;
(5) it is the national policy that areawide waste treatment management planning processes be developed and implemented to assure adequate control of sources of pollutants in each State;
(6) it is the national policy that a major research and demonstration effort be made to develop technology necessary to eliminate the discharge of pollutants into the navigable waters, waters of the contiguous zone, and the oceans; and
(7) it is the national policy that programs for the control of nonpoint sources of pollution be developed and implemented in an expeditious manner so as to enable the goals of this Act to be met through the control of both point and nonpoint sources of pollution.

[101(a)(7) added by PL 100-41]

. . . (d) Except as otherwise expressly provided in this Act, the Administrator of the Environmental Protection Agency (hereinafter in this Act called "Administrator") shall administer this Act.

SEC. 102 [33 U.S.C. 1252] Comprehensive Programs for Water Pollution Control

(a) The Administrator shall, after careful investigation, and in cooperation with other Federal agencies, State water pollution control agencies, interstate agencies, and the municipalities and industries involved, prepare or develop comprehensive programs for preventing, reducing, or eliminating the pollution of the navigable waters and ground waters and improving the sanitary condition of surface and underground waters. In the development of such comprehensive programs due regard shall be given to the improvements which are necessary to conserve such waters for the protection and propagation of fish and aquatic life and wildlife, recreational purposes, and the withdrawal of such waters for public water supply, agricultural, industrial, and other purposes. For the purpose of this section, the Administrator is authorized to make joint investigations with any such agencies of the condition of any waters in any State or States, and of the discharges of any sewage, industrial wastes, or substance which may adversely affect such waters.

(b) (1) In the survey of planning of any reservoir by the Corps of Engineers, Bureau of Reclamation, or other Federal agency, consideration shall be given to inclusion of storage for regulation of streamflow, except that any such storage and water releases shall not be provided as a substitute for adequate treatment or other methods of controlling waste at the source.

(2) The need for and the value of storage or regulation of streamflow (other than for water quality) including but not limited to navigation, salt water intrusion, recreation, esthetics, and fish and wildlife, shall be determined by the Corps of Engineers, Bureau of Reclamation, or other Federal agencies.

(3) The need for, the value of, and the impact of, storage for water quality control shall be determined by the Administrator, and his views on these matters shall be set forth in any report or presentation to Congress proposing authorization or construction of any reservoir including such storage.

(4) The value of such storage shall be taken into account in determining the economic value of the entire project of which it is a part, and costs shall be allocated to the purpose of regulation of streamflow in a manner which will insure that all project purposes, share equitably in the benefits of multiple-purpose construction.

(5) Costs of regulation of streamflow features incorporated in any Federal reservoir or other impoundment under the provisions of this Act

shall be determined and the beneficiaries identified and if the benefits are widespread or national in scope, the costs of such features shall be nonreimbursable.

(6) No license granted by the Federal Power Commission for a hydroelectric power project shall include storage for regulation of stream flow for the purpose of water quality control unless the Administrator shall recommend its inclusion and such reservoir storage capacity shall not exceed such proportion of the total storage required for the water quality control plan as the drainage area of such reservoir bears to the drainage area of the river basin or basins involved in such water quality control plan.

(c) (1) The Administrator shall, at the request of the Governor of a State, or a majority of the Governors when more than one State is involved, make a grant to pay not to exceed 50 per centum of the administrative expenses of a planning agency for a period not to exceed three years, which period shall begin after the date of enactment of the Federal Water Pollution Control Act Amendments of 1972, if such agency provides for adequate representation of appropriate State, interstate, local, or (when appropriate) international interests in the basin or portion thereof involved and is capable of developing an effective, comprehensive water quality control plan for a basin or portion thereof.

(2) Each planning agency receiving a grant under this subsection shall develop a comprehensive pollution control plan for the basin or portion thereof which—

(A) is consistent with any applicable water quality standards, effluent and other limitations, and thermal discharge regulations established pursuant to current law within the basin;

(B) recommends such treatment works as will provide the most effective and economical means of collection, storage, treatment, and elimination of pollutants and recommends means to encourage both municipal and industrial use of such works;

(C) recommends maintenance and improvement of water quality within the basin or portion thereof and recommends methods of adequately financing those facilities as may be necessary to implement the plan; and

(D) as appropriate, is developed in cooperation with, and is consistent with any comprehensive plan prepared by the Water Resources Council, any areawide waste management plans developed pursuant to section 208 of this Act, and any State plan developed pursuant to section 303(e) of this Act.

(3) For the purposes of this subsection the term "basin" includes, but is not limited to, rivers and their tributaries, streams, coastal waters, sounds, estuaries, bays, lakes, and portions thereof, as well as the lands drained thereby.

(d) The Administrator, after consultation with the States, and River Basin Commissions established under the Water Resources Planning Act, shall submit a report to Congress on or before July 1, 1978, which analyzes the relationship between programs under this Act, and the programs by which State and Federal agencies allocate quantities of water. Such report shall include recommendations concerning the policy in section 101(g) of the Act to improve coordination of efforts to reduce and eliminate pollution in concert with programs for managing water resources.

Source: "Federal Water Pollution Control Act, as Amended by the Clean Water Act of 1977." Available online. URL: http://www.epa.gov/npdes/pubs/cwatxt.txt. Accessed May 19, 2009.

The Safe Drinking Water Act of 1974 as Amended in 2001 (excerpts)

President Gerald R. Ford signed the Safe Drinking Water Act (SDWA) into law in 1974 to protect the nation's drinking water supply. Since then it has been amended numerous times. The excerpts below reflect the post–September 11, 2001, amendments to SDWA, which take into account terrorist attacks on drinking water systems. The title, chapter, subchapter, and part B refer to the law within the U.S. Code.

Title 42, Chapter 6A, Subchapter XII, Part B
300g. Coverage

Subject to sections 300g–4 and 300g–5 of this title, national primary drinking water regulations under this part shall apply to each public water system in each State; except that such regulations shall not apply to a public water system—

(1) which consists only of distribution and storage facilities (and does not have any collection and treatment facilities);

(2) which obtains all of its water from, but is not owned or operated by, a public water system to which such regulations apply;
(3) which does not sell water to any person; and
(4) which is not a carrier which conveys passengers in interstate commerce.

Sec. 1412 (b) STANDARDS.—
IDENTIFICATION OF CONTAMINANTS FOR LISTING.—

(A) GENERAL AUTHORITY.—The Administrator shall, in accordance with the procedures established by this subsection, publish a maximum contaminant level goal and promulgate a national primary drinking water regulation for a contaminant (other than a contaminant referred to in paragraph (2) for which a national primary drinking water regulation has been promulgated as of the date of enactment of the Safe Drinking Water Act Amendments of 1996) if the Administrator determines that—

(i) the contaminant may have an adverse effect on the health of persons;
(ii) the contaminant is known to occur or there is a substantial likelihood that the contaminant will occur in public water systems with a frequency and at levels of public health concern; and
(iii) in the sole judgment of the Administrator, regulation of such contaminant presents a meaningful opportunity for health risk reduction for persons served by public water systems.

(B) REGULATION OF UNREGULATED CONTAMINANTS.—
(i) LISTING OF CONTAMINANTS FOR CONSIDERATION.—
(I) Not later than 18 months after the date of enactment of the Safe Drinking Water Act Amendments of 1996 and every 5 years thereafter, the Administrator, after consultation with the scientific community, including the Science Advisory Board, after notice and opportunity for public comment, and after considering the occurrence data base established under section 1445(g), shall publish a list of contaminants which, at the time of publication, are not subject to any proposed or promulgated national primary drinking water regulation, which are known or anticipated to occur in public water systems, and which may require regulation under this title.
(II) The unregulated contaminants considered under subclause (I) shall include, but not be limited to, substances referred to in section 101(14) of the Comprehensive Environmental

Response, Compensation, and Liability Act of 1980, and substances registered as pesticides under the Federal Insecticide, Fungicide, and Rodenticide Act.

(III) The Administrator's decision whether or not to select an unregulated contaminant for a list under this clause shall not be subject to judicial review. . . .

(C) Priorities.—In selecting unregulated contaminants for consideration under subparagraph (B), the Administrator shall select contaminants that present the greatest public health concern. The Administrator, in making such selection, shall take into consideration, among other factors of public health concern, the effect of such contaminants upon subgroups that comprise a meaningful portion of the general population (such as infants, children, pregnant women, the elderly, individuals with a history of serious illness, or other subpopulations) that are identifiable as being at greater risk of adverse health effects due to exposure to contaminants in drinking water than the general population.

(D) Urgent threats to public health.—The Administrator may promulgate an interim national primary drinking water regulation for a contaminant without making a determination for the contaminant under paragraph (4)(C), or completing the analysis under paragraph (3)(C), to address an urgent threat to public health as determined by the Administrator after consultation with and written response to any comments provided by the Secretary of Health and Human Services, acting through the director of the Centers for Disease Control and Prevention or the director of the National Institutes of Health. A determination for any contaminant in accordance with paragraph (4)(C) subject to an interim regulation under this subparagraph shall be issued, and a completed analysis meeting the requirements of paragraph (3)(C) shall be published, not later than 3 years after the date on which the regulation is promulgated and the regulation shall be repromulgated, or revised if appropriate, not later than 5 years after that date.

(F) Health advisories and other actions.—The Administrator may publish health advisories (which are not regulations) or take other appropriate actions for contaminants not subject to any national primary drinking water regulation.

300i. Emergency powers

Notwithstanding any other provision of this subchapter the Administrator, upon receipt of information that a contaminant which is present in or is likely to enter a public water system or an underground source of drinking water, or that there is a threatened or potential terrorist attack (or other intentional act designed to disrupt the provision of safe drinking water or to impact adversely the safety of drinking water supplied to communities and individuals), which may present an imminent and substantial endangerment to the health of persons, and that appropriate State and local authorities have not acted to protect the health of such persons, may take such actions as he may deem necessary in order to protect the health of such persons. To the extent he determines it to be practicable in light of such imminent endangerment, he shall consult with the State and local authorities in order to confirm the correctness of the information on which action proposed to be taken under this subsection is based and to ascertain the action which such authorities are or will be taking. The action which the Administrator may take may include (but shall not be limited to)

(1) issuing such orders as may be necessary to protect the health of persons who are or may be users of such system (including travelers), including orders requiring the provision of alternative water supplies by persons who caused or contributed to the endangerment, and

(2) commencing a civil action for appropriate relief, including a restraining order or permanent or temporary injunction.

300i-1. Tampering with public water systems

Tampering

Any person who tampers with a public water system shall be imprisoned for not more than 20 years, or fined in accordance with title 18, or both.

(b) Attempt or threat

Any person who attempts to tamper, or makes a threat to tamper, with a public drinking water system be imprisoned for not more than 10 years, or fined in accordance with title 18, or both.

(c) Civil penalty

The Administrator may bring a civil action in the appropriate United States district court (as determined under the provisions of title 28) against any person who tampers, attempts to tamper, or makes a threat to tamper with a public water system. The court may impose on such person a civil penalty of not more than $1,000,000 for such tampering or not more than $100,000 for such attempt or threat.

(d) "Tamper" defined

For purposes of this section, the term "tamper" means—

(1) to introduce a contaminant into a public water system with the intention of harming persons; or

(2) to otherwise interfere with the operation of a public water system with the intention of harming persons.

300i-2. Terrorist and other intentional acts

(a) Vulnerability assessments

(1) Each community water system serving a population of greater than 3,300 persons shall conduct an assessment of the vulnerability of its system to a terrorist attack or other intentional acts intended to substantially disrupt the ability of the system to provide a safe and reliable supply of drinking water. The vulnerability assessment shall include, but not be limited to, a review of pipes and constructed conveyances, physical barriers, water collection, pretreatment, treatment, storage and distribution facilities, electronic, computer or other automated systems which are utilized by the public water system, the use, storage, or handling of various chemicals, and the operation and maintenance of such system. The Administrator, not later than August 1, 2002, after consultation with appropriate departments and agencies of the Federal Government and with State and local governments, shall provide baseline information to community water systems required to conduct vulnerability assessments regarding which kinds of terrorist attacks or other intentional acts are the probable threats to—

(A) substantially disrupt the ability of the system to provide a safe and reliable supply of drinking water; or

(B) otherwise present significant public health concern. . . .

6 (b) Emergency response plan

Each community water system serving a population greater than 3,300 shall prepare or revise, where necessary, an emergency response plan that incorporates the results of vulnerability assessments that have been completed. Each such community water system shall certify to the Administrator, as soon as reasonably possible after the enactment of this section, but not later than 6 months after the completion of the vulnerability assessment under subsection (a) of this section, that the system has completed such plan. The emergency response plan shall include, but not be limited to, plans, procedures, and identification of equipment that can be implemented or utilized in the event of a terrorist or other intentional attack on the public water system. The emergency response plan shall also include actions, procedures, and identification of equipment which can obviate or significantly lessen the impact of terrorist attacks or other intentional actions on the public health and the safety and supply of drinking water provided to communities and individuals. Community water systems shall, to the extent possible, coordinate with existing Local Emergency Planning Committees established under the Emergency Planning and Community Right-to-Know Act (42 U.S.C. 11001 et seq.) when preparing or revising an emergency response plan under this subsection.

Source: "Subchapter XII—Safety of Public Water Systems." U.S. Code Collection. Cornell University Law School. National Oceanic and Atmospheric Administration. Available online. URL: http://www.csc.noaa.gov/cmfp/reference/reference_frame.htm. Accessed May 20, 2009.

The Great Lakes Water Quality Agreement of 1978 (excerpts)

It soon became evident that the first Great Lakes Water Quality Agreement was too limited, both in its standards and its geographical scope. What was needed was a new agreement that would cover all five of the lakes, as well as pollution such as toxic chemicals. This second Great Lakes Water Quality Agreement was later amended by protocols that went into effect in 1987 and 1993.

ARTICLE II

Purpose

The purpose of the Parties is to restore and maintain the chemical, physical, and biological integrity of the waters of the Great Lakes Basin Ecosystem. In order to achieve this purpose, the Parties agree to make a maximum effort to develop programs, practices and technology necessary for a better understanding of the Great Lakes Basin Ecosystem and to eliminate or reduce to the maximum extent practicable the discharge of pollutants to the Great Lakes System.

Consistent with the provisions of this Agreement, it is the policy of the Parties that:

(a) The discharge of toxic substances in toxic amounts be prohibited and the discharge of any or all persistent toxic substances be virtually eliminated;

(b) Financial assistance to construct publicly owned waste treatment works be provided by a combination of local, state, provincial, and federal participation; and

(c) Coordinated planning processes and best management practices be developed and implemented by the respective jurisdictions to ensure adequate control of all sources of pollutants.

ARTICLE III

General Objectives

The Parties adopt the following General Objectives for the Great Lakes System. These waters should be:

(a) Free from substances that directly or indirectly enter the waters as a result of human activity and that will settle to form putrescent or otherwise objectionable sludge deposits, or that will adversely affect aquatic life or waterfowl;

(b) Free from floating materials such as debris, oil, scum, and other immiscible substances resulting from human activities in amounts that are unsightly or deleterious;

(c) Free from materials and heat directly or indirectly entering the water as a result of human activity that alone, or in combination with other materials, will produce color, odor, taste, or other conditions in such a degree as to interfere with beneficial uses;

(d) Free from materials and heat directly or indirectly entering the water as a result of human activity that alone, or in combination with other materials, will produce conditions that are toxic or harmful to human, animal, or aquatic life; and

(e) Free from nutrients directly or indirectly entering the waters as a result of human activity in amounts that create growths of aquatic life that interfere with beneficial uses.

ARTICLE IV

Specific Objectives

1. The Parties adopt the Specific Objectives for the boundary waters of the Great Lakes System as set forth in Annex 1, subject to the following:

(a) The Specific Objectives adopted pursuant to this Article represent the minimum levels of water quality desired in the boundary waters of the Great Lakes System and are not intended to preclude the establishment of more stringent requirements.

(b) The determination of the achievement of Specific Objectives shall be based on statistically valid sampling data.

(c) Notwithstanding the adoption of Specific Objectives, all reasonable and practicable measures shall be taken to maintain or improve the existing water quality in those areas of the boundary waters of the Great Lakes System where such water quality is better than that prescribed by the Specific Objectives, and in those areas having outstanding natural resource value.

(d) The responsible regulatory agencies shall not consider flow augmentation as a substitute for adequate treatment to meet the Specific Objectives.

(e) The Parties recognize that in certain areas of inshore waters natural phenomena exist which, despite the best efforts of the Parties, will prevent the achievement of some of the Specific Objectives. As early as possible, these areas should be identified explicitly by the appropriate jurisdictions and reported to the International Joint Commission.

(f) Limited use zones in the vicinity of present and future municipal, industrial and tributary point source discharges shall be designated by the responsible regulatory agencies within which some of the Specific Objectives may not apply. Establishment of these zones shall not be considered a substitute for adequate treatment or control of discharges at their source. The size shall be minimized to the greatest possible degree, being no larger than that attainable by all reasonable and practicable effluent treatment measures. The boundary of a limited use zone shall not transect the International Boundary. Principles for the designation of limited use zones are set out in Annex 2.

2. The Specific Objectives for the boundary waters of the Great Lakes System or for particular portions thereof shall be kept under review by the Parties and by the International Joint Commission, which shall make appropriate recommendations.

3. The Parties shall consult on:

(a) The establishment of Specific Objectives to protect beneficial uses from the combined effects of pollutants; and

(b) The control of pollutant loading rates for each lake basin to protect the integrity of the ecosystem over the long term.

ARTICLE V

Standards, Other Regulatory Requirements, and Research

1. Water quality standards and other regulatory requirements of the Parties shall be consistent with the achievement of the General and Specific

Objectives. The Parties shall use their best efforts to ensure that water quality standards and other regulatory requirements of the State and Provincial Governments shall similarly be consistent with the achievement of these Objectives. Flow augmentation shall not be considered as a substitute for adequate treatment to meet water quality standards or other regulatory requirements.

2. The Parties shall use their best efforts to ensure that:
   (a) The principal research funding agencies in both countries orient the research programs of their organizations in response to research priorities identified by the Science Advisory Board and recommended by the Commission; and
   (b) Mechanisms be developed for appropriate cost-effective international cooperation.

ARTICLE XV

Supersession

This Agreement supersedes the Great Lakes Water Quality Agreement of April 15, 1972, and shall be referred to as the "Great Lakes Water Quality Agreement of 1978."

ANNEX I

Specific Objectives

These Objectives are based on available information on cause/effect relationships between pollutants and receptors to protect the recognized most sensitive use in all waters. These Objectives may be amended, or new Objectives may be added, by mutual consent of the Parties.

I. CHEMICAL
A. Persistent Toxic Substances

1. Organic
(a) Pesticides

Aldrin/Dieldrin

The sum of the concentrations of aldrin and dieldrin in water should not exceed 0.001 microgram per liter. The sum of concentrations of aldrin and dieldrin in the edible portion of fish should not exceed 0.3 microgram per gram (wet weight basis) for the protection of human consumers of fish.

Chlordane

The concentration of chlordane in water should not exceed 0.06 microgram per liter for the protection of aquatic life.

DDT and Metabolites

The sum of the concentrations of DDT and its metabolites in water should not exceed 0.003 microgram per liter. The sum of the concentrations of DDT and its metabolites in whole fish should not exceed 1.0 microgram per gram (wet weight basis) for the protection of fish-consuming aquatic birds.

Endrin

The concentration of endrin in water should not exceed 0.002 microgram per liter. The concentration of endrin in the edible portion of fish should not exceed 0.3 microgram per gram (wet weight basis) for the protection of human consumers of fish.

Heptachlor/Hepiachlor Epoxide

The sum of the concentrations of heptachlor and heptachlor epoxide in water should not exceed 0.001 microgram per liter. The sum of the concentrations of heptachlor and heptachlor epoxide in edible portions of fish should not exceed 0.3 microgram per gram (wet weight basis) for the protection of human consumers of fish.

Lindane

The concentration of lindane in water should not exceed 0.01 microgram per liter for the protection of aquatic life. The concentration of lindane in edible portions of fish should not exceed 0.3 microgram per gram (wet weight basis) for the protection of human consumers of fish.

Methoxychlor

The concentration of methoxychlor in water should not exceed 0.04 microgram per liter for the protection of aquatic life.

Mirex

For the protection of aquatic organisms and fish-consuming birds and animals, mirex and its degradation products should be substantially absent from water and aquatic organisms. Substantially absent here means less than detection levels as determined by the best scientific methodology available.

Toxaphene

The concentration of toxaphene in water should not exceed 0.008 microgram per liter for the protection of aquatic life.

(b) Other Compounds

Phthalic Acid Esters

The concentration of dibutyl phthalate and di(2-ethylhexyl) phthalate in water should not exceed 4.0 microgram per liter and 0.6 microgram per liter, respectively, for the protection of aquatic life. Other phthalic acid esters should not exceed 0.2 microgram per liter in waters for the protection of aquatic life.

Polychlorinated Biphenyls (PCBs)

The concentration of total polychlorinated biphenyls in fish tissues (whole fish, calculated on a wet weight basis), should not exceed 0.1 microgram per gram for the protection of birds and animals which consume fish.

Unspecified Organic Compounds

For other organic contaminants, for which Specific Objectives have not been defined, but which can be demonstrated to be persistent and are likely to be toxic, the concentrations of such compounds in water or aquatic organisms should be substantially absent, i.e., less than detection levels as determined by the best scientific methodology available.

2. Inorganic

(a) Metals

Arsenic

The concentrations of total arsenic in an unfiltered water sample should not exceed 50 micrograms per liter to protect raw material for public water supplies.

Cadmium

The concentration of total cadmium in an unfiltered water sample should not exceed 0.2 microgram per liter to protect aquatic life.

Chromium

The concentration of total chromium in an unfiltered water sample should not exceed 50 micrograms per liter to protect raw waters for public water supplies.

Copper

The concentration of total copper in an unfiltered water sample should not exceed 5 micrograms per liter to protect aquatic life.

Iron

The concentration of total iron in an unfiltered water sample should not exceed 300 micrograms per liter to protect aquatic life.

Lead

The concentration of total lead in an unfiltered water sample should not exceed 10 micrograms per liter in Lake Superior, 20 micrograms per liter in Lake Huron and 25 micrograms per liter in all remaining Great Lakes to protect aquatic life.

Mercury

The concentration of total mercury in a filtered water sample should not exceed 0.2 microgram per liter nor should the concentration of total mercury in whole fish exceed 0.5 microgram per gram (wet weight basis) to protect aquatic life and fish-consuming birds.

Nickel

The concentration of total nickel in an unfiltered water sample should not exceed 25 micrograms per liter to protect aquatic life.

Selenium

The concentration of total selenium in an unfiltered water sample should not exceed 10 micrograms per liter to protect raw water for public water supplies.

Zinc

The concentration of total zinc in an unfiltered water sample should not exceed 30 micrograms per liter to protect aquatic life.

(b) Other Inorganic Substances

Fluoride

The concentration of total fluoride in an unfiltered water sample should not exceed 1200 micrograms per liter to protect raw waters for public water supplies.

Total Dissolved Solids

In Lake Erie, Lake Ontario and the International Section of the St. Lawrence River, the level of total dissolved solids should not exceed 200 milligrams per liter. In the St. Clair River, Lake St. Clair, the Detroit River and the Niagara River, the level should be consistent with maintaining the levels of total dissolved solids in Lake Erie and Lake Ontario at not to exceed 200 milligrams per liter. In the remaining boundary waters, pending further study, the level of total dissolved solids should not exceed present levels.

B. Non-Persistent Toxic Substances

1. Organic Substances

(a) Pesticides

Diazinon

The concentration of diazinon in an unfiltered water sample should not exceed 0.08 microgram per liter for the protection of aquatic life.

Guthion

The concentration of guthion in an unfiltered water sample should not exceed 0.005 microgram per liter for the protection of aquatic life.

Parathion

The concentration of parathion in an unfiltered water sample should not exceed 0.008 microgram per liter for the protection of aquatic life.

Other Pesticides

The concentration of unspecified, non-persistent pesticides should not exceed 0.05 of the median lethal concentration on a 96-hour test for any sensitive local species.

(b) Other Substances

Unspecified Non-Persistent Toxic Substances and Complex Effluents

Unspecified non-persistent toxic substances and complex effluents of municipal, industrial, or other origin should not be present in concentrations which exceed 0.05 of the median lethal concentration in a 96 hour test for any sensitive local species to protect aquatic life.

Oil and Petrochemicals

Oil and petrochemicals should not be present in concentrations that:

(1) can be detected as visible film, sheen, or discoloration on the surface;

(2) can be detected by odor;

(3) can cause tainting of edible aquatic organisms;

(4) can form deposits on shorelines and bottom sediments that are detectable by sight or odor, or are deleterious to resident aquatic organisms.

2. Inorganic Substances

Ammonia

The concentration of unionized ammonia (NH_3) should not exceed 20 micrograms per liter for the protection of aquatic life. Concentrations of total ammonia should not exceed 500 micrograms per liter for the protection of public water supplies.

Hydrogen Sulfide

The concentration of undissociated hydrogen sulfide should not exceed 2 micrograms per liter to protect aquatic life.

C. Other Substances

1. Dissolved oxygen

In the connecting channels and in the upper waters of the Lakes, the dissolved oxygen level should not be less than 6.0 milligrams per liter at any time; in hypolimnetic waters, it should not be less than necessary for the support of fish life, particularly cold water species.

2. pH

Values of pH should not be outside the range of 6.5 to 9.0, should discharge change the pH at the boundary of a limited use zone more than 0.5 units from that of the ambient waters.

3. Nutrients

(a) Phosphorus

The concentration should be limited to the extent necessary to prevent nuisance growths of algae, weeds and slimes that are or may become injurious to any beneficial water use. (Specific phosphorus control requirements are set out in Annex 3.)

4. Tainting Substances

(a) Raw public water supply sources should be essentially free from objectionable taste and odor for aesthetic reasons.

(b) Levels of phenolic compounds should not exceed 1 microgram per liter in public water supplies to protect against taste and odor in domestic water.

(c) Substances entering the water as the result of human activity that cause tainting of edible aquatic organisms should not be present in concentrations which will lower the acceptability of these organisms as determined by organoleptic tests.

II. PHYSICAL

A. Asbestos

Asbestos should be kept at the lowest practical levels and in any event should be controlled to the extent necessary to prevent harmful effects on human health.

B. Temperature

There should be no change in temperature that would adversely affect any local or general use of the waters.

C. Settleable and Suspended Solids, and Light Transmission

For the protection of aquatic life, waters should be free from substances attributable to municipal, industrial or other charges resulting from human activity that will settle to form putrescent or otherwise objectionable sludge deposits or that will alter the value of the Secchi disc depth by more than 10 percent.

III. MICROBIOLOGICAL

Waters used for body contact recreation activities should be substantially free from bacteria, fungi, or viruses that may produce enteric disorders or eye, ear, nose, throat and skin infections or other human diseases and infections.

IV. RADIOLOGICAL

The level of radioactivity in waters outside of any defined source control area should not result in a TED50 (total equivalent dose integrated over 50 years as calculated in accordance with the methodology established by the International Commission on Radiological Protection) greater than 1 millirem to the whole body from a daily ingestion of 2.2 liters of lake water for one year. For dose commitments between 1 and 5 millirem at the periphery of the source control area, source investigation and corrective action are recommended if releases are not as low as reasonably achievable. For dose commitments greater than 5 millirem, the responsible regulatory authorities shall determine appropriate corrective action.

Source: "1978 Agreement Between the United States and Canada on Great Lakes Water Quality." Available online. International Freshwater Treaties Database. Oregon State University. URL: http://ocid.nacse.org/tfdd/tfdddocs/352ENG.htm. Accessed May 20, 2009.

The Great Lakes Charter 1985 (excerpts)

On February 11, 1985, one of the most historic international environmental agreements, or international treaty of any kind, for that matter, was signed by the governors of the U.S. states of Minnesota, Wisconsin, Michigan, Illinois, Indiana, Ohio, Pennsylvania, and New York and the premiers of the Canadian provinces of Ontario and Québec—all of whose territories border the Great Lakes or, in the case of Québec, the St. Lawrence River. The charter was historic for two reasons. First, it was accomplished at the state and provincial level, and second, it recognized the Great Lakes region as a unitary ecosystem that disregarded a political border. Principle IV of the Charter helps ensure that major diversions from the basin will not take place.

THE GOVERNORS AND PREMIERS OF THE GREAT LAKES STATES AND PROVINCES JOINTLY FIND AND DECLARE THAT:

The water resources of the Great Lakes Basin are precious public natural resources, shared and held in trust by the Great Lakes States and Provinces.

The Great Lakes are valuable regional, national and international resources for which the federal governments of the United States and Canada and the International Joint Commission have, in partnership with the States and Provinces, and important, continuing an abiding role and responsibility.

The waters of the Great Lakes Basin are interconnected and part of a single hydrologic system. The multiple uses of these resources for municipal, industrial and agricultural water supply; mining; navigation; hydroelectric power and energy production; recreation; and the maintenance of fish and wildlife habitat and a balanced ecosystem are interdependent.

Studies conducted by the International Joint Commission, the Great Lakes States and Provinces, and other agencies have found that without careful and prudent management, the future development of diversions and consumptive uses of the water resources of the Great Lakes Basin may have significant adverse impacts on the environment, economy, and welfare of the Great Lakes region.

As trustees of the Basin's natural resources, the Great Lakes States and Provinces have a shared duty to protect, conserve, and manage the renewable but finite waters of the Great Lakes Basin for the use, benefit, and enjoyment of all their citizens, including generations yet to come. The most effective means of protecting, conserving, and managing the water resources of the Great Lakes is through the joint pursuit of unified and cooperative principles, policies and programs mutually agreed upon, enacted and adhered to by each and every Great Lakes state and province.

Management of the water resources of the Basin is subject to the jurisdiction, rights and responsibilities of the signatory states and provinces. Effective management of the water resources of the Great Lakes requires the exercise of such jurisdiction, rights, and responsibilities in the interest of all the people of the Great Lakes Region, acting in a continuing spirit of comity and mutual cooperation. The Great Lakes states and provinces reaffirm the mutual rights and obligations of all Basin jurisdictions to use, conserve, and protect Basin water resources, as expressed in the Boundary Waters Treaty of 1909, the Great Lakes Water Quality Agreement of 1978, and the principles of other applicable international agreements.

PURPOSE

THE PURPOSES OF THIS CHARTER are to conserve the levels and flows of the Great Lakes and their tributary and connecting waters; to protect and conserve the environmental balance of the Great Lakes Basin ecosystem; to provide for cooperative programs and management of the water resources of the Great Lakes Basin by the signatory States and Provinces; to make secure and protect present developments within the region; and to provide a secure foundation for future investment and development within the region.

PRINCIPLES FOR THE MANAGEMENT OF GREAT LAKES WATER RESOURCES

IN ORDER TO ACHIEVE THE PURPOSES OF THIS CHARTER, THE GOVERNORS AND PREMIERS OF THE GREAT LAKES STATES AND PROVINCES AGREE TO THE FOLLOWING PRINCIPLES:

Principle I
Integrity of the Great Lakes Basin

The planning and management of the water resources of the Great Lakes Basin should recognize and be founded upon the integrity of the natural resources and ecosystem of the Great Lakes Basin. The water resources of the Basin transcend political boundaries within the Basin, and should be recognized and treated as a single hydrologic system. In managing Great Lakes Basin waters, the natural resources and ecosystem of the Basin should be considered as a unified whole.

Principle II
Cooperation Among Jurisdictions

The signatory states and provinces recognize and commit to a spirit of cooperation among local, state, and provincial agencies, the federal governments of Canada and the United States, and the International Joint Commission in the study, monitoring, planning, and conservation of the water resources of the Great Lakes Basin.

Principle III
Protection of the Water Resources of the Great Lakes

The signatory States and Provinces agree that new or increased diversions and consumptive uses of Great Lakes Basin water resources are of serious concern. In recognition of their shared responsibility to conserve and protect the water resources of the Great Lakes Basin for the use, benefit, and enjoyment of all their citizens, the states and provinces agree to seek (where necessary) and to implement legislation establishing programs to manage and regulate the diversion and consumptive use of Basin water resources. It is the intent of the signatory states and provinces that diversions of Basin water resources will not be allowed if individually or cumulatively they would have any significant adverse impacts on lake levels, in-basin uses, and the Great Lakes Ecosystem.

Principle IV
Prior Notice and Consultation

It is the intent of the signatory states and provinces that no Great Lakes state or province will approve or permit any major new or increased diversion or consumptive use of the water resources of the Great Lakes Basin without notifying and consulting with and seeking the consent and concurrence of all affected Great Lakes States and Provinces.

Principle V
Cooperative Programs and Practices

The Governors and Premiers of the Great Lakes states and provinces commit to pursue the development and maintenance of a common base of data and information regarding the use and management of the Basin water resources, to the establishment of a systematic arrangement for the exchange of water data and information, to the creation of a Water Resources Management Committee, to the development of a Great Lakes Water Resources Management Program, and to additional and concerted and coordinated research efforts to provide improved information for future water planning and management decisions.

Source: "The Great Lakes Charter: Principles for the Management of Great Lakes Water Resources." Available online. URL: http://www.cglg.org/projects/water/docs/GreatLakesCharter.pdf. Accessed May 21, 2009.

The Great Lakes Compact, 2008 (excerpts)

The Great Lakes Compact essentially takes up where the Great Lakes Charter left off, except that it is a compact between the eight U.S. states that are riparian to the lakes and the federal government. Its main focus is to ensure that water is not diverted from the lakes or their tributaries in large amounts. Critics contend, however, that the compact does not go far enough; for example, it does not protect the region from transnationals or their subsidiaries that are engaged in bottling water.

ARTICLE 4
WATER MANAGEMENT AND REGULATION

Section 4.1. Water Resources Inventory, Registration and Reporting.

1. Within five years of the effective date of this Compact, each Party shall develop and maintain a Water resources inventory for the collection, interpretation, storage, retrieval exchange, and dissemination of information concerning the Water resources of the Party, including, but not limited to, information on the location, type, quantity, and use of those resources and the location, type, and quantity of Withdrawals, Diversions and Consump-

tive Uses. To the extent feasible, the Water resources inventory shall be developed in cooperation with local, state, federal, tribal and other private agencies and entities, as well as the Council. Each Party's agencies shall cooperate with that Party in the development and maintenance of the inventory.

2. The Council shall assist each Party to develop a common base of data regarding the management of the Water Resources of the Basin and to establish systematic arrangements for the exchange of those data with other states and provinces.

3. To develop and maintain a compatible base of Water use information, within five years of the effective date of this Compact any Person who Withdraws Water in an amount of 100,000 gallons per day or greater average in any 30-day period (including Consumptive Uses) from all sources, or Diverts Water of any amount, shall register the Withdrawal or Diversion by a date set by the Council unless the Person has previously registered in accordance with an existing state program. The Person shall register the Withdrawal or Diversion with the Originating Party using a form prescribed by the Originating Party that shall include, at a minimum and without limitation: the name and address of the registrant and date of registration; the locations and sources of the Withdrawal or Diversion; the capacity of the Withdrawal or Diversion per day and the amount Withdrawn or Diverted from each source; the uses made of the Water; places of use and places of discharge; and, such other information as the Originating Party may require. All registrations shall include an estimate of the volume of the Withdrawal or Diversion in terms of gallons per day average in any 30-day period.

4. All registrants shall annually report the monthly volumes of the Withdrawal, Consumptive Use and Diversion in gallons to the Originating Party and any other information requested by the Originating Party.

5. Each Party shall annually report the information gathered pursuant to this Section to a Great Lakes—St. Lawrence River Water use data base repository and aggregated information shall be made publicly available, consistent with the confidentiality requirements in Section 8.3.

6. Information gathered by the Parties pursuant to this Section shall be used to improve the sources and applications of scientific information

regarding the Waters of the Basin and the impacts of the Withdrawals and Diversions from various locations and Water sources on the Basin Ecosystem, and to better understand the role of groundwater in the Basin. The Council and the Parties shall coordinate the collection and application of scientific information to further develop a mechanism by which individual and Cumulative Impacts of Withdrawals, Consumptive Uses and Diversions shall be assessed.

Section 4.2. Water Conservation and Efficiency Programs.
1. The Council commits to identify, in cooperation with the provinces, Basin-wide Water conservation and efficiency objectives to assist the Parties in developing their Water conservation and efficiency program. These objectives are based on the goals of:

a. Ensuring improvement of the Waters and Water Dependent Natural Resources;
b. Protecting and restoring the hydrologic and ecosystem integrity of the Basin;
c. Retaining the quantity of surface water and groundwater in the Basin;
d. Ensuring sustainable use of Waters of the Basin; and,
e. Promoting the efficiency of use and reducing losses and waste of Water.

2. Within two years of the effective date of this Compact, each Party shall develop its own Water conservation and efficiency goals and objectives consistent with the Basin-wide goals and objectives, and shall develop and implement a Water conservation and efficiency program, either voluntary or mandatory, within its jurisdiction based on the Party's goals and objectives. Each Party shall annually assess its programs in meeting the Party's goals and objectives, report to the Council and the Regional Body and make this annual assessment available to the public.

3. Beginning five years after the effective date of this Compact, and every five years thereafter, the Council, in cooperation with the provinces, shall review and modify as appropriate the Basinwide objectives, and the Parties shall have regard for any such modifications in implementing their programs. This assessment will be based on examining new technologies, new patterns of Water use, new resource demands and threats, and Cumulative Impact assessment under Section 4.15.

4. Within two years of the effective date of this Compact, the Parties commit to promote Environmentally Sound and Economically Feasible Water Conservation Measures such as:

a. Measures that promote efficient use of Water;
b. Identification and sharing of best management practices and state of the art conservation and efficiency technologies;
c. Application of sound planning principles;
d. Demand-side and supply-side Measures or incentives; and,
e. Development, transfer and application of science and research.

5. Each Party shall implement in accordance with paragraph 2 above a voluntary or mandatory Water conservation program for all, including existing, Basin Water users. Conservation programs need to adjust to new demands and the potential impacts of cumulative effects and climate.

Section 4.3. Party Powers and Duties.

1. Each Party, within its jurisdiction, shall manage and regulate New or Increased Withdrawals, Consumptive Uses and Diversions, including Exceptions, in accordance with this Compact.

2. Each Party shall require an Applicant to submit an Application in such manner and with such accompanying information as the Party shall prescribe.

3. No Party may approve a Proposal if the Party determines that the Proposal is inconsistent with this Compact or the Standard of Review and Decision or any implementing rules or regulations promulgated thereunder. The Party may approve, approve with modifications or disapprove any Proposal depending on the Proposal's consistency with this Compact and the Standard of Review and Decision.

4. Each Party shall monitor the implementation of any approved Proposal to ensure consistency with the approval and may take all necessary enforcement actions.

5. No Party shall approve a Proposal subject to Council or Regional Review, or both, pursuant to this Compact unless it shall have been first submitted to and reviewed by either the Council or Regional Body, or both, and approved by the Council, as applicable. Sufficient opportunity shall be provided for comment on the Proposal's consistency with this Compact and the

Standard of Review and Decision. All such comments shall become part of the Party's formal record of decision, and the Party shall take into consideration any such comments received.

Section 4.4. Requirement for Originating Party Approval.
No Proposal subject to management and regulation under this Compact shall hereafter be undertaken by any Person unless it shall have been approved by the Originating Party.

Section 4.5. Regional Review.

1. **General.**
   a. It is the intention of the Parties to participate in Regional Review of Proposals with the Provinces, as described in this Compact and the Agreement.
   b. Unless the Applicant or the Originating Party otherwise requests, it shall be the goal of the Regional Body to conclude its review no later than 90 days after notice under Section 4.5.2 of such Proposal is received from the Originating Party.
   c. Proposals for Exceptions subject to Regional Review shall be submitted by the Originating Party to the Regional Body for Regional Review, and where applicable, to the Council for concurrent review.
   d. The Parties agree that the protection of the integrity of the Great Lakes—St. Lawrence River Basin Ecosystem shall be the overarching principle for reviewing Proposals subject to Regional Review, recognizing uncertainties with respect to demands that may be placed on Basin Water, including groundwater, levels and flows of the Great Lakes and the St. Lawrence River, future changes in environmental conditions, the reliability of existing data and the extent to which Diversions may harm the integrity of the Basin Ecosystem.
   e. The Originating Party shall have lead responsibility for coordinating information for resolution of issues related to evaluation of a Proposal, and shall consult with the Applicant throughout the Regional Review Process.
   f. A majority of the members of the Regional Body may request Regional Review of a regionally significant or potentially precedent setting Proposal. Such Regional Review must be conducted, to the extent possible, within the time frames set forth in this Section. Any such Regional Review shall be undertaken only after consulting the Applicant.

2. Notice from Originating Party to the Regional Body.

a. The Originating Party shall determine if a Proposal is subject to Regional Review. If so, the Originating Party shall provide timely notice to the Regional Body and the public.

b. Such notice shall not be given unless and until all information, documents and the Originating Party's Technical Review needed to evaluate whether the Proposal meets the Standard of Review and Decision have been provided.

c. An Originating Party may:

i. Provide notice to the Regional Body of an Application, even if notification is not required; or,

ii. Request Regional Review of an application, even if Regional Review is not required. Any such Regional Review shall be undertaken only after consulting the Applicant.

d. An Originating Party may provide preliminary notice of a potential Proposal.

3. Public Participation.

a. To ensure adequate public participation, the Regional Body shall adopt procedures for the review of Proposals that are subject to Regional Review in accordance with this Article.

b. The Regional Body shall provide notice to the public of a Proposal undergoing Regional Review. Such notice shall indicate that the public has an opportunity to comment in writing to the Regional Body on whether the Proposal meets the Standard of Review and Decision.

c. The Regional Body shall hold a public meeting in the state or province of the Originating Party in order to receive public comment on the issue of whether the Proposal under consideration meets the Standard of Review and Decision.

d. The Regional Body shall consider the comments received before issuing a Declaration of Finding.

e. The Regional Body shall forward the comments it receives to the Originating Party.

4. Technical Review.

a. The Originating Party shall provide the Regional Body with its Technical Review of the Proposal under consideration.

b. The Originating Party's Technical Review shall thoroughly analyze the Proposal and provide an evaluation of the Proposal sufficient for a determination of whether the Proposal meets the Standard of Review and Decision.

c. Any member of the Regional Body may conduct their own Technical Review of any Proposal subject to Regional Review.

d. At the request of the majority of its members, the Regional Body shall make such arrangements as it considers appropriate for an independent Technical Review of a Proposal.

e. All Parties shall exercise their best efforts to ensure that a Technical Review undertaken under Sections 4.5.4.c and 4.5.4.d does not unnecessarily delay the decision by the Originating Party on the Application. Unless the Applicant or the Originating Party otherwise requests, all Technical Reviews shall be completed no later than 60 days after the date the notice of the Proposal was given to the Regional Body.

5. **Declaration of Finding.**

a. The Regional Body shall meet to consider a Proposal. The Applicant shall be provided with an opportunity to present the Proposal to the Regional Body at such time.

b. The Regional Body, having considered the notice, the Originating Party's Technical Review, any other independent Technical Review that is made, any comments or objections including the analysis of comments made by the public, First Nations and federally recognized Tribes, and any other information that is provided under this Compact shall issue a Declaration of Finding that the Proposal under consideration:

   i. Meets the Standard of Review and Decision;

   ii. Does not meet the Standard of Review and Decision; or,

   iii. Would meet the Standard of Review and Decision if certain conditions were met.

c. An Originating Party may decline to participate in a Declaration of Finding made by the Regional Body.

d. The Parties recognize and affirm that it is preferable for all members of the Regional Body to agree whether the Proposal meets the Standard of Review and Decision.

e. If the members of the Regional Body who participate in the Declaration of Finding all agree, they shall issue a written Declaration of Finding with consensus.

f. In the event that the members cannot agree, the Regional Body shall make every reasonable effort to achieve consensus within 25 days.

g. Should consensus not be achieved, the Regional Body may issue a Declaration of Finding that presents different points of view and indicates each Party's conclusions.
h. The Regional Body shall release the Declarations of Finding to the public.
i. The Originating Party and the Council shall consider the Declaration of Finding before making a decision on the Proposal.

Section 4.6. Proposals Subject to Prior Notice.
1. Beginning no later than five years of the effective date of this Compact, the Originating Party shall provide all Parties and the Provinces with detailed and timely notice and an opportunity to comment within 90 days on any Proposal for a New or Increased Consumptive Use of 5 million gallons per day or greater average in any 90-day period. Comments shall address whether or not the Proposal is consistent with the Standard of Review and Decision. The Originating Party shall provide a response to any such comment received from another Party.

2. A Party may provide notice, an opportunity to comment and a response to comments even if this is not required under paragraph 1 of this Section. Any provision of such notice and opportunity to comment shall be undertaken only after consulting the Applicant.

Section 4.7. Council Actions.
1. Proposals for Exceptions subject to Council Review shall be submitted by the Originating Party to the Council for Council Review, and where applicable, to the Regional Body for concurrent review.

2. The Council shall review and take action on Proposals in accordance with this Compact and the Standard of Review and Decision. The Council shall not take action on a Proposal subject to Regional Review pursuant to this Compact unless the Proposal shall have been first submitted to and reviewed by the Regional Body. The Council shall consider any findings resulting from such review.

Section 4.8. Prohibition of New or Increased Diversions.
All New or Increased Diversions are prohibited, except as provided for in this Article.

Section 4.9. Exceptions to the Prohibition of Diversions.
1. Straddling Communities. A Proposal to transfer Water to an area within a Straddling Community but outside the Basin or outside the source Great Lake

Watershed shall be excepted from the prohibition against Diversions and be managed and regulated by the Originating Party provided that, regardless of the volume of Water transferred, all the Water so transferred shall be used solely for Public Water Supply Purposes within the Straddling Community, and:

a. All Water Withdrawn from the Basin shall be returned, either naturally or after use, to the Source Watershed less an allowance for Consumptive Use. No surface water or groundwater from outside the Basin may be used to satisfy any portion of this criterion except if it:
   i. Is part of a water supply or wastewater treatment system that combines water from inside and outside of the Basin;
   ii. Is treated to meet applicable water quality discharge standards and to prevent the introduction of invasive species into the Basin;
   iii. Maximizes the portion of water returned to the Source Watershed as Basin Water and minimizes the surface water or groundwater from outside the Basin;
b. If the Proposal results from a New or Increased Withdrawal of 100,000 gallons per day or greater average over any 90-day period, the Proposal shall also meet the Exception Standard; and,
c. If the Proposal results in a New or Increased Consumptive Use of 5 million gallons per day or greater average over any 90-day period, the Proposal shall also undergo Regional Review.

2. Intra-Basin Transfer. A Proposal for an Intra-Basin Transfer that would be considered a Diversion under this Compact, and not already excepted pursuant to paragraph 1 of this Section, shall be excepted from the prohibition against Diversions, provided that:

a. If the Proposal results from a New or Increased Withdrawal less than 100,000 gallons per day average over any 90-day period, the Proposal shall be subject to management and regulation at the discretion of the Originating Party.
b. If the Proposal results from a New or Increased Withdrawal 100,000 gallons per day or greater average over any 90-day period and if the Consumptive Use resulting from the Withdrawal is less than 5 million gallons per day average over any 90-day period:
   i. The Proposal shall meet the Exception Standard and be subject to management and regulation by the Originating Party, except that the Water may be returned to another Great Lake watershed rather than the Source Watershed;
   ii. The Applicant shall demonstrate that there is no feasible, cost effective, and environmentally sound water supply alternative

within the Great Lake watershed to which the Water will be transferred, including conservation of existing water supplies; and,

iii. The Originating Party shall provide notice to the other Parties prior to making any decision with respect to the Proposal.

c. If the Proposal results in a New or Increased Consumptive Use of 5 million gallons per day or greater average over any 90-day period:
   i. The Proposal shall be subject to management and regulation by the Originating Party and shall meet the Exception Standard, ensuring that Water Withdrawn shall be returned to the Source Watershed;
   ii. The Applicant shall demonstrate that there is no feasible, cost effective, and environmentally sound water supply alternative within the Great Lake watershed to which the Water will be transferred, including conservation of existing water supplies;
   iii. The Proposal undergoes Regional Review; and,
   iv. The Proposal is approved by the Council. Council approval shall be given unless one or more Council Members vote to disapprove.

3. Straddling Counties. A Proposal to transfer Water to a Community within a Straddling County that would be considered a Diversion under this Compact shall be excepted from the prohibition against Diversions, provided that it satisfies all of the following conditions:

a. The Water shall be used solely for the Public Water Supply Purposes of the Community within a Straddling County that is without adequate supplies of potable water;
b. The Proposal meets the Exception Standard, maximizing the portion of water returned to the Source Watershed as Basin Water and minimizing the surface water or groundwater from outside the Basin;
c. The Proposal shall be subject to management and regulation by the Originating Party, regardless of its size;
d. There is no reasonable water supply alternative within the basin in which the community is located, including conservation of existing water supplies;
e. Caution shall be used in determining whether or not the Proposal meets the conditions for this Exception. This Exception should not be authorized unless it can be shown that it will not endanger the integrity of the Basin Ecosystem;
f. The Proposal undergoes Regional Review; and,

g. The Proposal is approved by the Council. Council approval shall be given unless one or more Council Members vote to disapprove. A Proposal must satisfy all of the conditions listed above. Further, substantive consideration will also be given to whether or not the Proposal can provide sufficient scientifically based evidence that the existing water supply is derived from groundwater that is hydrologically interconnected to Waters of the Basin.

4. Exception Standard. Proposals subject to management and regulation in this Section shall be declared to meet this Exception Standard and may be approved as appropriate only when the following criteria are met:

a. The need for all or part of the proposed Exception cannot be reasonably avoided through the efficient use and conservation of existing water supplies;
b. The Exception will be limited to quantities that are considered reasonable for the purposes for which it is proposed;
c. All Water Withdrawn shall be returned, either naturally or after use, to the Source Watershed less an allowance for Consumptive Use. No surface water or groundwater from the outside the Basin may be used to satisfy any portion of this criterion except if it:
    i. Is part of a water supply or wastewater treatment system that combines water from inside and outside of the Basin;
    ii. Is treated to meet applicable water quality discharge standards and to prevent the introduction of invasive species into the Basin;
d. The Exception will be implemented so as to ensure that it will result in no significant individual or cumulative adverse impacts to the quantity or quality of the Waters and Water Dependent Natural Resources of the Basin with consideration given to the potential Cumulative Impacts of any precedent-setting consequences associated with the Proposal;
e. The Exception will be implemented so as to incorporate Environmentally Sound and Economically Feasible Water Conservation Measures to minimize Water Withdrawals or Consumptive Use;
f. The Exception will be implemented so as to ensure that it is in compliance with all applicable municipal, state and federal laws as well as regional interstate and international agreements, including the Boundary Waters Treaty of 1909; and,
g. All other applicable criteria in Section 4.9 have also been met.

Section 4.10. Management and Regulation of New or Increased Withdrawals and Consumptive Uses.

1. Within five years of the effective date of this Compact, each Party shall create a program for the management and regulation of New or Increased Withdrawals and Consumptive Uses by adopting and implementing Measures consistent with the Decision-Making Standard. Each Party, through a considered process, shall set and may modify threshold levels for the regulation of New or Increased Withdrawals in order to assure an effective and efficient Water management program that will ensure that uses overall are reasonable, that Withdrawals overall will not result in significant impacts to the Waters and Water Dependent Natural Resources of the Basin, determined on the basis of significant impacts to the physical, chemical, and biological integrity of Source Watersheds, and that all other objectives of the Compact are achieved. Each Party may determine the scope and thresholds of its program, including which New or Increased Withdrawals and Consumptive Uses will be subject to the program.

2. Any Party that fails to set threshold levels that comply with Section 4.10.1 any time before 10 years after the effective date of this Compact shall apply a threshold level for management and regulation of all New or Increased Withdrawals of 100,000 gallons per day or greater average in any 90 day period.

3. The Parties intend programs for New or Increased Withdrawals and Consumptive Uses to evolve as may be necessary to protect Basin Waters. Pursuant to Section 3.4, the Council, in cooperation with the provinces, shall periodically assess the Water management programs of the Parties. Such assessments may produce recommendations for the strengthening of the programs, including without limitation, establishing lower thresholds for management and regulation in accordance with the Decision-Making Standard.

Section 4.11. Decision-Making Standard.

Proposals subject to management and regulation in Section 4.10 shall be declared to meet this Decision-Making Standard and may be approved as appropriate only when the following criteria are met:

1. All Water Withdrawn shall be returned, either naturally or after use, to the Source Watershed less an allowance for Consumptive Use;

2. The Withdrawal or Consumptive Use will be implemented so as to ensure that the Proposal will result in no significant individual or cumulative

adverse impacts to the quantity or quality of the Waters and Water Dependent Natural Resources and the applicable Source Watershed;

3. The Withdrawal or Consumptive Use will be implemented so as to incorporate Environmentally Sound and Economically Feasible Water Conservation Measures;

4. The Withdrawal or Consumptive Use will be implemented so as to ensure that it is in compliance with all applicable municipal, state and federal laws as well as regional interstate and international agreements, including the Boundary Waters Treaty of 1909;

5. The proposed use is reasonable, based upon a consideration of the following factors:

a. Whether the proposed Withdrawal or Consumptive Use is planned in a fashion that provides for efficient use of the water, and will avoid or minimize the waste of Water;
b. If the Proposal is for an increased Withdrawal or Consumptive use, whether efficient use is made of existing water supplies;
c. The balance between economic development, social development and environmental protection of the proposed Withdrawal and use and other existing or planned withdrawals and water uses sharing the water source;
d. The supply potential of the water source, considering quantity, quality, and reliability and safe yield of hydrologically interconnected water sources;
e. The probable degree and duration of any adverse impacts caused or expected to be caused by the proposed Withdrawal and use under foreseeable conditions, to other lawful consumptive or non-consumptive uses of water or to the quantity or quality of the Waters and Water Dependent Natural Resources of the Basin, and the proposed plans and arrangements for avoidance or mitigation of such impacts; and,
f. If a Proposal includes restoration of hydrologic conditions and functions of the Source Watershed, the Party may consider that.

Section 4.12. Applicability.

1. Minimum Standard. This Standard of Review and Decision shall be used as a minimum standard. Parties may impose a more restrictive decision-

making standard for Withdrawals under their authority. It is also acknowledged that although a Proposal meets the Standard of Review and Decision it may not be approved under the laws of the Originating Party that has implemented more restrictive Measures.

2. Baseline.

a. To establish a baseline for determining a New or Increased Diversion, Consumptive Use or Withdrawal, each Party shall develop either or both of the following lists for their jurisdiction:
   i. A list of existing Withdrawal approvals as of the effective date of the Compact;
   ii. A list of the capacity of existing systems as of the effective date of this Compact. The capacity of the existing systems should be presented in terms of Withdrawal capacity, treatment capacity, distribution capacity, or other capacity limiting factors. The capacity of the existing systems must represent the state of the systems. Existing capacity determinations shall be based upon approval limits or the most restrictive capacity information.

b. For all purposes of this Compact, volumes of Diversions, Consumptive Uses, or Withdrawals of Water set forth in the list(s) prepared by each Party in accordance with this Section, shall constitute the baseline volume.

c. The list(s) shall be furnished to the Regional Body and the Council within one year of the effective date of this Compact.

3. Timing of Additional Applications. Applications for New or Increased Withdrawals, Consumptive Uses or Exceptions shall be considered cumulatively within ten years of any application.

4. Change of Ownership. Unless a new owner proposes a project that shall result in a Proposal for a New or Increased Diversion or Consumptive Use subject to Regional Review or Council approval, the change of ownership in and of itself shall not require Regional Review or Council approval.

5. Groundwater. The Basin surface water divide shall be used for the purpose of managing and regulating New or Increased Diversions, Consumptive Uses or Withdrawals of surface water and groundwater.

6. Withdrawal Systems. The total volume of surface water and groundwater resources that supply a common distribution system shall determine the volume of a Withdrawal, Consumptive Use or Diversion.

7. Connecting Channels. The watershed of each Great Lake shall include its upstream and downstream connecting channels.

8. Transmission in Water Lines. Transmission of Water within a line that extends outside the Basin as it conveys Water from one point to another within the Basin shall not be considered a Diversion if none of the Water is used outside the Basin.

9. Hydrologic Units. The Lake Michigan and Lake Huron watersheds shall be considered to be a single hydrologic unit and watershed.

10. Bulk Water Transfer. A Proposal to Withdraw Water and to remove it from the Basin in any container greater than 5.7 gallons shall be treated under this Compact in the same manner as a Proposal for a Diversion. Each Party shall have the discretion, within its jurisdiction, to determine the treatment of Proposals to Withdraw Water and to remove it from the Basin in any container of 5.7 gallons or less.

Section 4.13. Exemptions.

Withdrawals from the Basin for the following purposes are exempt from the requirements of Article 4.

1. To supply vehicles, including vessels and aircraft, whether for the needs of the persons or animals being transported or for ballast or other needs related to the operation of the vehicles.

2. To use in a non-commercial project on a short-term basis for firefighting, humanitarian, or emergency response purposes.

Section 4.14. U.S. Supreme Court Decree: *Wisconsin et al. v. Illinois et al.*

1. Notwithstanding any terms of this Compact to the contrary, with the exception of Paragraph 5 of this Section, current, New or Increased Withdrawals, Consumptive Uses and Diversions of Basin Water by the State of Illinois shall be governed by the terms of the United States Supreme Court decree in *Wisconsin et al. v. Illinois et al.* and shall not be subject to

the terms of this Compact nor any rules or regulations promulgated pursuant to this Compact. This means that, with the exception of Paragraph 5 of this Section, for purposes of this Compact, current, New or Increased Withdrawals, Consumptive Uses and Diversions of Basin Water within the State of Illinois shall be allowed unless prohibited by the terms of the United States Supreme Court decree in *Wisconsin et al. v. Illinois et al.*

2. The Parties acknowledge that the United States Supreme Court decree in *Wisconsin et al. v. Illinois et al.* shall continue in full force and effect, that this Compact shall not modify any terms thereof, and that this Compact shall grant the parties no additional rights, obligations, remedies or defenses thereto. The Parties specifically acknowledge that this Compact shall not prohibit or limit the State of Illinois in any manner from seeking additional Basin Water as allowed under the terms of the United States Supreme Court decree in *Wisconsin et al. v. Illinois et al.*, any other party from objecting to any request by the State of Illinois for additional Basin Water under the terms of said decree, or any party from seeking any other type of modification to said decree. If an application is made by any party to the Supreme Court of the United States to modify said decree, the Parties to this Compact who are also parties to the decree shall seek formal input from the Canadian provinces of Ontario and Québec, with respect to the proposed modification, use best efforts to facilitate the appropriate participation of said provinces in the proceedings to modify the decree, and shall not unreasonably impede or restrict such participation.

3. With the exception of Paragraph 5 of this Section, because current, New or Increased Withdrawals, Consumptive Uses and Diversions of Basin Water by the State of Illinois are not subject to the terms of this Compact, the State of Illinois is prohibited from using any term of this Compact, including Section 4.9, to seek New or Increased Withdrawals, Consumptive Uses or Diversions of Basin Water.

4. With the exception of Paragraph 5 of this Section, because Sections 4.3, 4.4, 4.5, 4.6, 4.7, 4.8, 4.9, 4.10, 4.11, 4.12 (Paragraphs 1, 2, 3, 4, 6 and 10 only), and 4.13 of this Compact all relate to current, New or Increased Withdrawals, Consumptive Uses and Diversions of Basin Waters, said provisions do not apply to the State of Illinois. All other provisions of this Compact not listed in the preceding sentence shall apply to the State of Illinois, including the Water Conservation Programs provision of Section 4.2.

5. In the event of a Proposal for a Diversion of Basin Water for use outside the territorial boundaries of the Parties to this Compact, decisions by the State of Illinois regarding such a Proposal would be subject to all terms of this Compact, except Paragraphs 1, 3 and 4 of this Section.

6. For purposes of the State of Illinois' participation in this Compact, the entirety of this Section 4.14 is necessary for the continued implementation of this Compact and, if severed, this Compact shall no longer be binding on or enforceable by or against the State of Illinois.

Section 4.15. Assessment of Cumulative Impacts.
1. The Parties in cooperation with the Provinces shall collectively conduct within the Basin, on a Lake watershed and St. Lawrence River Basin basis, a periodic assessment of the Cumulative Impacts of Withdrawals, Diversions and Consumptive Uses from the Waters of the Basin, every 5 years or each time the incremental Basin Water losses reach 50 million gallons per day average in any 90-day period in excess of the quantity at the time of the most recent assessment, whichever comes first, or at the request of one or more of the Parties. The assessment shall form the basis for a review of the Standard of Review and Decision, Council and Party regulations and their application. This assessment shall:

a. Utilize the most current and appropriate guidelines for such a review, which may include but not be limited to Council on Environmental Quality and Environment Canada guidelines;
b. Give substantive consideration to climate change or other significant threats to Basin Waters and take into account the current state of scientific knowledge, or uncertainty, and appropriate Measures to exercise caution in cases of uncertainty if serious damage may result;
c. Consider adaptive management principles and approaches, recognizing, considering and providing adjustments for the uncertainties in, and evolution of science concerning the Basin's water resources, watersheds and ecosystems, including potential changes to Basin-wide processes, such as lake level cycles and climate.

2. The Parties have the responsibility of conducting this Cumulative Impact assessment. Applicants are not required to participate in this assessment.

3. Unless required by other statutes, Applicants are not required to conduct a separate cumulative impact assessment in connection with an Application but shall submit information about the potential impacts of

a Proposal to the quantity or quality of the Waters and Water Dependent Natural Resources of the applicable Source Watershed. An Applicant may, however, provide an analysis of how their Proposal meets the no significant adverse Cumulative Impact provision of the Standard of Review and Decision.

ARTICLE 5
TRIBAL CONSULTATION

Section 5.1. Consultation with Tribes.

1. In addition to all other opportunities to comment pursuant to Section 6.2, appropriate consultations shall occur with federally recognized Tribes in the Originating Party for all Proposals subject to Council or Regional Review pursuant to this Compact. Such consultations shall be organized in the manner suitable to the individual Proposal and the laws and policies of the Originating Party.

2. All federally recognized Tribes within the Basin shall receive reasonable notice indicating that they have an opportunity to comment in writing to the Council or the Regional Body, or both, and other relevant organizations on whether the Proposal meets the requirements of the Standard of Review and Decision when a Proposal is subject to Regional Review or Council approval. Any notice from the Council shall inform the Tribes of any meeting or hearing that is to be held under Section 6.2 and invite them to attend. The Parties and the Council shall consider the comments received under this Section before approving, approving with modifications or disapproving any Proposal subject to Council or Regional Review.

3. In addition to the specific consultation mechanisms described above, the Council shall seek to establish mutually-agreed upon mechanisms or processes to facilitate dialogue with, and input from federally recognized Tribes on matters to be dealt with by the Council; and, the Council shall seek to establish mechanisms and processes with federally recognized Tribes designed to facilitate ongoing scientific and technical interaction and data exchange regarding matters falling within the scope of this Compact. This may include participation of tribal representatives on advisory committees established under this Compact or such other processes that are mutually agreed upon with federally recognized Tribes individually or through duly authorized intertribal agencies or bodies.

ARTICLE 6
PUBLIC PARTICIPATION

Section 6.1. Meetings, Public Hearings and Records.
1. The Parties recognize the importance and necessity of public participation in promoting management of the Water Resources of the Basin. Consequently, all meetings of the Council shall be open to the public, except with respect to issues of personnel.

2. The minutes of the Council shall be a public record open to inspection at its offices during regular business hours.

Section 6.2. Public Participation.
It is the intent of the Council to conduct public participation processes concurrently and jointly with processes undertaken by the Parties and through Regional Review. To ensure adequate public participation, each Party or the Council shall ensure procedures for the review of Proposals subject to the Standard of Review and Decision consistent with the following requirements:

1. Provide public notification of receipt of all Applications and a reasonable opportunity for the public to submit comments before Applications are acted upon.

2. Assure public accessibility to all documents relevant to an Application, including public comment received.

3. Provide guidance on standards for determining whether to conduct a public meeting or hearing for an Application, time and place of such a meeting(s) or hearing(s), and procedures for conducting of the same.

4. Provide the record of decision for public inspection including comments, objections, responses and approvals, approvals with conditions and disapprovals.

Source: "Great Lakes–St. Lawrence River Basin Water Resources Compact." Available online. URL: http://www.glelc.org/files/public_law_110_342.pdf. Accessed May 21, 2009.

5

International Documents

The documents in this chapter exemplify both the boundaryless nature of issues concerning water and the spirit of cooperation when it comes to the resource, even among countries that are otherwise hostile toward one another. Some of the treaties relate to specific water issues, while others reveal the importance of access to freshwater in maintaining peace. The documents are organized into the following sections:

Bolivia

Egypt, Ethiopia, and Sudan

Israel, Jordan, and the Palestinian Territories

India, Pakistan, and Bangladesh

The documents are arranged in chronological order within each section. Those that have been excerpted are identified as such; all others are reproduced in full.

BOLIVIA

Treaty with Peru Concerning Lake Titicaca, 1957

Bolivia and Peru first exchanged notes concerning the waters of Lake Titicaca in 1935. In the 1950s, the two countries signed three treaties. The last, signed in 1957, dealt with using the lake's waters to produce hydroelectricity.

Article 1

The Governments of Peru and Bolivia, in view of the recommendations made by the joint Peruvian-Bolivian Commission and by virtue of the fact that the two countries have joint, indivisible and exclusive ownership over the waters of Lake Titicaca, resolve to adopt a definite plan for a preliminary economic study concerning the joint utilization of the said waters in such manner as not to fundamentally alter the navigability or fishing facilities thereof or substantially affect the volume of water diverted from the Lake for industrial, irrigation or other purposes.

Article 2

The preliminary basic values of water drawn from the Lake in accordance with article 1 shall be as follows:

1. Kinetic energy of the water—$US 0.001 per kwh consumed;
2. Water used for irrigation purposes—$US 0.001 per cubic metre consumed.

These basic values, which are fixed on a preliminary basis and shall be divided equally between the two countries, shall be taken into consideration in carrying out the economic studies for the project to which this Agreement refers.

Article 3

The payments or other compensation to be made for the losses or the reduction in economic benefits sustained by either of the two countries as a result of hydroelectric development or use of the waters for irrigation and other purposes shall be the subject of an agreement to be concluded after the above-mentioned economic studies are completed.

Article 4

The preliminary economic study on the utilization by Peru and Bolivia of the waters of the Lake Titicaca shall contain, in a special introductory chapter, an estimate of the electricity consumption in both countries so that the construction of one or more hydroelectric stations capable of meeting the demand efficiently and equitably can be considered in the initial stage of development. It shall also include an agricultural and economic study of the areas where there is likely to be a market for the water for irrigation purposes after it has yielded its kinetic energy.

Article 5

The two Governments agree that they may jointly or separately initiate negotiations with responsible bodies or firms of world-wide reputation regarding the contract for the preliminary economic studies mentioned in this Agreement. Each country shall promptly give notice of such negotiations to the other through their Ministries of Foreign Affairs so that a meeting of the joint Sub-Commission for the utilization of the Waters of Lake Titicaca may be called, the said Sub-Commission being authorized to study and recommend acceptance and signature of the contract by the two Governments.

When the preliminary economic studies provided for in the said contract have been completed, the joint Sub-Commission shall submit them to the joint Peruvian-Bolivian Commission for consideration and approval.

Article 6

When the studies referred to in article 5 have been approved by the joint Peruvian-Bolivian Commission, both Governments shall invite tenders in the world market for the final studies and for the financing, in whole or in part, of the project.

Article 7

This Agreement shall come into force upon the exchange of the instruments of ratification, which shall take place as soon as possible at Lima.

Source: "Agreement between Bolivia and Peru Concerning a Preliminary Economic Study of the Joint Utilization of the Waters of Lake Titicaca." International Freshwater Treaties Database. Oregon State University. Available online. URL: http://ocid.nacse.org/tfdd/tfdddocs/98ENG.htm. Accessed June 5, 2009.

Joint Management Treaty of the Pilcomayo River, 1995 (excerpts)

In 1996 and 1997, tension was elevated between the Pilcomayo River riparian nations of Argentina and Paraguay on one side and Bolivia on the other, over the issue of pollution of the Pilcomayo by Bolivian mines. Warfare almost broke out, but an agreement was eventually reached with the help of the international agency Green Cross. The 1995 treaty established a precedent of cooperation among the three countries over the sharing and management of the Pilcomayo's waters.

Constituent Agreement of the Trinational Commission for the Development of the Pilcomayo River Basin

The Governments of the Republic of Argentina, the Republic of Bolivia, and the Republic of Paraguay;

Considering the necessity of establishing a permanent legal-technical mechanism responsible for the central administration of the Pilcomayo River Basin that impels the sustainable development of its zone of influence, optimizes the utilization of its natural resources, generates employment, attracts investments, and permits rational and fair usage of water resources;

Taking into account the Treaty of the Plata River Basin that foresaw the "rational utilization of water resources, especially through the regulation of the watercourses and their multiple and equitable uses," and in completion of the Joint Declaration signed by the presidents of Argentina, Bolivia, and Paraguay on April 26, 1994 in the city of Formosa in the Republic of Argentina;

Decide to approve the present Constituent Agreement of the Trinational Commission for the Development of the Pilcomayo River Basin, which will be governed by the following STATUTE.

Article I
The Commission

The Parties convening to establish a Trinational Commission for the Development of the Pilcomayo River Basin, hereafter called "the Commission"

Article IV
Domains and Functions

The Commission will be responsible for completing the proposed objectives in the Formosa Declaration, signed on April 26, 1994, by the presidents of Argentina, Bolivia, and Paraguay.

The Commission, consequently, will be responsible for the study and execution of joint projects in the Pilcomayo River that affect the development of the Basin.

For the fulfillment of this responsibility, the Commission will have as functions:

a. To continue the studies and tasks necessary to obtain the multiple, rational, and harmonious development of the resources of the river, for flood control, retention of sediment, and regulation of water volume.
b. To prepare the General Integral Management Plan of the Basin, with the corresponding evaluation of the investments necessary for its execution. In the preparation of said General Plan, priority will be given to the projects that tend to accomplish the objectives of regional development.
c. To prepare technical and legal documents to summon bids that conform to the legal norms enforced in each country, with the purpose of carrying out studies, projects, and works linked to the development of the Basin.

d. To carry out environmental-impact studies linked to the activities mentioned in the present Statute.
e. To approve the planning of the design of bridges, ducts, and other structures that cross the river and can affect the uses and hydraulic functionality of the river, such as its navigation.
f. To promote the development of services and infrastructure in the region.
g. To plan the development of hydraulic energy.
h. To facilitate the activities that promote tourism.
i. To determine in which zones the extraction of resources will be prohibited because of its effect on hydric behavior and fluvial morphology [the scientific study of the river formations].
j. To propose norms that apply to the discharge of various types of contaminated substances into the river.
k. To monitor and systematically analyze the quality of the water, communicating to the Parties the infractions that will be verified.
l. To propose norms that regulate the activities of commercial fishing and sports in the river.
m. To coordinate the adoption of adequate measures to avoid alterations in the ecological equilibrium, including the control of pests and other factors that contaminate the river.
n. To cooperate and assist in studies about endemics, pandemics, and epidemics of hydric origin.
o. To carry out studies about agricultural irrigation within the regional boundaries, in order to promote irrigation-system projects.
p. To establish protected areas with the objective of preserving the wildlife and sites of historic interest.
q. To compile and update the information necessary to create and maintain a bank of hydrological, meteorological, and geotechnical data.
r. To drive and coordinate the installation and operation of meteorological, hydrologic, and hydrographic measurement stations and networks, and to carry out appraisal projects.
s. To establish and operate a bank of cartographic data about the Basin.
t. To analyze and study the possibilities of equipping navigable sections once the river is regulated.
u. The further functions of the Parties will be entrusted to them within their domain.
v. In conformity with Article II, clause b, the Council of Delegates will regulate the functions that will be in the domain of the Executive Leadership.

Article XIII
Collaboration of Official Organizations of the Countries

The Commission will receive, for its diligence and for the completion of its work, the fullest collaboration of the official technical and administrative organizations of the three countries.

Article XIV
Solution to Controversies

Any disagreement that arises within the Commission in relation to its functions will be elevated by the Commission to the Member States so the Member States may procure a solution to the matter through direct negotiations.

Source: "Agreement Constituting the Trilateral Commission for the Development of the Riverbed Rio Pilcomayo." International Freshwater Treaties Database. Oregon State University. Translated by Donna Scism. Available online. URL: http://ocid.nacse.org/tfdd/tfdddocs/255SPA.pdf. Accessed June 5, 2009.

Law 2029, 1999 (excerpts)

On October 29, 1999, the Bolivian government passed Law 2029, which allowed for the privatization of water. On November 1, 1999, Aguas del Tunari, the Bolivian subsidiary of an international consortium, took over the water company of Cochabamba. The contract between the government and Aguas del Tunari was signed before Law 2029 was passed, that law providing an ex post facto legal basis for the contract. The law was eventually retracted. Note that the term "Service Provider for Water and Sanitary Drainage," or "EPSA," is used often in these excerpts (and throughout the law in general). Article 8, subsection k, provides a definition. In the Cochabamba incident, Aguas del Tunari was the EPSA.

Part I
General
Chapter I: The Purpose and Scope of the Law

ARTICLE 1—(Aim). This law aims to establish the rules governing the provision and use of drinking water and sewage systems and institutional framework that governs the procedure for granting concessions and licenses for the provision of services, rights and obligations of providers and users, establishing the principles for setting prices, Rates, Taxes and Fees, and determination of violations and penalties.

ARTICLE 2—(Scope). They are subject to this Law throughout the country, all natural or legal persons, public or private, whatever its form of incorporation, to provide, or Users are linked to some drinking water and sewerage services.

ARTICLE 3—(Basic Sanitation). The Sanitation Sector Services includes: drinking water, sanitary sewer, sanitary disposal of excreta, solid waste and storm drainage.

ARTICLE 4—(Scope of the Law). This Act applies to basic drinking water and sewage systems, and creates the Superintendency of Sanitation.

ARTICLE 5—(Principles). The principles governing the provision of drinking water and sewage systems are: a) universal access to services; b) quality and continuity of services, consistent with human development policies; c) efficient use and allocation of resources for the provision and use of services; d) recognition of the economic value of services that must be paid by beneficiaries according to their socio-economic and social equity; e) sustainability of services; f) fair treatment to all providers and service users within the same category, and g) environmental protection.

ARTICLE 6—(Sectoral Regulatory System). Services for Water and Sanitary Sewer Sanitation Sector are incorporated into the Sectoral Regulation System (SIRESE) and subject to the provisions contained in Law No. 1600, Law System sectoral regulators, to October 28, 1994, its regulations and this Act and its regulations.

ARTICLE 7—(Public Utility). Works for the provision of potable water and sewage systems are in the public interest, are of public utility and are under state protection.

Chapter II: Definitions

ARTICLE 8—(Definitions). For the purposes of this Law, establishes the following definitions: a) Water: Water fit for human consumption in accordance with the requirements established by existing legislation. b) Wastewater: liquid waste from the discharge of water used domestically or otherwise. c) Treated Wastewater: Wastewater treatment systems processed to meet quality requirements in relation to the class of the receiving body to be discharged. g) Provision: Act by which the administrative Superintendency of Basic Sanitation, of the Bolivian State, grants the right to an

EPSA for providing drinking water and sanitary sewer. k) Service Provider for Water and Sanitary Drainage (EPSA): Legal person, public or private, that provides one or more of the drinking water and sewage systems, and has one of the following forms of incorporation: i. municipal public enterprise, dependent on one or more municipal governments; ii. mixed corporation; iii. private enterprise; iv. cooperative utilities; v. civil partnership; vi. indigenous and peasant communities, under article 171 of the Constitution of the State; vii. any other organization that has a legal structure recognized by Law, except for municipal governments. l) License: administrative act by which the Sanitation Superintendent certifies that an EPSA or a municipal government that provides drinking water services and sewerage services in direct compliance with the requirements for approval of tariffs or fees and is eligible to access government programs and projects of the sector. m) Price: the amount charged by the provider of services to users for connections, reconnections, meter installation and operational concepts similar. n) Water Resources: Water in the state that is found in nature. o) Water Service: Public service that includes one or more of the collection, transmission, processing and storage of water to make drinking water and distribution system to the users through networks of pipes or alternative media. p) Sanitary Sewer Service: Public service that includes one or more of the collection, treatment and disposal of wastewater into bodies. r) Price: Price per unit charged by the user by an EPSA any drinking water or sanitary sewer. w) non Concession Zone: Human settlements whose population is dispersed or concentrated, if not exceed 10,000 inhabitants, is not financially self-sustaining. x) Concession Zone: Principal population concentrated in the living more than 10,000, or an association of human settlements or pooling of municipal governments for the provision of potable water or sanitary sewer whose combined population is less than 10,000 inhabitants and where the provision of services is financially self-sustaining. Grant was accepted in the population under 10,000 inhabitants that prove to be self-sustaining.

ARTICLE 9—(National Competition). Policies, rules and regulation of drinking water and sewage systems are of national competence. The awards, the regulation of drinking water and sanitary sewer easements and have the same are the responsibility of the Superintendencia de Saneamiento Básico.

Part III

Entities of the Service Providers

Chapter I: Rights and Responsibilities

ARTICLE 17—(The Form of Services). The provision of drinking water and sanitary sewer is the responsibility of municipal governments under the provisions of this Act, its regulations and other provisions of law. This liability may be enforced in either directly or through third parties, depending on whether or not an area [is a concession zone]. In [concession] areas providing drinking water and sanitary sewer will be made mandatory through EPSA.

In [non concession areas], municipal governments can provide drinking water or sewage either directly or through an EPSA. The National Government will encourage the pooling of people for the provision of potable water and sewerage services through EPSA.

Article 19—(Part Private). Private entities may participate in the provision of drinking water and sewage systems, by granting, in accordance with regulations. The State shall encourage private sector participation in the provision of Water Services Potable and sanitary sewer.

Article 24—(Rights of Service Providers). The EPSA have the following rights: a) Fees charged pursuant to this Act and its regulations; b) charge for services provided to users, with the approval of the Superintendency of Basic Sanitation, according to regulations; c) suspend services for the reasons stated in this Act and its regulations; d) to collect fines for the user, according to regulations, and e) the other set by regulation or by the concession contract, as appropriate.

Article 29—(Concessions of Potable Water and Sewage Systems). The EPSA to provide drinking water or sewerage services in areas Concesa have to request the relevant provision of the service concession to the Superintendencia de Saneamiento Básico. No natural person or legal entity whether public or private association or nonprofit corporation, cooperative, municipal or otherwise, may provide drinking water or sewerage services in areas Concesa without proper Grant issued by the Superintendence of Sanitation.

Apart from the requirement to obtain concessions only to the EPSA and municipal governments to provide drinking water or sewerage services in a direct way, in areas not Concesa.

Concessions for the provision of drinking water and sanitary sewer are granted, modified, renewed or revoked by the Superintendency of Basic Sanitation, the name of the State, by administrative order, pursuant to procedures established by regulation. Concession contracts must contain at least the rights and obligations of the concessionaires of the holders set out in this Act and its regulations. Concessions for water service and sanitary sewer services, should be allowed in together. The granting of drinking water and sanitary sewer will be awarded for a maximum of forty (40) years, according to regulation.

Source: "Law No. 2029: Services Act of Drinking Water and Sanitary Sewer, 29 October 1999." Translation available online. URL: http://translate.google.com/translate?hl=en&sl=es&u=http://www.congreso.gov.bo/leyes/2029.htm&ei=SR8oSqLtOlHOMpOovYQF&sa=X&oi=translate&resnum=1&ct=result&prev=/search%3Fq%3DLey%2B2029%26hl%3Den. Accessed June 4, 2009.

EGYPT, ETHIOPIA, AND SUDAN

Treaty between Egypt and Sudan, 1959

In 1959, Sudan signed a water treaty with Egypt—actually the United Arab Republic—that not only increased the allotments of Nile River water for both countries but paved the way for the construction of the Aswan High Dam. The dam's reservoir, ultimately named Lake Nasser, would lie in both countries. The United Arab Republic consisted of Egypt and Syria, with the former dominating. That union lasted from 1958 until 1961, though Egypt retained official use of the name until 1971. The word "milliard" in the text of this treaty is another term for "billion."

AGREEMENT BETWEEN THE REPUBLIC OF THE SUDAN AND THE UNITED ARAB REPUBLIC FOR THE FULL UTILIZATION OF THE NILE WATERS. SIGNED AT CAIRO, ON 8 NOVEMBER 1959

As the River Nile needs projects, for its full control and for increasing its yield for the full utilization of its waters by the Republic of the Sudan and the United Arab Republic on technical working arrangements other than those now applied:

And as these works require for their execution and administration, full agreement and cooperation between the two Republics in order to regulate their benefits and utilize the Nile waters in a manner which secures the present and future requirements of the two countries:

And as the Nile waters Agreement concluded in 1929 provided only for the partial use of the Nile waters and did not extend to include a complete control of the River waters, the two Republics have agreed on the following:

First

THE PRESENT ACQUIRED RIGHTS

1. That the amount of the Nile waters used by the United Arab Republic unto this Agreement is signed shall be her acquired right before obtaining the benefits of the Nile Control Projects and the projects which will increase its yield and which projects are referred to in this Agreement; The total of this acquired right is 48 Milliards of cubic meters per year as measured at Aswan.

2. That the amount of the waters used at present by the Republic of Sudan shall be her acquired right before obtaining the benefits of the projects referred to above. The total amount of this acquired right is 4 Milliards of cubic meters per year as measured at Aswan.

Second

THE NILE CONTROL PROJECTS AND THE DIVISION OF THEIR BENEFITS BETWEEN THE TWO REPUBLICS

1. In order to regulate the River waters and control their flow into the sea, the two Republics agree that the United Arab Republic constructs the Sudd el Aali at Aswan as the first link of a series of projects on the Nile for over-year storage.

2. In order to enable the Sudan to utilize its share of the water, the two Republics agree that the Republic of Sudan shall construct the Roseires Dam on the Blue Nile and any other works which the Republic of the Sudan considers essential for the utilization of its share.

3. The net benefit from the Sudd el Aali Reservoir shall be calculated on the basis of the average natural River yield of water at Aswan in the years of this century which is estimated at about 84 Milliards of cubic meters per year. The acquired rights of the two Republics referred to in Article "First" as measured at Aswan, and the average of losses of over-year storage of the Sudd El Aali Reservoir shall be deducted from this yield, and the balance shall be the net benefit which shall be divided between the two Republics.

4. The net benefit from the Sudd el Aali Reservoir mentioned in the previous item, shall be divided between the two Republics at the ratio of 14 1/2 for the Sudan and 7 1/2 for the United Arab Republic so long as the average river yield remains in future within the limits of the average yield referred to in the previous paragraph. This means that, if the average yield remains the same as the average of the previous years of this century which is estimated

at 84 Milliards, and if the losses of over-year storage remain equal to the present estimate of 10 Milliards the net benefit of the Sudd el Aali Reservoir shall be 22 Milliards of which the share of the Republic of the Sudan shall be 14 1/2 Milliards and the share of the United Arab Republic shall be 7 1/2 Milliards. By adding these shares to their acquired rights, the total share from the net yield of the Nile after the full operation of the Sudd el Aali Reservoir shall be 18 1/2 Milliards for the Republic of the Sudan and 55 1/2 Milliards for the United Arab Republic.

But if the average yield increases, the resulting net benefit from this increase shall be divided between the two Republics, in equal shares.

5. As the net benefit from the Sudd el Aali (referred to in item 3 Article Second) is calculated on the basis of the average natural yield of the river at Aswan in the years of this century after the deduction therefrom of the acquired rights of the two Republics and the average losses of over-year storage at the Sudd el Aali Reservoir, it is agreed that this net benefit shall be the subject of revision by the two parties at reasonable intervals to be agreed upon after starting the full operation of the Sudd el Aali Reservoir.

6. The United Arab Republic agrees to pay to the Sudan Republic 15 Million Egyptian Pounds as full compensation for the damage resulting to the Sudanese existing properties as a result of the storage in the Sudd el Aali Reservoir up to a reduced level of 182 meters (survey datum). The payment of this compensation shall be affected in accordance with the annexed agreement between the two parties.

7. The Republic of the Sudan undertakes to arrange before July 1963, the transfer of the population of Halfa and all other Sudanese inhabitants whose lands shall be submerged by the stored water.

8. It is understood that when the Sudd el Aali is fully operated for over-year storage, the United Arab Republic will not require storing any water at Gebel Aulia Dam. And the two contracting parties will in due course, discuss all matters related to this renunciation.

Third

PROJECTS FOR THE UTILIZATION OF LOST WATERS IN THE NILE BASIN

In view of the fact that at present, considerable volumes of the Nile Basin Waters are lost in the swamps of Bahr El Jebel, Bahr El Zeraf, Bahr el Ghazal and the Sobat River, and as it is essential that efforts should be exerted in order to prevent these losses and to increase the yield of the River for use in agricultural expansion in the two Republics, the two Republics agree to the following:

1. The Republic of the Sudan in agreement with the United Arab Republic shall construct projects for the increase of the River yield by preventing losses of waters of the Nile Basin in the swamps of Bahr El Jebel, Bahr el Zeraf, Bahr el Ghazal and its tributaries, the Sobat River and its tributaries and the White Nile Basin. The net yield of these projects shall be divided equally between the two Republics and each of them shall also contribute equally to the costs.

The Republic of the Sudan shall finance the above-mentioned projects out of its own funds and the United Arab Republic shall pay its share in the costs in the same ratio of 50% allotted for her in the yield of these projects.

2. If the United Arab Republic, on account of the progress in its planned agricultural expansion should find it necessary to start on any of the increase of the Nile yield projects, referred to in the previous paragraph, after its approval by the two Governments and at a time when the Sudan Republic does not need such project, the United Arab Republic shall notify the Sudan Republic of the time convenient for the former to start the execution of the project. And each of the two Republics shall, within two years after such notification, present a date-phased programme for the utilization of its share of the waters saved by the project, and each of the said programmes shall bind the two parties. The United Arab Republic shall at the expiry of the two years, start the execution of the projects, at its own expense. And when the Republic of Sudan is ready to utilize its share according to the agreed programme, it shall pay to the United Arab Republic a share of all the expenses in the same ratio as the Sudan's share in benefit is to the total benefit of the project; provided that the share of either Republic shall not exceed one half of the total benefit of the project.

Fourth

TECHNICAL COOPERATION BETWEEN THE TWO REPUBLICS

1. In order to ensure the technical cooperation between the Governments of the two Republics, to continue the research and study necessary for the Nile control projects and the increase of its yield and to continue

the hydrological survey of it upper reaches, the two Republics agree that immediately after the signing of this Agreement a Permanent Joint Technical Commission shall be formed of an equal number of members from both parties; and its functions shall be:

a) The drawing of the basic outlines of projects for the increase of the Nile yield, and for the supervision of the studies necessary for the finalising of projects, before presentation of the same to the Governments of the two Republics for approval.
b) The supervision of the execution of the projects approved by the two Governments.
c) The drawing up of the working arrangements for any works to be constructed on the Nile, within the boundaries of the Sudan, and also for those to be constructed outside the boundaries of the Sudan, by agreement with the authorities concerned in the countries in which such works are constructed.
d) The supervision of the application of all the working arrangements mentioned in (c) above in connection with works constructed within the boundaries of Sudan and also in connection with the Sudd el Aali Reservoir and Aswan Dam, through official engineers delegated for the purpose by the two Republics; and the supervision of the working of the upper Nile projects, as provided in the agreements concluded with the countries in which such projects are constructed.
e) As it is probable that a series of low years may occur, and a succession of low levels in the Sudd el Aali Reservoir may result to such an extent as not to permit in any one year the drawing of the full requirements of the two Republics, the Technical Commission is charged with the task of devising a fair arrangement for the two Republics to follow. And the recommendations of the Commission shall be presented to the two Governments for approval.

2. In order to enable the Commission to exercise the functions enumerated in the above item, and in order to ensure the continuation of the Nile gauging and to keep observations on all its upper reaches, these duties shall be carried out under the technical supervision of the Commission by the engineers of the Sudan Republic and the engineers of the United Arab Republic in the Sudan and in the United Arab Republic and in Uganda [*sic*].

3. The two Governments shall form the Joint Technical Commission, by a joint decree, and shall provide it with its necessary funds from their

budgets. The Commission may, according to the requirements of work, hold its meetings in Cairo or in Khartoum. The Commission shall, subject to the approval of the two Governments, lay down regulations for the Organisation of its meetings and its technical, administrative and financial activities.

Fifth

GENERAL PROVISIONS

1. If it becomes necessary to hold any negotiations concerning the Nile waters, with any riparian state, outside the boundaries of the two Republics, the Governments of the Sudan Republic and the United Arab Republic shall agree on a unified view after the subject is studied by the said Technical Commission. The said unified view shall be the basis of any negotiations by the Commission with the said states.

If the negotiations result in an agreement to construct any works on the river, outside the boundaries of the two Republics, the Joint Technical Commission shall after consulting the authorities in the Governments of the States concerned, draw all the technical execution details and the working and maintenance arrangements. And the Commission shall, after the sanction of the same by the Governments concerned, supervise the carrying out of the said technical agreements.

2. As the riparian states, other than the two Republics, claim a share in the Nile waters, the two Republics have agreed that they shall jointly consider and reach one unified view regarding the said claims. And if the said consideration results in the acceptance of allotting an amount of the Nile water to one or the other of the savbid states, the accepted amount shall be deducted from the shares of the two Republics in equal parts, as calculated at Aswan.

The Technical Commission mentioned in this agreement shall make the necessary arrangements with the states concerned, in order to ensure that their water consumption shall not exceed the amounts agreed upon.

Sixth

TRANSITIONAL PERIOD BEFORE BENEFITING FROM THE COMPLETE SUDD EL AALI RESERVOIR

As the benefiting of the two Republics from their appointed shares in the net benefit of the Sudd el Aali Reservoir shall not start before the construction and the full utilization of the Reservoir, the two parties shall agree on their agricultural expansion programmes in the transitional period from now up to the completion of the Sudd el Aali, without prejudice to their present water requirements.

Seventh

This Agreement shall come into force after its sanction by the two contracting parties, provided that either party shall notify the other party of the date of its sanction, through the diplomatic channels.

Eighth

Annex (1) and Annex (2, A and B) attached to this Agreement shall be considered as an integral part of this Agreement.
Written in Cairo in two Arabic original copies this 7th day of Gumada El Oula 1379, the 8th day of November 1959.

For the Republic of Sudan
(Signed) Lewa Mohammed
TALAAT FARID

For the United Arab Republic:
(*Signed*) Zakaria
MOHIE EL DIN

ANNEX 1

A SPECIAL PROVISION FOR THE WATER LOAN REQUIRED BY THE UNITED ARAB REPUBLIC

The Republic of the Sudan agrees in principle to give a water loan from the Sudan's share in the Sudd el Aali waters, to the United Arab Republic, in order to enable the latter to proceed with her planned programmes for Agricultural Expansion.

The request of the United Arab Republic for this loan shall be made after it revises its programmes within five years from the date of the signing of this agreement. And if the revision by United Arab Republic reveals her need for this loan, the Republic of the Sudan shall give it out of its own share a loan exceeding one and a half Milliards, provided that the utilisation of this loan shall cease in November, 1977.

ANNEX 2

A

To the Head of the Delegation of the Republic of Sudan

With reference to Article (Second) Paragraph 6 of the Agreement signed this day concerning the full utilization of the River Nile Waters, compensation amounting to 15 Million Egyptian Pounds in sterling or in a third currency agreed upon by the two parties, and calculated on the basis of a fixed rate of $2.87156 to the Egyptian Pound, shall be paid by the Government of the United Arab Republic, as agreed upon, in installments in the following manner:

£3 million on the first of January, 1960
£4 million on the first of January, 1961
£4 million on the first of January, 1962
£4 million on the first of January, 1963

I shall be grateful if you confirm your agreement to the above.

With highest consideration
Head of the United Arab Republic Delegation
(*Signed*) Zakaria MOHIE EL DIN

B

To the Head of United Arab Republic Delegation
I have the honour to acknowledge receipt of your letter dated today and stipulating the following: [*See Annex 2, A*]

I have the honour to confirm the agreement of the Government of the Republic of the Sudan to the contents of the said letter.

With highest consideration.
Head of the Delegation of the Republic of Sudan:
(Signed) Lewa Mohammed TALAAT FARID
Translation by the Government of the United Arab Republic.

Source: "Agreement between the Republic of the Sudan and the United Arab Republic for the Full Utilization of the Nile Waters." International Freshwater Treaties Database. Oregon State University. Available online. URL: http://ocid.nacse.org/tfdd/tfdddocs/110ENG.pdf. Accessed May 22, 2009.

Treaty of Cooperation between Egypt and Ethiopia, 1993

After years of saber rattling between Egypt and Ethiopia over Nile River water, President Hosni Mubarak of Egypt and Meles Zenawi, president of Ethiopia's then-transitional government, signed an accord on July 1, 1993, in which they pledged mutual cooperation, especially with regard to the conservation and development of the Nile. Egypt, as the downriver country, benefited the most from the accord, and especially from Article 5.

Framework for General Cooperation Between The Arab Republic of Egypt And Ethiopia

The Arab Republic of Egypt and Ethiopia,
Determined to consolidate the ties of friendship, to enhance cooperation between the two countries and to establish a broad base of common interests,

Desirous of the realization of their full economic and resource potentials,

Recognizing the importance of the traditional ties existing between the two countries that have been consolidated during their long history of close relations and linked by the Nile River with its basin as a center of mutual interest,

Reaffirming their commitment to the UN and OAU Charters, Principles of International Law, as well as the Lagos Plan of Action,

Hereby agree on the following framework for cooperation:

Article 1

The two parties reaffirm their commitment to the principles of good neighbourliness, peaceful settlement of disputes, and non-interference in the internal affairs of states.

Article 2

The two parties are committed to the consolidation of mutual trust and understanding between the two countries.

Article 3

The two parties recognize the importance of their cooperation as an essential means to promote their economic and political interests as well as stability of the region.

Article 4

The two parties agree that the issue of the use of the Nile waters shall be worked out in detail through discussions by experts from both sides, on the basis of the rules and principles of international law.

Article 5

Each party shall refrain from engaging in any activity related to the Nile waters that may cause appreciable harm to the interests of the other party.

Article 6

The two parties agree on the necessity of the conservation and protection of the Nile waters. In this regard, they undertake to consult and cooperate in projects that are mutually advantageous, such as projects that would enhance the volume of flow and reduce the loss of Nile waters through comprehensive and integrated development schemes.

Article 7

The two parties will create [an] appropriate mechanism for periodic consultations on matters of mutual concern, including the Nile waters, in a manner that would enable them to work together for peace and stability in the region.

Article 8

The two parties shall endeavour towards a framework for effective cooperation among countries of the Nile basin for the promotion of common interest in the development of the basin.

This framework for cooperation is made in two originals in the Arabic and English languages, both texts being equally authentic.

Done at Cairo, this 1st day of the month of July 1993.

Source: "Framework for General Cooperation Between the Arab Republic of Egypt and Ethiopia." Available online. International Freshwater Treaties Database. Oregon State University. URL: http://ocid.nacse.org/tfdd/tfdddocs/359ENG.pdf. Accessed May 22, 2009.

The Nile Basin Initiative Act of 2002

The Nile Basin Initiative (NBI) was launched in Dar es Salaam in 1999 and three years later the NBI received legal status in Entebbe, Uganda, where its headquarters are located.

An Act to confer legal status in Uganda on the Nile Basin Initiative, and otherwise give the force of law in Uganda to the signed Agreed Minute No. 7 of the 9th Annual Meeting of the Nile Basin States held in Cairo, Egypt, on 14th February 2002; and to provide for other connected or incidental matters.

WHEREAS, currently, there is no regional or international Treaty or Agreement among the riparian States of the Nile River Basin, namely, Burundi, Democratic Republic of the Congo, Egypt, Eritrea, Ethiopia, Kenya, Rwanda, Sudan, Tanzania and Uganda, on cooperation in the utilization of the waters of the Nile River Basin;

AND WHEREAS the Government of the Republic of Uganda and the Governments of the other Nile Basin States at the meeting of their Council of Ministers held in Dar es Salaam, Tanzania, on 22nd February, 1999 established a "Transitional Institutional Mechanism of the Nile Basin Initiative (NBI)", pending the conclusion of a "Cooperative Framework Agreement", based on a vision "to achieve sustainable socioeconomic development through equitable utilization of, and benefit from, the common Nile Basin water resources";

AND WHEREAS the Government of Uganda and the Governments of the other States at the 9th annual meeting of their Council of Ministers held in Cairo, Egypt, on 14th February, 2002, adopted "Agreed Minute No. 7" of that meeting to, among other things, "invest the NBI, on a transitional basis, with legal personality to perform all of the functions entrusted to it, including the power to sue and be sued, and to acquire or dispose of movable and immovable property";

AND WHEREAS following the ratification by the Republic of Uganda of the Agreed Minute No. 7, it is necessary to give legal effect in Uganda to the provisions of that Agreed Minute No. 7;

DATE of ASSENT: 11th October 2002.
Date of Commencement: 1st November 2002.

Now, THEREFORE, BE IT ENACTED by Parliament as follows:

1. This Act may be cited as the Nile Basin Initiative Act, 2002.

2. In this Act, unless the context otherwise requires—

"Agreed Minute No. 7" means that part or portion of the minutes of the annual meeting of the Council of Ministers of the Nile Basin States held in Cairo, Egypt, on 14th February, 2002 and set out in the Schedule to this Act;

"Minister" means the Minister responsible for Water;

"Nile Basin Initiative" or "NBI" means the transitional arrangement established by the Nile Basin States at the meeting of their Council of Ministers held in Dar-es-Salaam, Tanzania, on 22nd February, 1999, to foster cooperation and sustainable development of the Nile River for the benefit of the inhabitants of those countries;

"Nile Basin States" means the States of Burundi, Democratic Republic of the Congo, Egypt, Eritrea, Ethiopia, Kenya, Rwanda, Sudan, Tanzania and Uganda; "Nile River Basin" means all the area forming the basin of the Nile River.

3. (1) The NBI shall have the capacity, within Uganda, of a body corporate with perpetual succession, and with power to acquire, hold, manage and dispose of movable and immovable corporate property, and to sue and be sued in its own name.

(2) The NBI shall have the capacity, within Uganda, to perform any of the functions conferred upon it by and under the Agreed Minute No. 7, and to do all things, including borrowing, that are, in the opinion of the Nile Basin States or the appropriate organ of the NBI, necessary or desirable for the performance of those functions.

(3) Subsection (2) of this section relates only to the capacity of the NBI as a body corporate, and nothing in that subsection shall be construed as authorising the disregard by the NBI of any law, or anything affecting any power of the NBI conferred by any law.

4. (1) There shall be charged on and paid out of the Consolidated Fund, without further appropriation other than this Act, all payments required to be made from time to time by the Government of Uganda under the terms or provisions of the Agreed Minute No. 7.

(2) Subject to article *159* of the Constitution of the Republic of Uganda, for the purposes of providing any sums required for making payments under this section, the Minister responsible for finance may, on behalf of the Government, make such arrangements as are necessary or raise loans by creation and issue of securities hearing such rates of interest and subject to such conditions as to repayment redemption or otherwise as the Minister thinks fit; and the charges and expenses incurred in connection with their issue shall be charged on and issued out of the Consolidated Fund.

(3) Any moneys received by the Government under the Agreed Minute No. 7 shall be paid into and form part of the Consolidated Fund, and shall be available in any manner in which the Consolidated Fund is available.

5. (1) The NBI, its staff and officials shall enjoy, within Uganda, such privileges and immunities as are necessary for their functions.

(2) The privileges and immunities referred to in subsection (1) of this section shall be in accordance with the provisions of the Diplomatic Privileges Act. 1965, Act No. 2 of *1965.*
SCHEDULE
AGREED MINUTE NO. 7

6. LEGAL STATUS OF THE NILE BASIN INITIATIVE (NBI)
The Ministers of Water affairs of the Nile Basin Countries, referring to the provisions of the signed Agreed Minutes of their meeting in Dar es Salaam on 22 February 1999 establishing a Transitional Institutional Mechanism of the Nile Basin Initiative (NBI), pending the conclusion of a Cooperative Framework Agreement to advance the Nile Basin Strategic Action Program in realization of the Shared Vision for the Nile Basin, "to achieve sustainable socio-economic development through equitable utilization of, and benefit from, the common Nile Basin water resources";

Invest the NBI, on a transitional basis, with legal personality to perform all of the functions entrusted to it, including the power to sue and be sued, and to acquire or dispose of movable and immovable property;

Recall that organs of the NBI include: the Council of Ministers of Water Affairs of the Nile Basin Countries (Nile-COM), which provides policy guidance and makes decisions on matters relating to the NM; the

Technical Advisory Committee (Nile-TAC), which renders technical advice and assistance to the Nile-COM; and the Nile Basin Secretariat (Nile-SEC), which renders administrative services to the Nile-COM and Nile-TAC;

Decide that NBI shall enjoy in the territory of each Nile Basin State the legal personality referred to above and such privileges and immunities as are necessary for the fulfilled of its functions;

Confirm that the headquarters of the NBI is situated at Entebbe, Uganda;

Confirm that the Executive Director of the Nile-SEC is the principal executive officer of the NBI;

Decide that the Executive Director, staff and officials of the NBI shall enjoy in the territory of each Nile Basin State such privileges and immunities as are necessary for the fulfillment of their functions;

Confirm that a draft budget for each financial year is to be prepared by the Executive Director and approved by the Nile-COM on the recommendation of the Nile-TAC; that the resources of the budget are derived from annual contributions of the Nile Basin States and such other sources as may be determined by the Nile-COM; and that the contributions of the Nile Basin States are based on the budget as approved by the Nile-COM;

Confirm that the accounts of the NBI relating to each fiscal year are to be audited the following fiscal year by an internationally recognized auditing firm selected on the basis of competitive bidding and submitted to Nile-TAC/COM for its review and approval; and that the NBI is to follow procurement and financial management practices that conform with international practices with the addition of any specific requirements of individual funding institutions.

Source: "The Nile Basin Initiative Act, 2002." Available online. URL: http://faolex.fao.org/docs/pdf/uga80648.pdf. Accessed May 25, 2009.

ISRAEL, JORDAN, AND THE PALESTINIAN TERRITORIES

Declaration of Principles of 1993 (excerpts)

Also known as the Oslo Accords, the Declaration of Principles was the first major peace agreement between Israel and the Palestinian Liberation Organization, granting a measure of autonomy to the Palestinian territories of the West Bank and the Gaza Strip. While not placed foremost among the principles, cooperation in the area of freshwater development was specifically referred to.

ARTICLE XI
ISRAELI-PALESTINIAN COOPERATION IN ECONOMIC FIELDS

Recognizing the mutual benefit of cooperation in promoting the development of the West Bank, the Gaza Strip and Israel, upon the entry into force of this Declaration of Principles, an Israeli-Palestinian Economic Cooperation Committee will be established in order to develop and implement in a cooperative manner the programs identified in the protocols attached as Annex III and Annex IV.

ANNEX III
PROTOCOL ON ISRAELI-PALESTINIAN COOPERATION IN ECONOMIC AND DEVELOPMENT PROGRAMS

The two sides agree to establish an Israeli-Palestinian continuing Committee for Economic Cooperation, focusing, among other things, on the following:

1. Cooperation in the field of water, including a Water Development Program prepared by experts from both sides, which will also specify the mode of cooperation in the management of water resources in the West Bank and Gaza Strip, and will include proposals for studies and plans on water rights of each party, as well as on the equitable utilization of joint water resources for implementation in and beyond the interim period.

ANNEX IV
PROTOCOL ON ISRAELI-PALESTINIAN COOPERATION CONCERNING REGIONAL DEVELOPMENT PROGRAMS

1. The two sides will cooperate in the context of the multilateral peace efforts in promoting a Development Program for the region, including the West Bank and the Gaza Strip, to be initiated by the G-7. The parties will request the G-7 to seek the participation in this program of other interested states, such as members of the Organisation for Economic Cooperation and Development, regional Arab states and institutions, as well as members of the private sector.

2. The Development Program will consist of two elements:

  a. an Economic Development Program for the 'West Bank and the Gaza Strip.
  b. a Regional Economic Development Program.

A. The Economic Development Program for the West Bank and the Gaza strip will consist of the following elements:

1. A Social Rehabilitation Program, including a Housing and Construction Program.
2. A Small and Medium Business Development Plan.
3. An Infrastructure Development Program (water, electricity, transportation and communications, etc.)
4. A Human Resources Plan.
5. Other programs.

B. The Regional Economic Development Program may consist of the following elements:

1. The establishment of a Middle East Development Fund, as a first step, and a Middle East Development Bank, as a second step.
2. The development of a joint Israeli-Palestinian-Jordanian Plan for coordinated exploitation of the Dead Sea area.
3. The Mediterranean Sea (Gaza)—Dead Sea Canal.
4. Regional Desalinization and other water development projects.
5. A regional plan for agricultural development, including a coordinated regional effort for the prevention of desertification.
6. Interconnection of electricity grids.
7. Regional cooperation for the transfer, distribution and industrial exploitation of gas, oil and other energy resources.
8. A Regional Tourism, Transportation and Telecommunications Development Plan.
9. Regional cooperation in other spheres.

3. The two sides will encourage the multilateral working groups, and will coordinate towards their success. The two parties will encourage intersessional activities, as well as pre-feasibility and feasibility studies, within the various multilateral working groups.

Source: "Declaration of Principles on Interim Self-Government Arrangements." Peace Agreements Digital Collection: Israel-PLO. United States Institute of Peace. Available online. URL: http://www.usip.org/library/pa/israel_plo/oslo_09131993.html. Accessed May 25, 2009.

Peace Treaty between Israel and Jordan, 1994 (excerpts)

On October 26, 1994, Israel and Jordan signed a historic peace treaty—technically the two countries had been at war since 1948. During the negotiations, water had been a major sticking point, and the issue was not resolved until an annex concerning water problems was added to the document. The abbreviation MCM stands for million cubic meters.

Article 6.—Water

With the view to achieving a comprehensive and lasting settlement of all the water problems between them:

1. The Parties agree mutually to recognize the rightful allocations of both of them in Jordan River and Yarmouk River waters and Araba/Arava groundwater in accordance with the agreed acceptable principles, quantities and quality as set out in Annex H, which shall be fully respected and complied with.

2. The Parties, recognizing the necessity to find a practical, just and agreed solution to their water problems and with the view that the subject of water can form the basis for the advancement of cooperation between them, jointly undertake to ensure that the management and development of their water resources do not, in any way, harm the water resources of the other Party.

3. The Parties recognize that their water resources are not sufficient to meet their needs. More water should be supplied for their use through various methods, including projects of regional and international cooperation.

4. In light of paragraph 3 of this Article, with the understanding that cooperation in water-related subjects would be to the benefit of both Parties, and will help alleviate their water shortages, and that water issues along their entire boundary must be dealt with in their totality, including the possibility of trans-boundary water transfers, the Parties agree to search for ways to alleviate water shortages and to cooperate in the following fields:

a. development of existing and new water resources, increasing the water availability, including cooperation on a regional basis as appropriate, and minimizing wastage of water resources through the chain of their uses;
b. prevention of contamination of water resources;
c. mutual assistance in the alleviation of water shortages;
d. transfer of information and joint research and development in water-related subjects, and review of the potentials for enhancement of water resources development and use.

5. The implementation of both Parties' undertakings under this Article is detailed in Annex II.

ANNEX II
WATER RELATED MATTERS

Pursuant to Article 6 of the Treaty, Israel and Jordan agreed on the following Articles on water related matters:

Article I.—Allocation

1. Water from the Yarmouk River
   a. Summer period—15th May to 15th October of each year. Israel pumps (12) MCM [million cubic meters] and Jordan gets the rest of the flow.
   b. Winter period—16th October to 16th May of each year. Israel pumps (13) MCM and Jordan is entitled to the rest of the flow subject to provisions outlined hereinbelow: Jordan concedes to Israel pumping an additional (20) MCM from the Yarmouk in winter in return for Israel conceding to transferring to Jordan during the summer period the quantity specified in paragraph (2.a) below from the Jordan River.
   c. In order that waste of water will be minimized, Israel and Jordan may use, downstream of point 121/Adassiya Diversion, excess flood water that is not usable and will evidently go to waste unused.

2. Water from the Jordan River
   a. Summer period—15th May to 15th October of each year. In return for the additional water that Jordan concedes to Israel in winter in accordance with paragraph (l.b) above, Israel concedes to transfer to Jordan in the summer period (20) MCM from the Jordan River directly upstream from Deganya gates on the river. Jordan shall pay the operation and maintenance cost of such transfer through existing systems (not including capital cost) and shall bear the total cost of any new transmission system. A separate protocol shall regulate this transfer.
   b. Winter period—16th October to 14th May of each year. Jordan is entitled to store for its use a minimum average of (20) MCM of the floods in the Jordan River south of its confluence with the Yarmouk (as outlined in Article II below). Excess floods that are not usable and that will otherwise be wasted can be utilized for the benefit of the two Parties including pumped storage off the course of the river.
   c. In addition to the above, Israel is entitled to maintain its current uses of the Jordan River waters between its confluence with the

Yarmouk, and its confluence with Tiral Zvi/Wadi Yabis. Jordan is entitled to an annual quantity equivalent to that of Israel, provided however, that Jordan's use will not harm the quantity or quality of the above Israeli uses. The Joint Water Committee (outlined in Article VII below) will survey existing uses for documentation and prevention of appreciable harm.

d. Jordan is entitled to an annual quantity of (10) MCM of desalinated water from the desalination of about (20) MCM of saline springs now diverted to the Jordan River. Israel will explore the possibility of financing the operation and maintenance cost of the supply to Jordan of this desalinated water (not including capital cost). Until the desalination facilities are operational, and upon the entry into force of the Treaty, Israel will supply Jordan (10) MCM of Jordan River water from the same location as in (2.a) above, outside the summer period and during dates Jordan selects, subject to the maximum capacity of transmission.

3. Additional Water

Israel and Jordan shall cooperate in finding sources for the supply to Jordan of an additional quantity of (50) MCM/year of water of drinkable standards. To this end, the Joint Water Committee will develop, within one year from the entry into force of the Treaty, a plan for the supply to Jordan of the above mentioned additional water. This plan will be forwarded to the respective governments for discussion and decision.

4. Operation and Maintenance

a. Operation and maintenance of the systems on Israeli territory that supply Jordan with water, and their electricity supply, shall be Israel's responsibility. The operation and maintenance of the new systems that serve only Jordan will be contracted at Jordan's expense to authorities or companies selected by Jordan.

b. Israel will guarantee easy unhindered access of personnel and equipment to such new systems for operation and maintenance. This subject will be further detailed in the agreements to be signed between Israel and the authorities or companies selected by Jordan.

Article II.—Storage

1. Israel and Jordan shall cooperate to build a diversion/storage dam on the Yarmouk River directly downstream of the point 121/Adassiya Diversion. The purpose is to improve the diversion efficiency into the King

Abdullah Canal of the water allocation of the Hashemite Kingdom of Jordan, and possibly for the diversion of Israel's allocation of the river water. Other purposes can be mutually agreed.

2. Israel and Jordan shall cooperate to build a system of water storage on the Jordan River, along their common boundary, between its confluence with the Yannouk River and its confluence with Tirat Zvi/Wadi Yabis, in order to implement the provision of paragraph (2.b) of Article I above. The storage system can also be made to accommodate more floods; Israel may use up to (3) MCM/year of added storage capacity.

3. Other storage reservoirs can be discussed and agreed upon mutually.

Article III.—Water Quality and Protection

1. Israel and Jordan each undertake to protect, within their own jurisdiction, the shared waters of the Jordan and Yarmouk Rivers, and Arava/Araba groundwater, against any pollution, contamination, harm or unauthorized withdrawals of each other's allocations.

2. For this purpose, Israel and Jordan will jointly monitor the quality of water along their boundary, by use of jointly established monitoring stations to be operated under the guidance of the Joint Water Committee.

3. Israel and Jordan will each prohibit the disposal of municipal and industrial wastewater into the courses of the Yarmouk and the Jordan Rivers before they are treated to standards allowing their unrestricted agricultural use. Implementation of this prohibition shall be completed within three years from the entry into force of the Treaty.

4. The quality of water supplied from one country to the other at any given location shall be equivalent to the quality of the water used from the same location by the supplying country.

5. Saline springs currently diverted to the Jordan River are earmarked for desalination within four years. Both countries shall cooperate to ensure that the resulting brine will not be disposed of in the Jordan River or in any of its tributaries.

6. Israel and Jordan will protect water systems each in its own territory, supplying water to the other, against any pollution, contamination, harm or unauthorized withdrawal of each other's allocations.

Article IV.—Groundwater in Emek Ha'arava/Wadi Araba

1. In accordance with the provisions of this Treaty, some wells drilled and used by Israel along with their associated systems fall on the Jordanian side of the borders. These wells and systems are under Jordan's sovereignty. Israel shall retain the use of these wells and systems in the quantity and quality detailed in an Appendix to this Annex, that shall be jointly prepared by 31st December, 1994. Neither country shall take, nor cause to be taken, any measure that may appreciably reduce the yields or quality of these wells and systems.

2. Throughout the period of Israel's use of these wells and systems, replacement of any well that may fail among them, shall be licensed by Jordan in accordance with the laws and regulations then in effect. For this purpose, the failed well shall be treated as though it was drilled under license from the competent Jordanian authority at the time of its drilling. Israel shall supply Jordan with the log of each of the wells and the technical information about it to be kept on record. The replacement well shall be connected to the Israeli electricity and water systems.

3. Israel may increase the abstraction rate from wells and systems in Jordan by up to (10) MCM/year about the yields referred to in paragraph 1 above, subject to a determination by the Joint Water Committee that this undertaking is hydrogeologically feasible and does not harm existing Jordanian uses. Such increase is to be carried out within five years from the entry into force of the Treaty.

4. Operation and Maintenance
    a. Operation and maintenance of the wells and systems on Jordanian territory that supply Israel with water, and their electricity supply shall be Jordan's responsibility. The operation and maintenance of these wells and systems will be contracted at Israel's expense to authorities or companies selected by Israel.
    b. Jordan will guarantee easy unhindered access of personnel and equipment to such wells and systems for operation and maintenance. This subject will be further detailed in the agreements to be signed between Jordan and the authorities or companies selected by Israel.

Article V.—Notification and Agreement

1. Artificial changes in or of the course of the Jordan and Yarmouk Rivers can only be made by mutual agreement.

2. Each country undertakes to notify the other, six months ahead of time, of any intended projects, which are likely to change the flow of either of the above rivers along their common boundary, or the quality of such flow. The subject will be discussed in the Joint Water Committee with the aim of preventing harm and mitigating adverse impacts such projects may cause.

Article VI.—Cooperation

1. Israel and Jordan undertake to exchange relevant data on water resources through the Joint Water Committee.

2. Israel and Jordan shall cooperate in developing plans for purposes of increasing water supplies and improving water use efficiency, within the context of bilateral, regional or international cooperation.

Article VII.—Joint Water Committee

1. For the purpose of the implementation of this Annex, the Parties will establish a Joint Water Committee comprised of three members from each country.

2. The Joint Water Committee will, with the approval of the respective governments, specify its work procedures, the frequency of its meetings, and the details of its scope of work. The Committee may invite experts and/or advisors as may be required.

3. The Committee may form, as it deems necessary, a number of specialized subcommittees and assign them technical tasks. In this context, it is agreed that these subcommittees will include a northern subcommittee and a southern subcommittee, for the management on the ground of the mutual water resources in these sectors.

Source: "Treaty of Peace between the State of Israel and the Hashemite Kingdom of Jordan." Published by the Ministry of Foreign Affairs of Israel. Available online. International Freshwater Treaties Database. Oregon State University. URL: http://ocid.nacse.org/tfdd/tfdddocs/168ENG.htm. Accessed May 23, 2009.

Israeli-Palestinian Interim Agreement, 1995 (excerpts)

On September 28, 1995, Israel and the Palestinian Authority, as per the 1993 Declaration of Principles (Oslo Accords), signed an interim agreement, with four annexes, that covered a wide range of topics in the areas of politics and civil and military affairs. Below are two portions of Annex Three—article 40 and schedule 11—that cover water and sewerage in the West Bank and the Gaza Strip. The abbreviation MCM stands for million cubic meters.

ARTICLE 40
Water and Sewage

On the basis of good-will, both sides have reached the following agreement in the sphere of Water and Sewage:

Principles

1. Israel recognizes the Palestinian water rights in the West Bank. These will be negotiated in the permanent status negotiations and settled in the Permanent Status Agreement relating to the various water resources.

2. Both sides recognize the necessity to develop additional water for various uses.

3. While respecting each side's powers and responsibilities in the sphere of water and sewage in their respective areas, both sides agree to coordinate the management of water and sewage resources and systems in the West Bank during the interim period, in accordance with the following principles:
   a. Maintaining existing quantities of utilization from the resources, taking into consideration the quantities of additional water for the Palestinians from the Eastern Aquifer and other agreed sources in the West Bank as detailed in this Article.
   b. Preventing the deterioration of water quality in water resources.
   c. Using the water resources in a manner which will ensure sustainable use in the future, in quantity and quality.
   d. Adjusting the utilization of the resources according to variable climatological and hydrological conditions.
   e. Taking all necessary measures to prevent any harm to water resources, including those utilized by the other side.
   f. Treating, reusing or properly disposing of all domestic, urban, industrial, and agricultural sewage.
   g. Existing water and sewage systems shall be operated, maintained and developed in a coordinated manner, as set out in this Article.
   h. Each side shall take all necessary measures to prevent any harm to the water and sewage systems in their respective areas.
   i. Each side shall ensure that the provisions of this Article are applied to all resources and systems, including those privately owned or operated, in their respective areas.

Transfer of Authority

4. The Israeli side shall transfer to the Palestinian side, and the Palestinian side shall assume, powers and responsibilities in the sphere of water and sewage in the West Bank related solely to Palestinians, that are currently held by the military government and its Civil Administration, except for the issues that will be negotiated in the permanent status negotiations, in accordance with the provisions of this Article.

5. The issue of ownership of water and sewage related infrastructure in the West Bank will be addressed in the permanent status negotiations.

Additional Water

6. Both sides have agreed that the future needs of the Palestinians in the West Bank are estimated to be between 70–80 mcm/year.

7. In this framework, and in order to meet the immediate needs of the Palestinians in fresh water for domestic use, both sides recognize the necessity to make available to the Palestinians during the interim period a total quantity of 28.6 mcm/year, as detailed below:

a. Israeli Commitment:
   1. Additional supply to Hebron and the Bethlehem area, including the construction of the required pipeline—1 mcm/year.
   2. Additional supply to Ramallah area—0.5 mcm/year.
   3. Additional supply to an agreed take-off point in the Salfit area—0.6 mcm/year.
   4. Additional supply to the Nablus area—1 mcm/year.
   5. The drilling of an additional well in the Jenin area—1.4 mcm/year.
   6. Additional supply to the Gaza Strip—5 mcm/year.
   7. The capital cost of items (1) and (5) above shall be borne by Israel.

b. Palestinian Responsibility:
   1. An additional well in the Nablus area—2.1 mcm/year.
   2. Additional supply to the Hebron, Bethlehem and Ramallah areas from the Eastern Aquifer or other agreed sources in the West Bank—17 mcm/year.
   3. A new pipeline to convey the 5 mcm/year from the existing Israeli water system to the Gaza Strip. In the future, this quantity will come from desalination in Israel.
   4. The connecting pipeline from the Salfit take-off point to Salfit.

5. The connection of the additional well in the Jenin area to the consumers.
6. The remainder of the estimated quantity of the Palestinian needs mentioned in paragraph 6 above, over the quantities mentioned in this paragraph (41.4–51.4 mcm/year), shall be developed by the Palestinians from the Eastern Aquifer and other agreed sources in the West Bank. The Palestinians will have the right to utilize this amount for their needs (domestic and agricultural).
8. The provisions of paragraphs 6–7 above shall not prejudice the provisions of paragraph 1 to this Article.
9. Israel shall assist the Council in the implementation of the provisions of paragraph 7 above, including the following:
   a. Making available all relevant data.
   b. Determining the appropriate locations for drilling of wells.
10. In order to enable the implementation of paragraph 7 above, both sides shall negotiate and finalize as soon as possible a Protocol concerning the above projects, in accordance with paragraphs 18–19 below.

The Joint Water Committee

11. In order to implement their undertakings under this Article, the two sides will establish, upon the signing of this Agreement, a permanent Joint Water Committee (JWC) for the interim period, under the auspices of the CAC [Joint Civil Affairs Coordination and Cooperation Committee].
12. The function of the JWC shall be to deal with all water and sewage related issues in the West Bank including, inter alia:
   a. Coordinated management of water resources.
   b. Coordinated management of water and sewage systems.
   c. Protection of water resources and water and sewage systems.
   d. Exchange of information relating to water and sewage laws and regulations.
   e. Overseeing the operation of the joint supervision and enforcement mechanism.
   f. Resolution of water and sewage related disputes.
   g. Cooperation in the field of water and sewage, as detailed in this Article.
   h. Arrangements for water supply from one side to the other.
   i. Monitoring systems. The existing regulations concerning measurement and monitoring shall remain in force until the JWC decides otherwise.

j. Other issues of mutual interest in the sphere of water and sewage.

13. The JWC shall be comprised of an equal number of representatives from each side.
14. All decisions of the JWC shall be reached by consensus, including the agenda, its procedures and other matters.
15. Detailed responsibilities and obligations of the JWC for the implementation of its functions are set out in Schedule 8.

Supervision and Enforcement Mechanism

16. Both sides recognize the necessity to establish a joint mechanism for supervision over and enforcement of their agreements in the field of water and sewage, in the West Bank.
17. For this purpose, both sides shall establish, upon the signing of this Agreement, Joint Supervision and Enforcement Teams (JSET), whose structure, role, and mode of operation is detailed in Schedule 9.

Water Purchases

18. Both sides have agreed that in the case of purchase of water by one side from the other, the purchaser shall pay the full real cost incurred by the supplier, including the cost of production at the source and the conveyance all the way to the point of delivery. Relevant provisions will be included in the Protocol referred to in paragraph 19 below.
19. The JWC will develop a Protocol relating to all aspects of the supply of water from one side to the other, including, inter alia, reliability of supply, quality of supplied water, schedule of delivery and off-set of debts.

Mutual Cooperation

20. Both sides will cooperate in the field of water and sewage, including, inter alia:
    a. Cooperation in the framework of the Israeli-Palestinian Continuing Committee for Economic Cooperation, in accordance with the provisions of Article XI and Annex III of the Declaration of Principles.
    b. Cooperation concerning regional development programs, in accordance with the provisions of Article XI and Annex IV of the Declaration of Principles.

c. Cooperation, within the framework of the joint Israeli-Palestinian-American Committee, on water production and development related projects agreed upon by the JWC.
d. Cooperation in the promotion and development of other agreed water-related and sewage-related joint projects, in existing or future multilateral forums.
e. Cooperation in water-related technology transfer, research and development, training, and setting of standards.
f. Cooperation in the development of mechanisms for dealing with water-related and sewage related natural and manmade emergencies and extreme conditions.
g. Cooperation in the exchange of available relevant water and sewage data, including:
    1. Measurements and maps related to water resources and uses.
    2. Reports, plans, studies, researches and project documents related to water and sewage.
    3. Data concerning the existing extractions, utilization and estimated potential of the Eastern, North-Eastern and Western Aquifers (attached as Schedule 10).

Protection of Water Resources and Water and Sewage Systems

21. Each side shall take all necessary measures to prevent any harm, pollution, or deterioration of water quality of the water resources.
22. Each side shall take all necessary measures for the physical protection of the water and sewage systems in their respective areas.
23. Each side shall take all necessary measures to prevent any pollution or contamination of the water and sewage systems, including those of the other side.
24. Each side shall reimburse the other for any unauthorized use of or sabotage to water and sewage systems situated in the areas under its responsibility which serve the other side.

The Gaza Strip

25. The existing agreements and arrangements between the sides concerning water resources and water and sewage systems in the Gaza Strip shall remain unchanged, as detailed in Schedule 11.

SCHEDULE 11
The Gaza Strip

Pursuant to Article 40, Paragraph 25:

1. All water and sewage (hereinafter referred to as "water") systems and resources in the Gaza Strip shall be operated, managed and developed (including drilling) by the Council, in a manner that shall prevent any harm to the water resources.

2. As an exception to paragraph 1, the existing water systems supplying water to the Settlements and the Military Installation Area, and the water systems and resources inside them shall continue to be operated and managed by Mekorot Water Co.

3. All pumping from water resources in the Settlements and the Military Installation Area shall be in accordance with existing quantities of drinking water and agricultural water. Without derogating from the powers and responsibilities of the Council, the Council shall not adversely affect these quantities. Israel shall provide the Council with all data concerning the number of wells in the Settlements and the quantities and quality of the water pumped from each well, on a monthly basis.

4. Without derogating from the powers and responsibilities of the Council, the Council shall enable the supply of water to the Gush Katif settlement area and Kfar Darom settlement by Mekorot, as well as the maintenance by Mekorot of the water systems supplying these locations.

5. The Council shall pay Mekorot for the cost of water supplied from Israel and for the real expenses incurred in supplying water to the Council.

6. All relations between the Council and Mekorot shall be dealt with in a commercial agreement.

7. The Council shall take the necessary measures to ensure the protection of all water systems in the Gaza Strip.

8. The two sides shall establish a subcommittee to deal with all issues of mutual interest including the exchange of all relevant data to the management and operation of the water resources and systems and mutual prevention of harm to water resources.

9. The subcommittee shall agree upon its agenda and upon the procedures and manner of its meetings, and may invite experts or advisers as it sees fit.

Source: "The Israeli-Palestinian Interim Agreement on the West Bank and the Gaza Strip Annex III." Available online. International Freshwater Treaties Database. Oregon State University. URL: http://ocid.nacse.org/tfdd/tfdddocs/229ENG.htm. Accessed May 25, 2009.

INDIA, PAKISTAN, AND BANGLADESH

Inter-Dominion Agreement between the Government of India and the Government of Pakistan, 1948

Following the 1947 partition that divided Pakistan and India, tensions over water increased in the state of Punjab, which lies in both countries. In April 1948, Indian East Punjab abruptly cut off water to Pakistani West Punjab, threatening its municipal and agricultural supplies. The crisis lasted for a month and was resolved by the Inter-Dominion Agreement, signed in New Delhi on May 4, 1948.

A dispute has arisen between the East and West Punjab Government regarding the supply by East Punjab of water to the Central Bari Doab and the Depalpur canals in West Punjab. The contention of the East Punjab Government is that under the Punjab Partition (Apportionment of Assets and liabilities) Order, 1947, and the Arbitral Award the proprietary rights in the waters of the rivers in East Punjab vest wholly in the East Punjab Government and that the West Punjab Government cannot claim any share of these waters as a right. The West Punjab Government disputes this contention, its view being that the point has conclusively been decided in its favour by implication by the Arbitral Award and that in accordance with international law and equity, West Punjab has a right to the waters of the East Punjab rivers.

2. The East Punjab Government has revived the flow of water into these canals on certain conditions of which two are disputed by West Punjab. One, which arises out of the contention in paragraph 1, is the right to the levey of seigniorage charges for water and the other is the question of the capital cost of the Madhopur Head Works and carrier channels to be taken into account.

3. The East and West Punjab Governments are anxious that this question should be settled in a spirit of goodwill and friendship. Without prejudice to its legal rights in the matter the East Punjab Government assured the West Punjab Government that it has no intention to withhold water from West Punjab without giving it time to tap alternative sources. The West Punjab Government on its part recognize the natural anxiety of the East Punjab Government to discharge the obligations to develop areas where water is scarce and which were underdeveloped in relation to parts of West Punjab.

4. Apart, therefore, from the question of law involved the Governments are anxious to approach the problem in a practical spirit of the East Punjab Government progressively diminishing its supply to these canals in order to give reasonable time to enable the West Punjab Government to tap alternative sources.

5. The West Punjab Government has agreed to deposit immediately in the Reserve Bank such *ad hoc* sum as may be specified by the Prime Minister of India. Out of this sum, that Government agrees to the immediate transfer to East Punjab Government of sums over which there are no dispute.

6. After an examination by each party of the legal issues, of the method of estimating the cost of water to be supplied by the East Punjab Government and of the technical survey of water resources and the means of using them for supply to these canals, the two Governments agree that further meetings between their representatives should take place.

7. The Dominion Governments of India and Pakistan accept above terms and express the hope that a friendly solution will be reached.

(Signed)	*(Signed)*
JAWAHARLAL NEHRU	GHULAM MOHD.
SWARAN SINGH	SHAUKAT HYAT KHAN
N. V. GADGIL	MUMTAZ DAULTANA

Source: "Inter-Dominion Agreement between the Government of India and the Government of Pakistan, on the Canal Water Dispute between [India and Pakistan]." Available online. International Freshwater Treaties Database. Oregon State University. URL: http://ocid.nacse.org/tfdd/tfdddocs/61ENG.htm. Accessed May 25, 2009.

Indus Waters Treaty, 1960

Although the 1948 Inter-Dominion Agreement between Pakistan and India solved the most immediate water-related problems in the divided state of Punjab, it soon became obvious that a permanent treaty for the Indus basin was needed. Amid much wrangling between the two countries, the World Bank (formally known as the International Bank for Reconstruction and Development) stepped in to broker a deal; it was also a cosignatory of the treaty, which was signed in Karachi, Pakistan, on September 19, 1960.

PREAMBLE

The Government of India and the Government of Pakistan, being equally desirous of attaining the most complete and satisfactory utilisation of the waters of the Indus system of rivers and recognizing the need, therefore, of fixing and delimiting, in a spirit of goodwill and friendship, the rights and obligations of each in relation to the other concerning the use of these waters and of making provision for the settlement, in a cooperative spirit, of all such questions as may hereafter arise in regard to the interpretation or application of the provisions agreed upon herein, have resolved to conclude a Treaty in furtherance of these objectives, and for this purpose named as their plenipotentiaries:

The Government of India:
Shri Jawaharlal Nehru,
Prime Minister of India,
and
The Government of Pakistan:
Field Marshal Mohammad Ayub Khan, H.P., H.J.,
President of Pakistan

who, having communicated to each other their respective Full Powers and found them in good and due form, have agreed upon the following Articles and Annexures:

Article I
DEFINITIONS

As used in this Treaty:

(1) The terms "Article" and "Annexure" mean respectively an Article of, an Annexure to, this Treaty.

Except as otherwise indicated, references to Paragraphs are to the paragraphs in the Article or in the Annexure in which the reference is made

(2) The term "Tributary" of a river means any surface channel, whether in continuous or intermittent flow and by whatever name called, whose waters in the natural course would fall into that river, e.g. a tributary, a torrent, a natural drainage an artificial drainage, a *nadi,* a *nallah,* a *nai,* a *khad,* a *cho.* The term also includes any subtributary or branch or subsidiary channel, by whatever name called, whose waters, in the natural course, would directly or otherwise flow into that surface channel.

(3) The term "The Indus," "The Jhelum," "The Chenab," "The Ravi," "The Beas" or "The Sutlej" means the named river (including Connecting Lakes, if any) and all its Tributaries: Provided however that

- (i) none of the rivers named above shall be deemed to be a Tributary
- (ii) the Chenab shall be deemed to include the river Panjnad; and
- (iii) the river Chandra and the river Bhaga shall be deemed to be Tributaries of the Chenab.

(4) The term "Main" added after Indus, Jhelum, Chenab, Sutlej, Beas or Ravi means the main stem of the named river excluding its Tributaries, but including all channels and creeks of the main stem of that river and such Connecting Lakes as form part of the main stem itself. The Jhelum Main shall be deemed to extend up to Verinag, and the Chenab Main up to the confluence of the river Chandra and the river Bhaga.

(5) The term "Eastern Rivers" means The Sutlej, The Beas and The Ravi taken together.

(6) The term "Western Rivers" means The Indus, The Jhelum and The Chenab taken together.

(7) The term "the Rivers" means all the rivers, The Sutlej, The Beas, The Ravi, The Indus, The Jhelum and The Chenab.

(8) The term "Connecting Lake" means any lake which receives water from, or yields water to, any of the Rivers; but any lake which occasionally and irregularly receives only the spill of any of the Rivers and returns only the whole or part of that spill is not a Connecting Lake.

(9) The term "Agricultural Use" means the use of water for irrigation, except for irrigation of household gardens and public recreational gardens.

(10) The term "Domestic Use" means the use of water for:

(a) drinking, washing, bathing, recreation, sanitation (including the conveyance and dilution of sewage and of industrial and other wastes), stock and poultry and other like purposes;
(b) household and municipal purposes (including use for household gardens and public recreational gardens); and
(c) industrial purposes (including mining, mining and other like purposes); but the term does not include Agricultural Use or use for the generation of hydroelectric power.

(11) The term "Non-Consumptive Use" means any control or use of water for navigation, floating of timber or other property, flood protection or flood control, fishing or fish culture, wild life or other like beneficial purposes, provided that exclusive of seepage and evaporation of water incidental to the control or use, the water (undiminished in volume within the practical range of measurement) remains in, or is returned to, the same river or its Tributaries; but the term does not include Agricultural Use or use for the generation of hydroelectric power.

(12) The term "Transition Period" means the period beginning and ending as provided in Article II (6).

(13) The term "Bank" means the International Bank for Reconstruction Development.

(14) The term "Commissioners" means either of the Commissioners appointed under the provisions of Article VIII (1) and the term "Commission" means the Permanent Indus Commission constituted in accordance with Article VIII (3).

(15) The term "interference with the waters" means

(a) Any act of withdrawal therefrom; or
(b) Any man-made obstruction to their flow which causes a change in the volume (within the practical range of measurement) of the daily flow of the waters. Provided however that an obstruction

which involves only an insignificant and incidental change in the volume of the daily flow, for example, fluctuations due to afflux caused by bridge piers or a temporary by-pass, etc., shall not be deemed to be an interference with the waters.

(16) The term "Effective Date" means the date on which this effect in accordance with the provisions of Article XII, that is, the first of April 1960

Article II
PROVISIONS REGARDING EASTERN RIVERS

(1) All the waters of the Eastern Rivers shall be available for the unrestricted use of India, except as otherwise expressly provided in this Article.

(2) Except for Domestic Use and Non-Consumptive Use, Pakistan shall be under an obligation to let flow, and shall not permit any interference with, the waters of the Sutlej Main and the Ravi Main in the reaches where these rivers flow in Pakistan and have not yet finally crossed into Pakistan. The points of final crossing are the following: (a) near the new Hasta Bund upstream of Suleimanke in the case of the Sutlej Main, and (b) about one and a half miles upstream of the syphoon for the B-R—B-D Link in the case of the Ravi Main.

(3) Except for Domestic Use, Non-Consumptive Use and Agricultural (as specified in Annexure B), Pakistan shall be under an obligation to let flow, and shall not permit any interference with, the waters (while flowing in Pakistan) of any Tributary which in its natural course joins the Sutlej Main or the Ravi Main before these rivers have finally crossed into Pakistan.

(4) All the waters, while flowing in Pakistan, of any Tributary which, in its natural course, joins the Sutlej Main or the Ravi Main after these rivers have finally crossed into Pakistan shall be available for the unrestricted use of Pakistan: Provided however that this provision shall not be construed as giving Pakistan any claim or right to any releases by India in any such Tributary. If Pakistan should deliver any of the waters of any such Tributary, which on the Effective Date joins the Ravi Main after this river has finally crossed into Pakistan, into a reach of the Ravi Main upstream of this crossing, India shall not make use of these waters; each Party agrees to establish such discharge observation stations and make such observations as may be necessary for the determination of the

component of water available for the use of Pakistan on account of the aforesaid deliveries by Pakistan, and Pakistan agrees to meet the cost of establishing the aforesaid discharge observation stations and making the aforesaid observations.

(5) There shall be a Transition Period during which, to the extent specified in Annexure H, India shall

(i) limit its withdrawals for Agricultural Use,
(ii) limit abstractions for storages, and
(iii) make deliveries to Pakistan from the Eastern Rivers.

(6) The Transition Period shall begin on 1st April 1960 and it shall end on 31st March 1970, or, if extended under the provisions of Part 8 of Annexure H on the date up to which it has been extended. In any event, whether or not the replacement referred to in Article IV (1) has been accomplished, the Transition Period shall end not later than 31st March 1973.

(7) If the Transition Period is extended beyond 31st March 1970, the provisions of Article V (5) shall apply.

(8) If the Transition Period is extended beyond 31st March 1970, the provisions of Paragraph (5) shall apply during the period of extension beyond 31st March 1970.

(9) During the Transition Period, Pakistan shall receive for unrestricted use the waters of the Eastern Rivers which are to be released by India in accordance with the provisions of Annexure H. After the end of the Transition Period, Pakistan shall, have no claim or right to releases by India of any of the waters of the Eastern Rivers. In case there are any releases, Pakistan shall enjoy the unrestricted use of the waters so released after they have finally crossed into Pakistan: Provided that in the event that Pakistan makes any use of these waters, Pakistan shall not acquire any right whatsoever, by prescription or otherwise, to a continuance of such releases or such use.

Article III
PROVISIONS REGARDING WESTERN RIVERS

(1) Pakistan shall receive for unrestricted use all those waters of the Western Rivers which India is under obligation to let flow under the provisions of Paragraph (2)

(2) India shall be under an obligation to let flow all the waters of the Western Rivers, and shall not permit any interference with these waters, except for the following uses, restricted (except as provided in item (c) (ii) of Paragraph 5 of Annexure C) in the case of each of the rivers, The Indus, The Jhelum and The Chenab, to the drainage basin thereof:

(a) Domestic Use;
(b) Non-Consumptive Use;
(c) Agricultural Use, as set out in Annexure C; and
(d) Generation of hydroelectric power, as set out in Annexure D.

(3) Pakistan shall have the unrestricted use of all waters originating from the source other than the Eastern Rivers which are delivered by Pakistan into The Ravi or The Sutlej, and India shall not make use of these waters. Each Party agrees to establish such discharge observation stations and make such observations as may be considered necessary by the Commission for the determination of the component of water available for the use of Pakistan on account of the aforesaid deliveries by Pakistan.

(4) Except as provided in Annexures D and E, India shall not store any water, or construct any storage works on, the Western Rivers.

Article IV
PROVISIONS REGARDING EASTERN RIVERS AND WESTERN RIVERS

(1) Pakistan shall use its best endeavours to construct and bring into operation, with due regard to expedition and economy, that part of a system of works which will accomplish the replacement, from the Western Rivers and other sources, of water supplies for irrigation canals in Pakistan which, on 15th August 1947, were dependent on water supplies from Eastern Rivers.

(2) Each Party agrees that any Non-Consumptive Use made by it shall be not made as not to materially change, on account of such use, the flow in any channel the prejudice of the uses on that channel by the other Party under the provisions of this Treaty. In executing any scheme of flood protection or flood each Party will avoid, as far as practicable, any material damage to the other Party, and any such scheme carried out by India on the Western Rivers shall not involve any use of water or any storage in addition to that provided under Article III.

(3) Nothing in this Treaty shall be construed as having the effect of preventing either Party from undertaking schemes of drainage, river training, conservation of soil against erosion and dredging, or from removal of stones, gravel or sand of the Rivers: Provided that

(a) in executing any of the schemes mentioned above, each Party will avoid, as far as practicable, any material damage to the other Party;
(b) any such scheme carried out by India on the Western Rivers shall not involve any use of water or any storage in addition to that provided under Article III;
(c) except as provided in Paragraph (5) and Article VII (1) (b), India not take, any action to increase the catchment area, beyond the area on the Effective Date of any natural or artificial drainage or drain which crosses into Pakistan, and shall not undertake such construction or remodelling of any drainage or drain which so crosses or falls into a drainage or drain which so crosses as might cause material damage in Pakistan or entail the construction of a new drain or enlargement of an existing drainage or drain in Pakistan; and
(d) should Pakistan desire to increase the catchment area, beyond the area on the Effective Date, of any natural or artificial drainage or drain, which reduces drainage waters from India, or, except in an emergency, to pour any waters into it in excess of the quantities received by it as on the Effective Date, Pakistan shall before undertaking any work for these purposes, increase the capacity of that drainage or drain to the extent necessary so as not to impair its efficacy for dealing with drainage waters received from India as on the Effective Date.

(4) Pakistan shall maintain in good order its portions of the drainages mentioned below with capacities not less than the capacities as on the Effective Date

(i) Hudiara Drain
(ii) Kasur Nala
(iii) Salimshah Drain
(iv) Fazilka Drain.

(5) If India finds it necessary that any of the drainages mentioned in Paragraph (4) should be deepened or widened in Pakistan, Pakistan agrees to

undertake to do so as a work of public interest, provided India agrees to pay the cost of the deepening or widening.

(6) Each Party will use its best endeavours to maintain the natural channels of the Rivers, as on the Effective Date, in such condition as will avoid, as far as practicable, any obstruction to the flow in these channels likely to cause material damage to the other Party.

(7) Neither Party will take any action which would have the effect of diverting the Ravi Main between Madhopur and Lahore, or the Sutlej Main between Harike and Suleimanke, from its natural channel between high banks.

(8) The use of the natural channels of the Rivers for the discharge of flood or other excess waters shall be free and not subject to limitation by either Party, and neither Party shall have any claim against the other in respect of any damage caused by such use. Each Party agrees to communicate to the other Party, as far in advance as practicable, any information it may have in regard to such extraordinary discharge of water from reservoirs and flood flows as may affect the other Party.

(9) Each Party declares its intention to operate its storage dams, barrages and irrigation canals in such manner, consistent with the normal operations of its hydraulic systems, as to avoid, as far as feasible, material damage to the other Party.

(10) Each Party declares its intention to prevent, as far as practicable undue pollution of the waters of the Rivers which might affect adversely uses similar in nature to those to which the waters were put on the Effective Date, and agrees to take all reasonable measures to ensure that, before any sewage or industrial waste is allowed to flow into the Rivers, it will be treated, where necessary, in such manner as not materially to affect those uses: Provided that the criterion of reasonableness shall be the customary practice in similar situations on the Rivers.

(11) The Parties agree to adopt, is far as feasible, appropriate measures for the recovery, and restoration to owners, of timber and other property floated or floating down the Rivers, subject to appropriate charges being paid by the owners.

(12) The use of water for industrial purposes under Articles II (2), II (3) and III (2) shall not exceed:

(a) in the case of an industrial process known on the Effective Date, such quantum of use as was customary in that process on the Effective Date
(b) in the case of an industrial process not known on the Effective Date
  (i) such quantum of use as was customary on the Effective Date in similar or in any way comparable industrial processes; or
  (ii) if there was no industrial process on the Effective Date similar or in any way comparable to the new process, such quantum of use as would not have a substantially adverse effect on the other Party.

(13) Such part of any water withdrawn for Domestic Use under the provisions of Articles II (3) and III (2) as is subsequently applied to Agricultural Use shall be accounted for as part of the Agricultural Use specified in Annexure B and Annexure G respectively; each Party will use its best endeavours to return to the same river (directly or through one of its Tributaries) all water withdrawn therefrom for industrial purposes and not consumed either in the industrial processes for which it was withdrawn or in some other Domestic Use.

(14) In the event that either Party should develop a use of the waters of the Rivers which is not in accordance with the provisions of this Treaty, that Party shall not acquire by reason of such use any right, by prescription or otherwise, to a continuance of such use.

(15) Except as otherwise required by the express provisions of this Treaty nothing in this Treaty shall be construed as affecting existing territorial rights over the waters of any of the Rivers or the beds or banks thereof, or as affecting existing property rights under municipal law over such waters or beds or banks.

Article V
FINANCIAL PROVISIONS

(1) In consideration of the fact that the purpose of part of the system of works referred to in Article IV (1) is the replacement, from the Western Rivers and other sources, of water supplies for irrigation canals in Pakistan which, on 15th August 1947 were dependent on water supplies from the

Eastern Rivers, India agrees to make a fixed contribution of Pounds Sterling 62,060,000 towards the costs of these works. The amount in Pounds Sterling of this contribution shall remain unchanged irrespective of any alteration in the par value of any currency.

(2) The sum of Pounds Sterling 62,060,000 specified in Paragraph (1) shall be paid in ten equal annual installments on the 1st of November of each year. The first of such annual installments shall be paid on 1st November 1960, or if the Treaty has not entered into force by that date, then within one month after the Treaty enters into force.

(3) Each of the installments specified in Paragraph (2) shall be paid to the Bank for the credit of the Indus Basin Development Fund to be established and administered by the Bank, and payment shall be made in Pounds Sterling, or in such other currency or currencies as may from time to time be agreed between India and the Bank.

(4) The payments provided for under the provisions of Paragraph (3) shall be made without deduction or set-off on account of any financial claims of India on Pakistan arising otherwise than under the provisions of this Treaty: Provided that this provision shall in no way absolve Pakistan from the necessity of paying in other ways debts to India which may be outstanding against Pakistan.

(5) If, at the request of Pakistan, the Transition Period is extended in accordance with the provisions of Article II (6) and of Part 8 of Annexure H, the Bank shall thereupon pay to India out of the Indus Basin Development Fund the appropriate amount specified in the Table below:

PERIOD OF AGGREGATE EXTENSION OF TRANSITION

Period	Payment to India
One year	£Stg. 3,125,000
Two years	£Stg. 6,406,250
Three years	£Stg. 9,850,000

(6) The provisions of Article IV (1) and Article V (1) shall not be construed as conferring upon India any right to participate in the decisions as to the system of works which Pakistan constructs pursuant to Article IV (1) or as constituting an assumption of any responsibility by India or as an agreement by India in regard to such works.

(7) Except for such payments as are specifically provided for in this Treaty neither Party shall be entitled to claim any payment for observance of the provisions of this Treaty or to make any charge for water received from it by the other Party.

Article VI
EXCHANGE OF DATA

(1) The following data with respect to the flow in, and utilisation of the waters of, the Rivers shall be exchanged regularly between the Parties:

(a) Daily (or as observed or estimated less frequently) gauge and discharge data relating to flow of the Rivers at all observation sites.
(b) Daily extractions for or releases from reservoirs.
(c) Daily withdrawals at the heads of all canals operated by government or by a government agency (hereinafter in this Article called canals), including link canals.
(d) Daily escapages from all canals, including link canals.
(e) Daily deliveries from link canals.

These data shall be transmitted monthly by each Party to the other as soon as the data for a calendar month have been collected and tabulated, but not later than three months after the end of the months to which they relate: Provided that such of the data specified above as are considered by either Party to be necessary for operational purposes shall be supplied daily or at less frequent intervals, as may be requested. Should one Party request the supply of any of these data by telegram, telephone, wireless, it shall reimburse the other Party for the cost of transmission.

(2) If, in addition to the data specified in Paragraph (1) of this Article, either Party requests the supply of any data relating to the hydrology of the Rivers, or to canal or reservoir operation connected with the Rivers, or to any provision of this Treaty, such data shall be supplied by the other Party to the extent that these are available.

Article VII
FUTURE COOPERATION

(1) The two Parties recognize that they have a common interest in the optimum development of the Rivers, and, to that end, they declare their intention to cooperate, by mutual agreement, to the fullest possible extent. In particular:

(a) Each Party, to the extent it considers practicable and on agreement by the other Party to pay the costs to be incurred, will, at the request of the other Party, set up or install such hydrologic observation stations within the drainage basins of the Rivers, and set up or install such meteorological observation stations relating thereto and carry out such observations thereat, as may be requested, and will supply the data so obtained.

(b) Each Party, to the extent it considers practicable and on agreement by the other Party to pay the costs to be incurred, will, at the request of the other Party carry out such new drainage works as may be required in connection with new drainage works of the other Party.

(c) At the request of either Party, the two Parties may, by mutual agreement, cooperate in undertaking engineering works on the Rivers.

The formal arrangements, in each case, shall be as agreed upon between the Parties.

(2) If either Party plans to construct any engineering work which would cause interference with the waters of any of the Rivers and which, in its opinion, would affect the other Party materially, it shall notify the other Party of its plans and shall supply such data relating to the work as may be available and as would enable the other Party to inform itself of the nature, magnitude and effect of the work. If a work would cause interference with the waters of any of the Rivers but would not, in the opinion of the Party planning it, affect the other Party materially, nevertheless the Party planning the work shall, on request, supply the other Party with such data: regarding the nature, magnitude and effect, if any, of the work as may be available.

Article VIII

PERMANENT INDUS COMMISSION

(1) India and Pakistan shall each create a permanent post of Commissioner for Indus Waters, and shall appoint to this post, as often as a vacancy occurs, a person who should ordinarily be a high-ranking engineer competent in the field of hydrology and water-use. Unless either Government should decide to take up any particular question directly with the other Government, each Commissioner will be the representative of his Government for all matters arising out of this Treaty, and will serve as the regular

channel of communication on all matters relating to the implementation of the Treaty, and, in particular, with respect to

(a) the furnishing or exchange of information or data provided for in the Treaty; and
(b) the giving of any notice or response to any notice provided for in the Treaty.

(2) The status of each Commissioner and his duties and responsibilities towards his Government will be determined by that Government.

(3) The two Commissioners shall together form the Permanent Indus Commission.

(4) The purpose and functions of the Commission shall be to establish and maintain cooperation between the Parties in the development of the waters of the Rivers and, in particular,

(a) to study and report to the two Governments on any problem relating to the development of the waters of the Rivers which may be jointly referred to the Commission by the two Governments: in the event that a reference is made by one Government alone, the Commissioner of the other Government shall obtain the authorization of his Government before he proceeds to act on the reference;
(b) to make every effort to settle promptly, in accordance with the provisions of Article IX (1), any question arising thereunder;
(c) to undertake, once in every five years, a general tour of inspection of the Rivers for ascertaining the facts connected with various developments and works on the Rivers;
(d) to undertake promptly, at the request of either Commissioner, a tour of inspection of such works or sites on the Rivers as may be considered necessary by him for ascertaining the facts connected with those works or sites; and
(e) to take, during the Transition Period, such steps as may be necessary for the implementation of the provisions of Annexure H.

(5) The Commission shall meet regularly at least once a year, alternately in India and Pakistan. This regular annual meeting shall be held in November or in such other month as may be agreed upon between the Commissioners. The Commission shall also meet when requested by either Commissioner.

(6) To enable the Commissioners to perform their functions in the Commission, each Government agrees to accord to the Commissioner of the other Government the same privileges and immunities as are accorded to representatives of member States to the principal and subsidiary organs of the United Nations under Sections 11, 12 and 13 of Article IV of the Convention on the Privileges and Immunities of the United Nations (dated 13th February, 1946) during the periods specified in those Sections. It is understood and agreed that these privileges and immunities are accorded to the Commissioners not for the personal benefit of the individuals themselves but in order to safeguard the independent exercise of their functions in connection with the Commission; consequently, the Government appointing the Commissioner not only has the right but is under a duty to waive the immunity of its Commissioner in any case where, in the opinion of the appointing Government, the immunity would impede course of justice and can be waived without prejudice to the purpose for which immunity is accorded.

(7) For the purposes of the inspections specified in Paragraph (4) (c) and (d) each Commissioner may be accompanied by two advisers or assistants to whom appropriate facilities will be accorded.

(8) The Commission shall submit to the Government of India and to the Government of Pakistan, before the first of June of every year, a report on its work for the year, ended on the preceding 31st of March, and may submit to the two Governments other reports at such times as it may think desirable.

(9) Each Government shall bear the expenses of its Commissioner and his ordinary staff. The cost of any special staff required in connection with the work mentioned in Article VII (1) shall be borne as provided therein.

(10) The Commission shall determine its own procedures.

Article IX
SETTLEMENT OF DIFFERENCES AND DISPUTES

(1) Any question which arises between the Parties concerning the interpretation or application of this Treaty or the existence of any fact which, if established, might constitute a breach of this Treaty shall first be examined by the Commission, which will endeavour to resolve the question by agreement.

(2) If the Commission does not reach agreement on any of the questions mentioned in Paragraph (1), then a difference will be deemed to have arisen, which shall be dealt with as follows:

(a) Any difference which, in the opinion of either Commissioner, falls within the provisions of Part I of Annexure F shall, at the request of either Commissioner, be dealt with by a Neutral Expert in accordance with the provisions of Part 2 of Annexure F;

(b) If the difference does not come within the provisions of Paragraph (2) (a), or if a Neutral Expert, in accordance with the provisions of Paragraph 7 of Annexure F has informed the Commission that, in his opinion, the difference, or a part thereof, should be treated as a dispute, then a dispute will be deemed to have arisen which shall be settled in accordance with the provisions of Paragraphs (3), (4) and (5)

Provided that, at the discretion of the Commission, any difference may either be dealt with by a Neutral Expert in accordance with the provisions of Part 2 of Annexure F or be deemed to be a dispute to be settled in accordance with the provisions of Paragraphs (3), (4) and (5), or may be settled in any other way agreed upon by the Commission.

(3) As soon as a dispute to be settled in accordance with this and the succeeding paragraphs of this Article has arisen, the Commission shall, at the request of either Commissioner, report the fact to the two Governments, as early as practicable, stating in its report the points on which the Commission is in agreement and the issues in dispute, the views of each Commissioner on these issues and his reasons therefor.

(4) Either Government may, following receipt of the report referred to in Paragraph (3), or if it comes to the conclusion that this report is being unduly delayed in the Commission, invite the other Government to resolve the dispute by agreement. In doing so it shall state the names of its negotiators and their readiness to meet with the negotiators to be appointed by the other Government at a time and place to be indicated by the other Government. To assist in these negotiations, the two Governments may agree to enlist the services of one or more mediators acceptable to them.

(5) A court of Arbitration shall be established to resolve the dispute in the manner provided by Annexure G

(a) upon agreement between the Parties to do so; or
(b) at the request of either Party, if, after negotiations have begun pursuant to Paragraph (4), in its opinion the dispute is not likely to be resolved by negotiation or mediation; or
(c) at the request of either Party, if, after the expiry of one month following receipt by the other Government of the invitation referred to in Paragraph (4), that Party comes to the conclusion that the other Government is unduly delaying the negotiations.

(6) The provisions of Paragraphs (3), (4) and (5) shall not apply to any difference while it is being dealt with by a Neutral Expert.

Article X
EMERGENCY PROVISION

If, at any time prior to 31st March 1965, Pakistan should represent to the Bank that, because of the outbreak of large-scale international hostilities arising out of causes beyond the control of Pakistan, it is unable to obtain from abroad the materials and equipment necessary for the completion, by 31st March 1973, of that part of the system of works referred to in Article IV (1) which related to the replacement referred to therein, (hereinafter referred to as the "replacement element") and if, after consideration of this representation in consultation with India, the Bank is of the opinion that

(a) these hostilities are on a scale of which the consequence is that Pakistan is unable to obtain in time such materials and equipment as must be procured from abroad for the completion, by 31st March 1973, of the replacement element, and
(b) since the Effective Date, Pakistan has taken all reasonable steps to obtain the said materials and equipment and, with such resources of materials and equipment as have been available to Pakistan both from within Pakistan and from abroad, has carried forward the construction of the replacement element with due diligence and all reasonable expedition, the Bank shall immediately notify each of the Parties accordingly. The Parties undertake, without prejudice to the provisions of Article XII (3) and (4), that on being so notified, they will forthwith consult together and enlist the good offices of the Bank in their consultation, with a view to reaching mutual agreement as to whether or not, in the light of all the circumstances then prevailing, any modifications of the provisions of

this Treaty are appropriate and advisable and, if so, the nature and the extent of the modifications.

Article XI
GENERAL PROVISIONS

(1) It is expressly understood that

(a) this Treaty governs the rights and obligations of each Party in relation to the other with respect only to the use of the waters of the Rivers and matters incidental thereto; and
(b) nothing contained in this Treaty, and nothing arising out of the execution thereof, shall be construed as constituting a recognition or waiver (whether tacit, by implication or otherwise) of any rights or claims whatsoever of either of the Parties other than those rights or claims which are expressly recognized or waived in this Treaty.

Each of the Parties agrees that it will not invoke this Treaty, anything contained therein, or anything arising out of the execution thereof, in support of any of its own rights or claims whatsoever or in disputing any of the rights or claims whatsoever of the other Party, other than those rights or claims which are expressly recognized or waived in this Treaty.

(2) Nothing in this Treaty shall be construed by the Parties as in any way establishing any general principle of law or any precedent.

(3) The rights and obligations of each Party under this Treaty shall remain affected by any provisions contained in, or by anything arising out of the execution of, any agreement establishing the Indus Basin Development Fund.

Article XII
FINAL PROVISIONS

(1) This Treaty consists of the Preamble, the Articles hereof and Annexures A to H hereto, and may be cited as "The Indus Waters Treaty 1960."

(2) This Treaty shall be ratified and the ratifications thereof shall be exchanged in New Delhi. It shall enter into force upon the exchange of ratifications, and will then take effect retrospectively from the first of April 1960.

(3) The provisions of this Treaty may from time to time be modified by a duly ratified treaty concluded for that purpose between the two Governments.

(4) The provisions of this Treaty, or the provisions of this Treaty as modified under the provisions of Paragraph (3), shall continue in force until terminated by a duly ratified treaty concluded for that purpose between the two Governments.

IN WITNESS WHEREOF the respective Plenipotentiaries have signed this Treaty and have hereunto affixed their seals.

DONE in triplicate in English at Karachi on this Nineteenth day of September 1960.

For the Government of India:
(*Signed*) Jawaharlal NEHRU

For the Government of Pakistan:
(*Signed*) Mohammad Ayub KHAN
Field Marshal, H.P., H.J.

For the International Bank for Reconstruction and Development,
for the purposes specified in Articles V and X and Annexures F, G and H
(*Signed*) W. A. B. ILIFF

Source: "Indus Waters Treaty 1960 between the Government of India, the Government of Pakistan and the International Bank for Reconstruction and Development." International Freshwater Treaties Database. Oregon State University. URL: http://ocid.nacse.org/tfdd/tfdddocs/114ENG.htm. Accessed May 26, 2009.

1977 Treaty between India and Bangladesh on the Sharing of Ganges Waters and the 1996 Reiteration of the Treaty (excerpts)

The construction of the Farakka Barrage on the Ganges by India reduced neighboring Bangladesh's supply of the river's water, and a water conflict ensued. The two countries signed a treaty to deal with the problem in 1977; the treaty was reiterated, with some changes, in 1996. The 1996 reiteration is excerpted.

THE GOVERNMENT OF THE PEOPLE'S REPUBLIC OF BANGLADESH AND THE GOVERNMENT OF THE REPUBLIC OF INDIA,

DETERMINED to promote and strengthen their relations of friendship and good neighbourliness,

INSPIRED by the common desire of promoting the well-being of their peoples,

BEING desirous of sharing by mutual agreement the waters of the international rivers flowing through the territories of the two countries and of making the optimum utilisation of the water resources of their region by joint efforts.

RECOGNISING that the need of making an interim arrangement for sharing of the Ganges waters at Farakka in a spirit of mutual accommodation and the need for a solution of the long-term problem of augmenting the flows of the Ganges are in the mutual interests of the peoples of the two countries.

BEING desirous of finding a fair solution of the question before them, without affecting the rights and entitlements of either country other than those covered by this Agreement, or establishing any general principles of law or precedent,

HAVE AGREED AS FOLLOWS:

A. Arrangements for sharing of the waters of the Ganges at Farakka.

ARTICLE I

The quantum of waters agreed to be released by India to Bangladesh will be at Farakka.

ARTICLE II

(i) The sharing between Bangladesh and India of the Ganges waters at Farakka from the 1st January to the 31st May every year will be with reference to the quantum shown in column 2 of the Schedule annexed hereto which is based on 75 percent availability calculated from the recorded flows of the Ganges at Farakka from 1948 to 1973.

(ii) India shall release to Bangladesh waters by 10-day periods in quantum shown in column 4 of the Schedule:

Provided that if the actual availability at Farakka of the Ganges waters during a 10-day period is higher or lower than the quantum shown in column 2 of the Schedule it shall be shared in the proportion applicable to that period;

Provided further that if during a particular 10-day period, the Ganges flows at Farakka come down to such a level that the share of Bangladesh is lower than 80 percent of the value shown in column 4, the release of waters to

Bangladesh during that 10-day period shall not fall below 80 percent of the value shown in column 4.

ARTICLE III

The waters released to Bangladesh at Farakka under Article I shall not be reduced below Farakka except for reasonable uses of waters, not exceeding 200 cusecs, by India between Farakka and the point on the Ganges where both its banks are in Bangladesh.

ARTICLE IV

A Committee consisting of the representatives nominated by the two Governments (hereinafter called the Joint Committee) shall be constituted. The Joint Committee shall set up suitable teams at Farakka and Hardinge Bridge to observe and record at Farakka the daily flows below Farakka Barrage and in the Feeder Canal, as well as at Hardinge Bridge.

ARTICLE V

The Joint Committee shall decide its own procedure and method of functioning.

ARTICLE VI

The Joint Committee shall submit to the two Governments all data collected by it and shall also submit a yearly report to both the Governments.

ARTICLE VII

The Joint Committee shall be responsible for implementing the arrangements contained in this part of the Agreement and examining any difficulty arising out of the implementation of the above arrangements and of the operation of Farakka Barrage. Any difference or dispute or arising in this regard, if not resolved by the Joint Committee, shall be referred to a panel of an equal number of Bangladeshi and Indian experts nominated by the two Governments. If the difference or dispute still remains unresolved, it shall be referred to the two Governments which shall meet urgently at the appropriate level to resolve it by mutual discussion and failing that by such other arrangements as they may mutually agree upon.

B. Long-Term Arrangements

ARTICLE VIII

The two Governments recognise the need to cooperate with each other in finding a solution to the long-term problem of augmenting the flows of the Ganges during the dry season.

ARTICLE IX

The Indo-Bangladesh Joint Rivers Commission established by the two Governments in 1972 shall carry out investigation and study of schemes relating to the augmentation of the dry season flows of the Ganges, proposed or to be proposed by either Government with a view to finding a solution which is economical and feasible. It shall submit its recommendations to the two Governments within a period of three years.

ARTICLE X

The two Governments shall consider and agree upon a scheme or schemes, taking into account the recommendations of the Joint Rivers Commission, and take necessary measures to implement it or them as speedily as possible.

ARTICLE XI

Any difficulty, difference or dispute arising from or with regard to this part of the Agreement, if not resolved by the Joint Rivers Commission, shall be referred to the two Governments which shall meet urgently at the appropriate level to resolve it by mutual discussion.

C. Review and Duration

ARTICLE XII

The provisions of this Agreement will be implemented by both parties in good faith. During the period for which the Agreement continues to be in force in accordance with Article XV of the Agreement, the quantum of waters agreed to be released to Bangladesh at Farakka in accordance with this Agreement shall not be reduced.

ARTICLE XIII

The Agreement will be reviewed by the two Governments at the expiry of three years from the date of coming into force of this Agreement. Further reviews shall take place six months before the expiry of this Agreement or as may be agreed upon between the two Governments.

ARTICLE XIV

The review or reviews referred to in Article XIII shall entail consideration of the working, impact, implementation and progress of the arrangements contained in parts A and B of this Agreement.

ARTICLE XV

This Agreement shall enter into force upon signature and shall remain in force for a period of 5 years from the date of its coming into force. It may be extended further for a specified period by mutual agreement in the light of the review or, reviews referred to in Article XIII,

IN WITNESS WHEREOF the undersigned, being duly authorized thereto by the respective Governments, have signed this Agreement.
Done in duplicate at Dacca on the 5th November, 1977 in the Bengali, Hindi and English languages. In the event of any conflict between the texts, the English text shall prevail.

Rear Admiral Musharraf Husain Khan
Chief of Naval Staff and Member, President's Council of Advisers
in-charge of the Ministry of Communications, Flood Control, Water
Resources and Power, Government of the People's Republic of Bangladesh.

Surjit Singh Barnala
Minister for Agriculture and Irrigation, Government of the Republic of India

SCHEDULE
[Vida Article II (i)]

Sharing of waters at Farakka between the lst January and the 31st May every year.

1	2	3	4
Period	**Flows reaching Farraka (based on 75% availability from observed date (1948-73)**	**Withdrawal by India at Farraka**	**Release to Bangladesh**
	Cusecs	Cusecs	Cusecs
January 1-10	98,500	40,000	58,500
January 11-20	89,750	38,500	51,250
January 21-31	82,500	35,000	47,500
February 1-10	79,250	33,000	46,250
February 11-20	74,000	31,500	42,500
February 21-28/29	70,000	30,750	39,250
March 1-10	65,250	26,750	38,500
March 11-20	63,560	25,000	38,000
March 21-30	61,000	25,000	36,000
April 1-10	59,000	24,000	35,000

April 11-20	55,500	20,750	34,750
April 21-30	55,000	20,500	34,500
May 1-10	56,500	21,500	35,000
May 11-20	59,250	24,000	35,250
May 21-31	65,500	26,750	38,750

*[Reprinted from the text provided by the Government of Bangladesh.]

[According to Article XV, the Agreement entered into force upon signature, November 5, 1977 and shall remain in force for five years.]

THE GOVERNMENT OF THE REPUBLIC OF INDIA AND THE GOVERNMENT OF THE PEOPLE'S REPUBLIC OF BANGLADESH.

DETERMINED to promote and strengthen their relations of friendship and good neighborliness.

INSPIRED by the common desire of promoting the well-being of their peoples.

BEING desirous of sharing by mutual agreement the waters of the international rivers flowing through the territories of the two countries and of making the optimum utilisation of the water resources of their region in the fields of flood management, irrigation, river basin development and generation of hydropower for the mutual benefit of the peoples of the two countries.

RECOGNISING that the need for making an arrangement for sharing of the Ganga/Ganges waters at Farakka in a spirit of mutual accommodation and the need for a solution to the long-term problem of augmenting the flows of the Ganga/Ganges are in the mutual interests of the peoples of the two countries.

BEING desirous of finding a fair and just solution without affecting the rights and entitlements of either country other than those covered by this Treaty, or establishing any general principles of law or precedent.

HAVE AGREED AS FOLLOWS:

ARTICLE I

The quantum of waters agreed to be released by India to Bangladesh will be at Farakka.

ARTICLE II

(i) The sharing between India and Bangladesh of the Ganga/Ganges waters at Farakka by ten day periods from the 1st January to the 31st May every year will be with reference to the formula at Annexure I and an indicative schedule giving the implications of the sharing arrangement under Annexure I is at Annexure II.

(ii) The indicative schedule at Annexure II, as referred to in sub para (I) above, is based on 40 years (1949–1988), 10-day period average availability of water at Farakka. Every effort would be made by the upper riparian to protect flows of water at Farakka as in the 40-years average availability as mentioned above.

(iii) In the event flow at Farakka falls below 50,000 usage in any 10-day period, the two governments will enter into immediate consultations to make adjustments on an emergency basis, in accordance with the principles of equity, fair play and to harm to either party.

ARTICLE III

The waters released to Bangladesh at Farakka under Article I shall not be reduced below Farakka except for reasonable uses of waters, not exceeding 200 cusecs, by India between Farakka and the point on the Ganga/Ganges where both its banks are in Bangladesh.

ARTICLE IV

A Committee consisting of representatives nominated by the two Governments in equal numbers (hereinafter called the Joint Committee) shall be constituted following the signing of this Treaty. The Joint Committee shall set up suitable teams at Farakka and Hardinge Bridge to observe and record at Farakka the daily flows below Farakka Barrage, in the Feeder Canal, and at the Navigation Lock, as well as at the Hardinge Bridge.

ARTICLE V

The Joint Committee shall decide its own procedure and method of functioning.

ARTICLE VI

The Joint Committee shall submit to the two Governments all data collected by it and shall also submit a yearly report to both the Governments. Following submission of the reports the two Governments will meet at appropriate levels to decide upon such further actions as may be needed.

ARTICLE VII

The Joint Committee shall be responsible for implementing the arrangements contained in this Treaty and examining any difficulty arising out of the implementation of the above arrangements and of the operation of Farakka Barrage. Any difference or dispute arising in this regard, if not resolved by the Joint Committee, shall be referred to the Indo-Bangladesh Joint Rivers Commission. If the difference or dispute still remains unresolved, it shall be referred to the two Governments which shall meet urgently at the appropriate level to resolve it by mutual discussion.

ARTICLE XI

For the period of this Treaty, in the absence of mutual agreement on adjustments following reviews as mentioned in Article X, India shall release downstream of Farakka Barrage, water at a rate not less than 90% (ninety per cent) of Bangladesh's share according to the formula referred to in Article II, until such time as mutually agreed flows are decided upon.

ARTICLE XII

This Treaty shall enter into force upon signature and shall remain in force for a period of thirty years and it shall be renewable on the basis of mutual consent.

IN WITNESS WHEREOF the undersigned, being duty authorised thereto by the respective Governments, have signed this Treaty.

DONE at New Delhi 12th December, 1996 in Hindi, Bangia and English languages. In the event of any conflict between the texts, the English text shall prevail.

(H. D. DEVE GOWDA)
PRIME MINISTER
REPUBLIC OF INDIA.

(SHEIKH BASINA)
PRIME MINISTER
PEOPLES REPUBLIC OF BANGLADESH.

ANNEXURE—I

AVAILABILITY AT FARAKKA	SHARE OF INDIA	SHARE OF BANGLADESH
70,000 cusecs or less	50%	50%
70,000–75,000 cusecs	Balance of flow	35,000 cusecs
75,000 cusecs or more	40,000 cusecs	Balance of flow

Subject to the condition that India and Bangladesh each shall receive guaranteed 35,000 cusecs of water in alternate three 10-day periods during the period March 1 to May 10.

ANNEXURE—II

(Sharing of waters at Farakka between January 01 and May 31 every year.)

If actual availability corresponds to average flows of the period 1949 to 1988, the implication of the formula in Annex-I for the share of each side is:

PERIOD	AVERAGE OF TOTAL FLOW 1949–88	INDIA'S SHARE	BANGLADESH'S SHARE
	(cusecs)	(cusecs)	(cusecs)
January			
1-10	107,516	40,000	57,516
11-20	97,673	40,000	57,673
21-31	90,154	40,000	50,154
February			
1-10	86,323	40,000	46,323
11-20	82,858	40,000	42,859
21-31	79,105	40,000	30,106
March			
1-10	74,419	30,410	35,000
11-20	68,931	33,931	35,000
21-31	64,688	35,000*	29,688
April			
1-10	63,180	28,180	35,000*
11-20	62,633	35,000*	27,633
21-31	60,992	25,992	35,000*
May			
1-10	67,351	35,000*	32,351
11-20	73,590	38,590	35,000
21-31	81,854	40,000	41,854

(* Three ten day periods during which 35,000 cusecs shall be provided).

Source: "Agreement between the Government of the People's Republic of Bangladesh and the Government of the Republic of India on Sharing of the Ganges Waters at Farakka and on Augmenting Its Flows." Available online. International Freshwater Treaties Database. Oregon State University. URL: http://ocid.nacse.org/tfdd/tfdddocs/158ENG.htm. Accessed May 26, 2009.

PART III

Research Tools

6

How to Research Freshwater Supply

There is no shortage of research information on the global freshwater supply crisis, and it increases every day. The main task is sifting through it all and organizing it in a manner that supports your thesis. In this respect, there are a number of aspects to the freshwater supply, any one of which can serve as a starting point for your own work: a general overview (such as this book); a particular country or basin; the scientific—that is, hydrological—viewpoint; and the political, economic, environmental, and health components, to name a few. If you live in a region that has little rainfall, you probably have an awareness of the local situation and are able to glean information from articles in local newspapers and the like.

DEFINING THE TOPIC

Even if you are interested in writing a general overview of the global freshwater supply, you will have to have a narrower focus than this book. Nevertheless, you can probably pick and choose elements of this book to discuss, while researching the problem in those parts of the world in which you are most interested: Russia and the former Soviet Union, Europe, Western Africa, sub-Saharan Africa, China, Southeast Asia, and Central and South America are all possibilities with different and overlapping problems. You might choose to narrow your focus, which is a better option if you are writing a short paper. Whether you are writing an overview or focusing on a specific country, region, or locale, however, it is good to keep in mind that the main problems are water quantity, water quality, and water distribution. All the other dilemmas having to do with water stem from these three basic issues.

- **Water quantity.** Numerous problems are caused by a lack of water. The first that comes to mind is conflict, though not necessarily armed conflict. Another problem is food. Where water is already scarce, for example, a drought of just a couple of years' duration can lead to famine—as occurred in Ethiopia and Eritrea in the 1980s. Particularly in developed nations—and the western United States is a good example of this—water quantity is tied to politics and economics, so it is good to keep these in mind when researching quantity issues. Politics also play a (markedly different) role in quantity issues in the Jordan River Basin, as well as the lower Nile Basin. Additionally, remember that a lack of water directly contributes to poor sanitation and hygiene. Those most affected by this are children and the elderly.

Other aspects of water quantity are the overdrawing of groundwater and the overdevelopment and/or draining of surface water to the detriment of an ecosystem.

- **Water quality.** What good is water if drinking it causes a person to become ill or if the taste and smell are so bad it is undrinkable? In this case, the water could be used for farming or for sanitation purposes, but if it is not potable, people's standard of living will be far below what it should be. As noted in this book, waterborne diseases lead to the deaths of millions of children worldwide each year. And waterborne disease is not the only problem when it comes to water quality. Water pollution exists in every country. Industrial, agricultural, and domestic waste harms rivers and lakes and alters or even destroys ecosystems.
- **Water distribution.** This is quickly becoming the most complicated of the three basic issues of the global water crisis. It is connected to the other two, though less to water quality. Lack of easy access to freshwater is among the major causes of a region or even a nation's underdevelopment. For many people in developing nations, the acquisition of water is a daily task that occupies too much of their time—and that task, by and large, falls on the shoulders of girls and women. Girls may miss all or part of the school day because they are hauling water. Thus, in developing countries, gender and cultural stereotypes play a role in water acquisition that governments have yet to address.

Another aspect of water distribution worth investigating is the growing influence of transnational corporations in the delivery of water to municipalities. As noted, sometimes this leads to major failures, as in Cochabamba, Bolivia, and Atlanta, Georgia, but not always. In fact, there are those in developing nations who would welcome a transnational to distribute water to their towns and villages because the government has failed to do so. If you are pursuing this topic, it is a good idea to keep in mind the ongoing debate over whether water should be classified as a human "need" or a human "right."

Once you have chosen which aspect of the global freshwater crisis you plan to write about—and it is best to choose a topic with which you feel comfortable or already have some knowledge and interest—you need to write an outline and undertake your research.

GATHERING RESEARCH MATERIAL

You might well use this book as a starting point. Chapters 9 and 10, the list of organizations and agencies and the annotated bibliography, respectively, should prove helpful, especially if you are writing about the water problem in the United States or one of the other areas discussed in this book. If you have chosen a different topic, the general section of the annotated bibliography will be helpful, as will chapter 9. It would also be wise to read the glossary to familiarize yourself with terms that you will probably come across in your reading.

You may well find it helpful to examine *The Water Atlas: A Unique Visual Analysis of the World's Most Critical Resource,* by Robin Clarke and Jannet King (see annotated bibliography). Each of its 31, two-page sections focuses on a different aspect of the water crisis. The sections themselves are grouped into six headings—"A Finite Resource," "Uses and Abuses," "Water Health," "Reshaping the Natural World," "Water Conflicts," and "Ways Forward"—that contain brief introductions. The sections include easy-to-understand, color-coded maps that highlight the particular aspect of the water crisis under discussion. They also contain statistics and graphs. In the back of the book are two tables of country statistics grouped under the headings "Needs and Resources" and "Uses and Abuses."

When reading the statistics in this or any other book or article, it is a good idea to take a look at the year of publication on the copyright page so you know exactly how old those statistics are. Also keep in mind that most figures are approximations, or averages. The hydrologic cycle is not fixed but subject to fluctuations, and climate change affects it in ways that have yet to be fully understood; also, the flow of water can be measured with certainty only over a short period of time. Recall the error made in 1922 in calculating the flow of the Colorado River for the various state allotments. The delegates did understand, however, that a river does not flow at the same rate every year and so made the allotments using 10-year averages. It is wise, then, to read all statistics as approximations or averages and keep in mind that they are most useful in comparing a country's resource and the future of that resource with other nations in its region.

If your topic is political and/or economic, try to get material that covers both sides of an argument, or as many points of view as possible. Remember that if you are working on a specific river basin, for example, there are at least as many points of view as there are riparian states or countries. Again, in the case of the Colorado River Basin, there were seven states, plus Mexico, plus the U.S. government. In the case of the Nile River Basin, Eritrea is not a riparian nation—that is, it does not border the Nile—but it is still a basin country. The politics become even more complicated when discussing groundwater. If all of this sounds daunting, just remember to keep your focus narrow.

Books

If you are writing about a particular current water crisis, then it makes sense to choose books that are up to date, while keeping in mind that older books may contain important background information that will help provide context. Still, things related to water change fairly quickly, including water development projects, population growth and population shifts, climate change, and treaties. The quickest way to discover whether a book will have the information you need and where to find it in the text is to check the index. If the book is popular enough in the field, you may even be able to find a synopsis online.

The water crisis is a phenomenon that has gained a lot of attention over the past few decades, but that attention is mostly from professionals. What this means is that there have been a lot of books and articles written about the problem, but the media has devoted little attention to it, except for the local media in areas that are in one way or another affected by the crisis. Even then, coverage will probably be mainly local. The upshot is that public libraries contain only a tiny percentage of what has been published on the subject. Without a doubt, college or university libraries will have far more material. If you are fortunate enough to live near a college or university that is home to the water resources research institute for your state, then all the better.

The Internet

Nowadays it is not so difficult to find magazine and newspaper articles on the Internet in addition to stories that originate online. Most newspapers and many magazines have been archiving their material since at least the early 1990s. If you use a regular search engine, however, you may find it time consuming to locate the information you are looking for so, again, a library will be most useful. Public libraries are accessible from home computers, but college and university libraries are only accessible from home if you have a connection with the school. All libraries, public or academic, will have online access to certain newspapers and journals such as the *New York Times* or *Smithsonian, Time,* and *Newsweek* magazines, but academic libraries (and possibly some public ones) will have access to numerous databases. Two of these, Lexis-Nexis and JStor, should provide you with ample material for whatever topic you are working on. Lexis-Nexis is particularly useful. It actually has multiple databases, and of these "Lexis-Nexis Academic" will best serve your needs. It is divided into nine subsections including "Major U.S. and World Publications," "Major World Publications (non-English)," "Newswire Services," "TV and Radio Broadcast Transcripts," and "Web Publications." If you are writing about water problems in the United States, another valuable database is Findlaw.com, which includes United States Supreme Court decisions dating back to 1893.

It is best to avoid Internet encyclopedias, such as Wikipedia, whose contributors are unknown and where the articles themselves are often under revision. Web sites that provide links to other sites are good places to begin, as you can then research the links as well.

The Web sites of the various organizations and agencies listed in chapter 9 of this book will be helpful in and of themselves, and most of these also have links to other useful Web sites (though not necessarily organizations). Finally, a keyword search in one of the common search engines should bring up a good number of Web sites—you might want to remember to narrow your search terms, though, to avoid wasting time.

Once you have your material, arrange it to adhere as closely as possible to your outline. Not only will doing this make writing your paper easier, but it will help you sift out whatever is superfluous. Moreover, you will be able to see more quickly if there are any gaps in your research. If there are, continue seeking out information until you are satisfied that you have covered your topic fully.

EVALUATING MATERIAL

Almost all books about freshwater will contain some statistics, and it will be your task to realize to what purpose they are being used. Remember that few water statistics are hard and fast numbers, but for most people it is easier to present them that way because they wish to make a general, usually comparative, point. People who are trying to argue a particular case (such as politicians) by and large treat water statistics as hard numbers.

When choosing what material to use, you should ask yourself a few questions. Doing so will help you to sort through things and you will find that as you do, you will strengthen your line of reasoning and decide how best to say what you have noted in your outline.

HOW IMPORTANT IS THIS MATERIAL TO MY PAPER?

Presumably, all your research material is important, but you will need to prioritize. It is this sorting process that will help you gather your thoughts about how you want to say what you want to say.

HOW RELIABLE IS THE AUTHOR?

Many environmentalists, ecologists, hydrologists, and writers have focused on the water crisis for their entire careers. Generally their work is highly regarded and often quoted. Most have no agenda beyond increasing the water supply, ensuring its quality, and expanding its distribution equitably. These are the people whose work you want to look at closely. Some politicians and economists also have made the global water crisis their priority—you can refer to chapter 8 for a brief list, or to the bibliography.

HOW UP TO DATE IS THE MATERIAL?

This is important because, as noted above, the global water issue, like water itself, is in flux. While the basic situation has not changed over the past 20 or 30 years—the arid regions are still dry, for example—national and regional infrastructures have. Some have gotten old and some have been repaired, while those recently built employ new technology. More important, as more and more people have paid attention to the water crisis, the politics and economics of the situation have changed, and will continue to do so. The last thing you want to do is discuss a situation that is so dated that it is on the verge of irrelevance or no longer even exists.

HOW READABLE IS THIS MATERIAL?

This seems so obvious that you might not think it a necessary question to pose, but you do not want to be trying to decipher the meaning of a text after you have begun writing yourself. While most of the material you gather will probably be written in a pretty straightforward manner, science- or economics-oriented books and articles can be heavy with jargon or assume knowledge on the part of the reader that you may not have. If you do not have a good grasp of your research material, then you will not be able to convey the message properly to your reader.

CITING SOURCES

As you come across material you intend to use, take down the information you will need to cite in your footnotes or endnotes and bibliography. Trying to find it later will be far more troublesome and time consuming.

Footnotes and Endnotes

Mark your notes in the text with consecutive numbers, then match those numbers to the citations at the end of the page (footnotes) or chapter (endnotes). Cite your sources as follows: the author's name, as normally written, followed by a period. The title of the source followed by a period: If the source is a book, italicize it; if it is an article, place it in quotation marks. If the source is an article, the name of the newspaper or magazine comes next, followed by the date of the issue. If the source is a book, the place of publication follows the title, followed by a colon. If the city of publication is well known, such as New York or London, there is no need for any further geographical information. If it is not well known, add the state or the country, followed by a comma. There is no need to add U.S.A. The year of publication, followed by the page number/s, comes last.

Article:[1] Ali Askouri. "A Culture Drowned: Sudan Dam Will Submerge Historically Rich Area, Destroy Nile Communities." *World Rivers Review* 19, no. 2 (April 2004).

Book:[2] Robin Clarke and Jannet King. *The Water Atlas: A Unique Analysis of the World's Most Critical Resource.* New York: The New Press, 2004, p. 22.

Bibliography

For ease of use, bibliographies are often arranged in subsections, such as "Books," "Periodicals," and "Web Sources." Citations are arranged in alphabetical order by author, this time with the last name followed by the first, and if given the person's middle name or initial. After this the formula is the same as that for footnotes, except that page numbers are not needed.

Article: Gertner, Jon. "The Future Is Drying Up." *New York Times Magazine*, (10/21/07).

Book: Ball, Philip. *Life's Matrix: A Biography of Water.* New York: Farrar, Straus and Giroux, 1999.

The Internet

Internet sources require a slightly different form of citation. If a byline is given, follow the rules of order above, depending on whether it is a footnote or a bibliography citation. The name of the article should always be in quotation marks, followed by the name of the Web site and the date the article was published (if provided). Include phrase "Available online." The URL comes next, followed by the date you accessed the page.

Footnote:[1] Derek Sands. "Analysis: Mideast Turns to Nukes for Water." *Energy Daily* (9/14/07). Available online. URL: http://www.energy-daily.com/reports/Analysis_Mideast_turns_to_nukes_for_water_999.html. Accessed May 29, 2009.

Bibliography: "About CERP: Brief Overview." Comprehensive Everglades Restoration Plan. Available online. URL: http://www.evergladesplan.org/about/about_cerp_brief.aspx. Accessed January 14, 2009.

FURTHER TIPS

Always keep in mind that everything having to do with the freshwater supply is ongoing and changing, and none of your research will be completely up to date. Therefore, your own project, no matter how narrowly you have focused the topic, will necessarily be a sort of report on the continuing process. This may help you to convey the enormity and immediacy of the crisis.

Text can be quoted directly as a means of strengthening your argument or if the quote, presumably by an expert, conveys an idea better than you could. If you quote material remember to cite it, but do not overload your work with direct quotes.

Put your research into perspective by not collecting so much information that you get bogged down. On the other hand, it is far better to have too much material from which to work than not enough. Arranging and prioritizing your material before you begin writing will help in this regard, but you do not have to use every piece of information you have gathered. Finally, use statistics sparingly. Not only can they be manipulated, but numbers tend to blur (and even remove) the human face from a crisis of nature that is all too human.

7

Facts and Figures

GENERAL

1 The Hydrologic Cycle

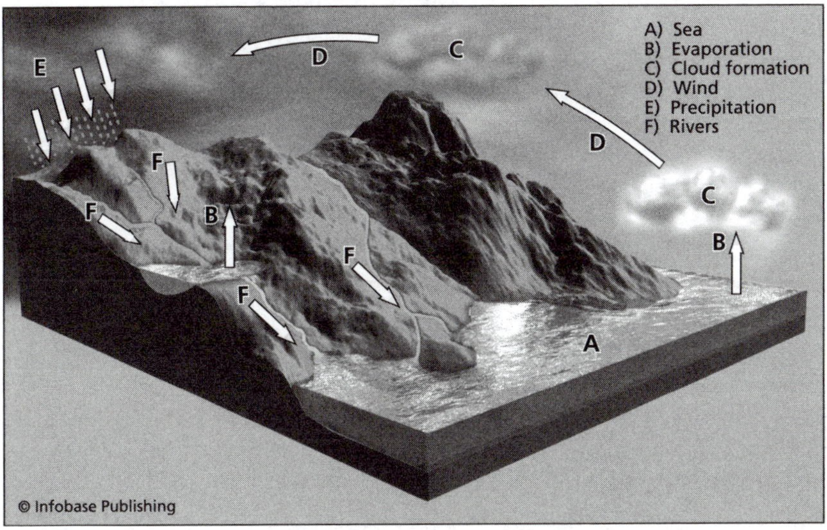

Earth's finite amount of water is kept in balance by the cycle of evaporation, cloud formation, precipitation, percolation, and flow.

UNITED STATES

2.1 The Colorado River Basin

The Colorado River Basin showing the major public works. The Hoover Dam and its reservoir, Lake Mead, are southeast of Las Vegas. The Glen Canyon Dam and Lake Powell are on the Utah-Arizona border, and the Parker Dam and Lake Havasu are on the California-Arizona border.

Source: Ed Marston, ed. *Western Water Made Simple.* Washington, D.C.: Island Press, 1987.

2.2 The Lower Colorado Basin

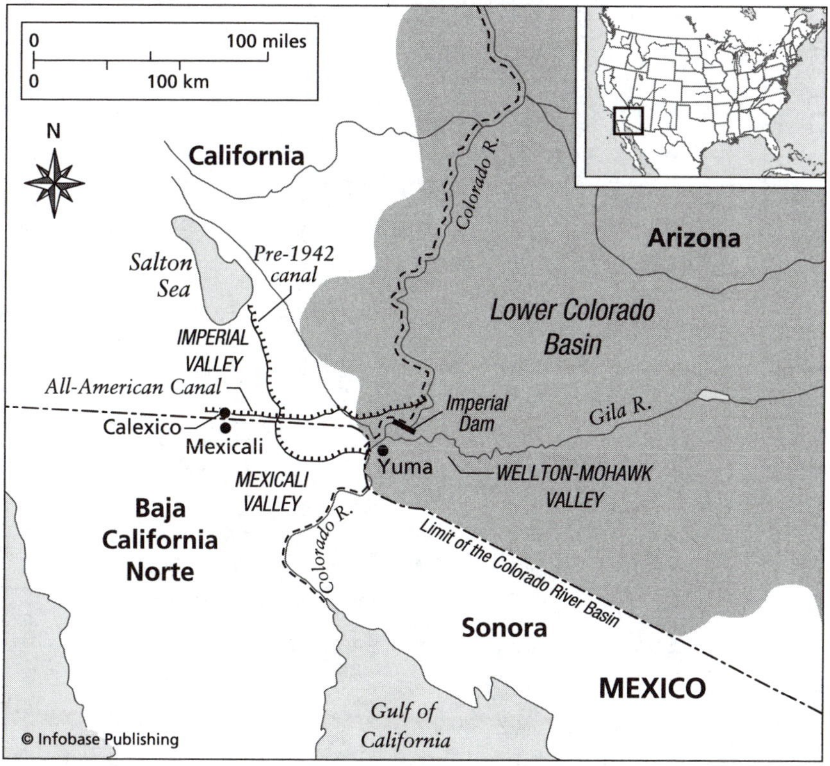

The All-American Canal and the Imperial Dam are two of the southernmost public water projects on the Colorado River. The canal lies just north of the Mexican border, while the dam is near Yuma, Arizona.

Source: Ed Marston, ed. *Western Water Made Simple.* Washington, D.C.: Island Press, 1987.

2.3 Areas of Concern in the Great Lakes and St. Lawrence River Basin

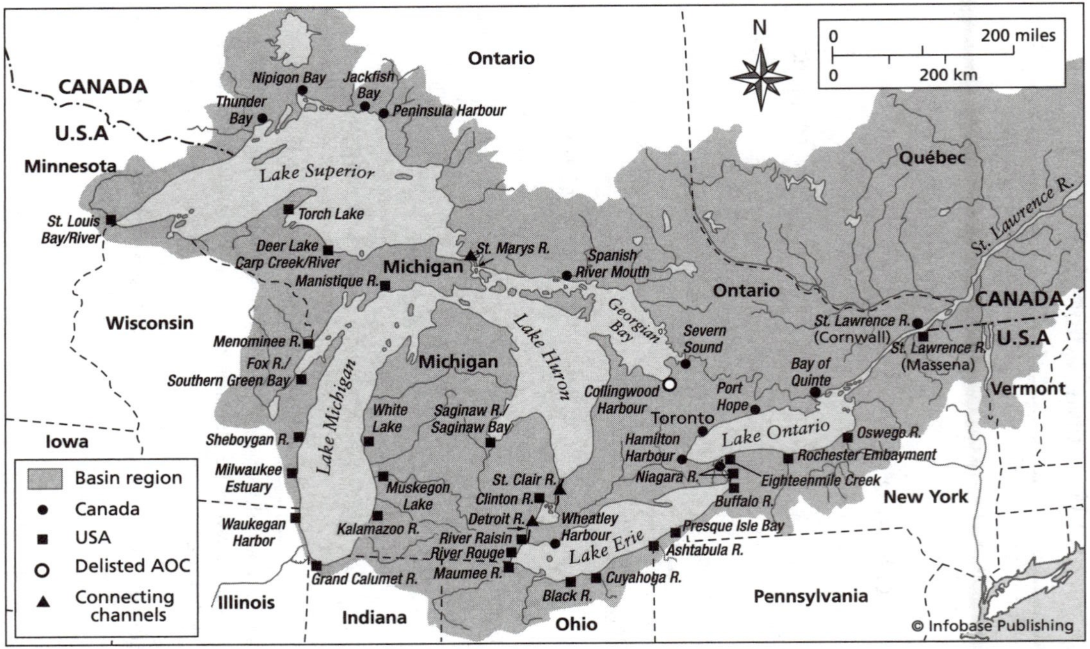

Of the 42 identified areas of concern (AOC) in the Great Lakes region, only one had been delisted by the beginning of the 21st century.

Source: Environment Canada. *Our Great Lakes* (1999). In Mark Sproule-Jones. *Restoration of the Great Lakes: Promises, Practices, Performances.* Vancouver, B.C.: University of British Columbia Press, 2002.

2.4 The Apalachicola-Chattahoochee-Flint Basin

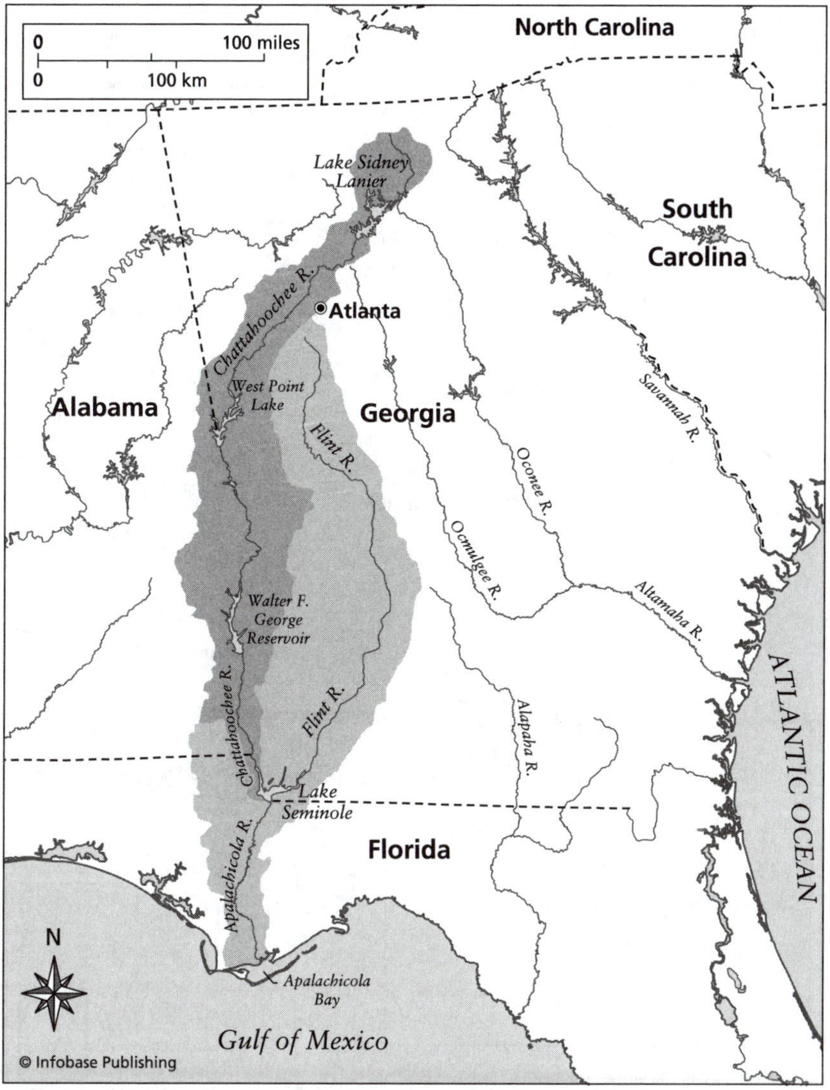

The Chattahoochee River begins in northeast Georgia and runs almost the entire length of the state; for most of its run it forms the border between Georgia and Alabama. Lake Lanier is at the top of the river. The Flint River also runs southward; the Apalachicola River begins where the Flint River joins the Chattahoochee at the Georgia–Florida border.

2.5 Water Storage in the Floridian Aquifer in Southern Florida

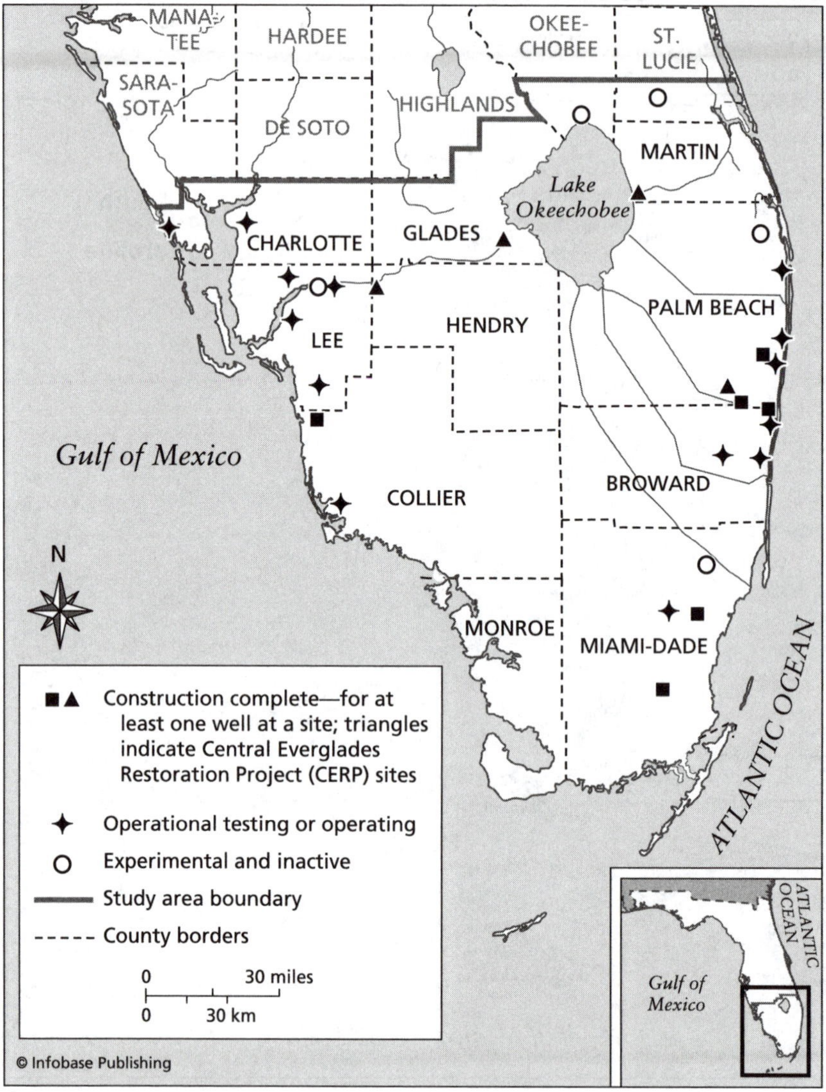

Location and status of water recovery sites in southern Florida, ca. 2004.

Source: U.S. Department of the Interior/U.S. Geological Survey. Fact Sheet 2004–3128 (November 2004). Available online. URL: http://pubs.usgs.gov/fs/2004/3128. Accessed June 8, 2009.

2.6 Water Recovery in the Floridian Aquifer in Southern Florida

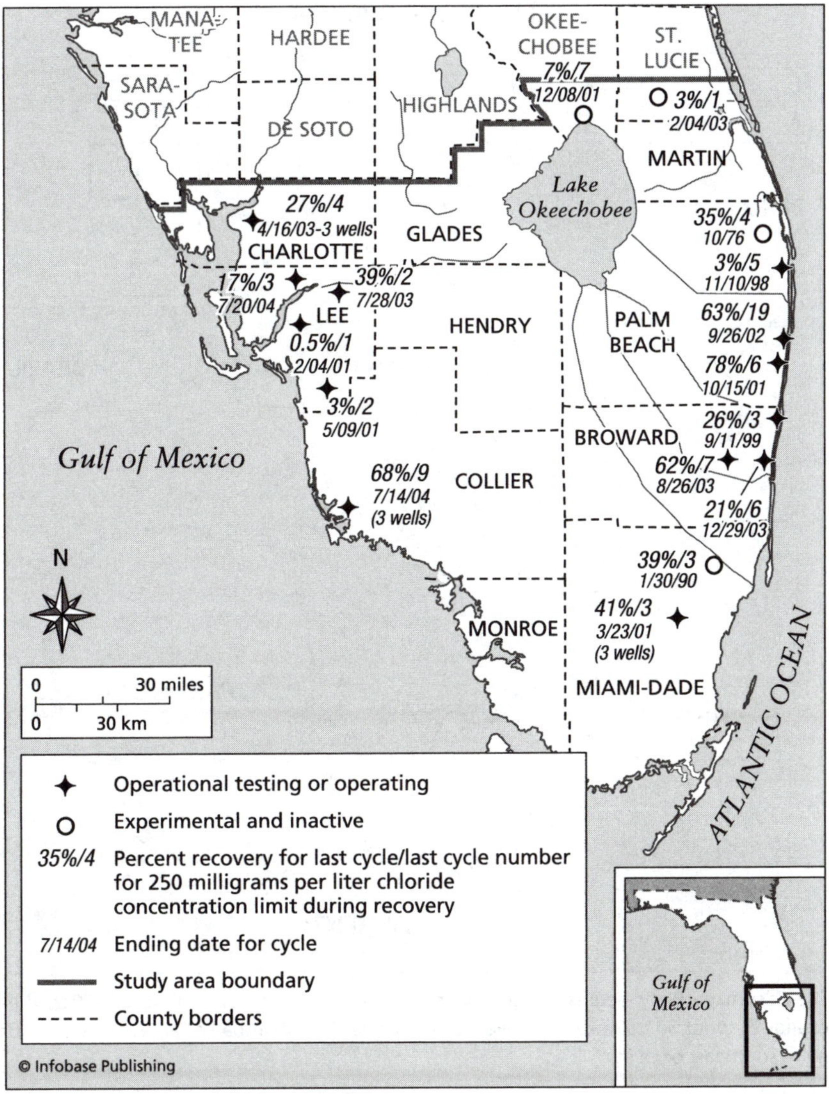

Water recovery efficiencies for aquifer storage and recovery sites in southern Florida, ca. 2004.

Source: U.S. Department of the Interior/U.S. Geological Survey. Fact Sheet 2004–3128. (November 2004). Available online. URL: http://pubs.usgs.gov/fs/2004/3128. Accessed June 8, 2009.

BOLIVIA

3 The Pilcomayo River

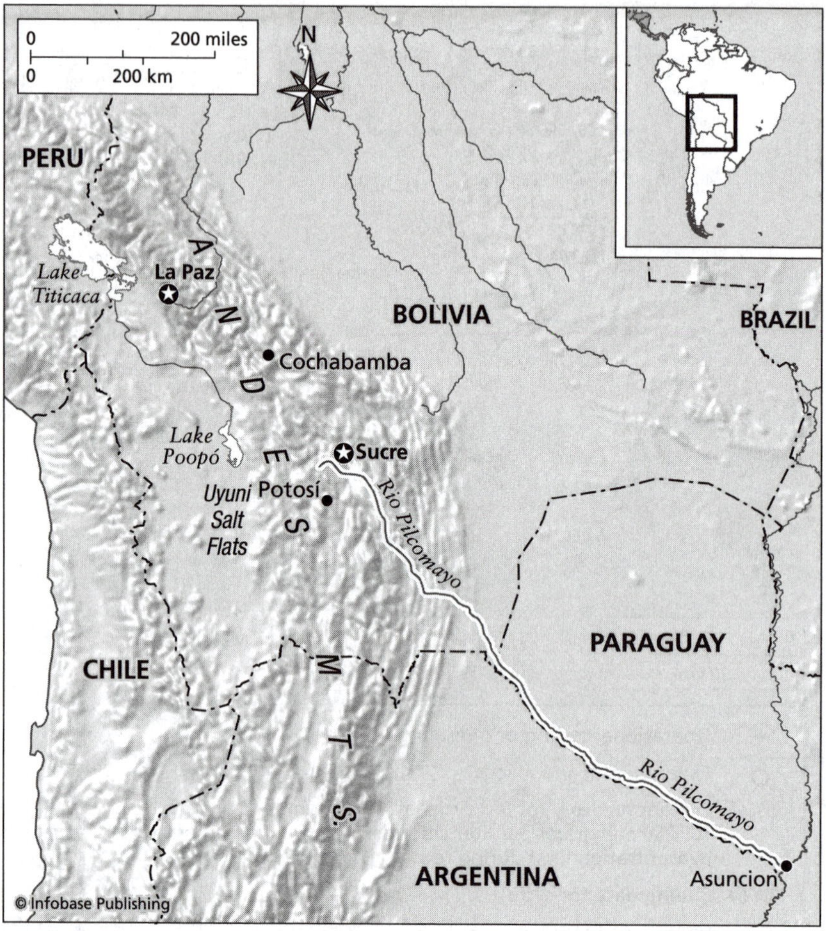

The Pilcomayo River begins in the Andes, northwest of the city of Potosí. La Paz, the capital, is about 50 miles east of Lake Titicaca, while Cochabamba lies approximately 300 miles southeast of La Paz.

EGYPT, ETHIOPIA, AND SUDAN

4 The Nile Basin Water Project at the Turn of the Century

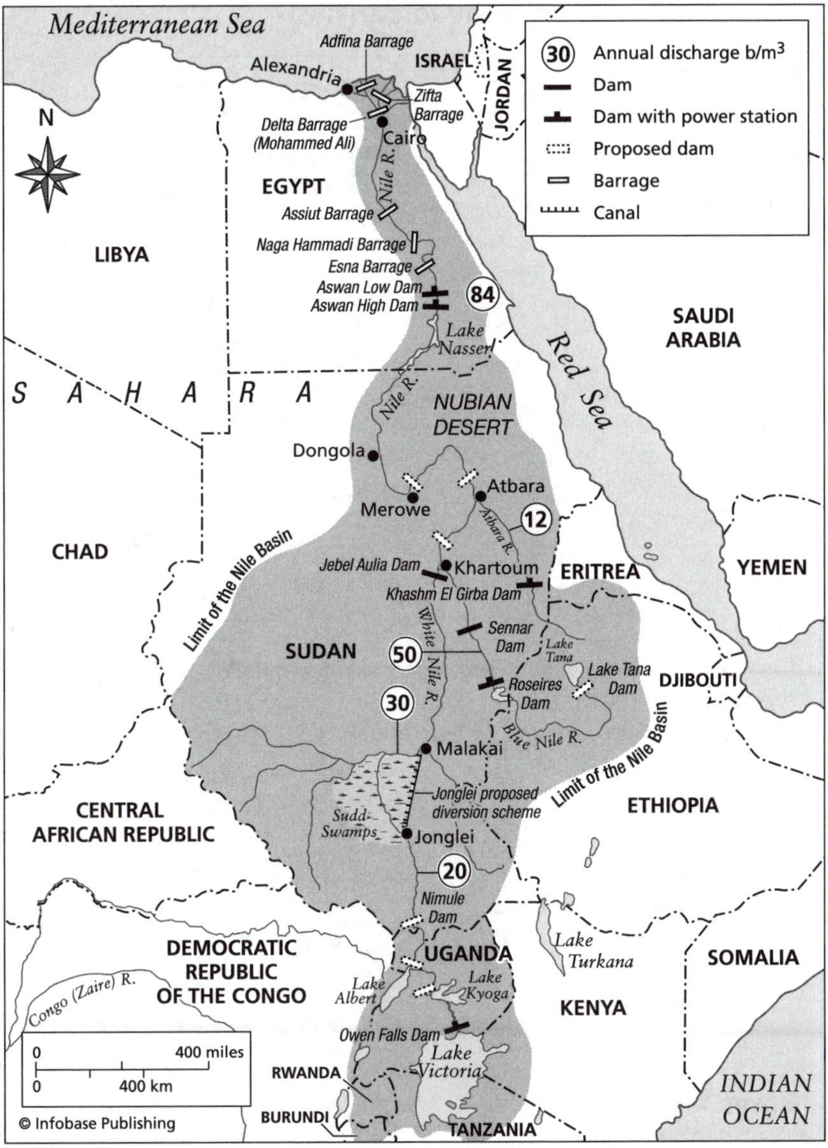

Proposed and actual dams and barrages and the Jonglei Canal, as of 1999.

Source: Arnon Soffer. *Rivers of Fire: The Conflict over Water in the Middle East.* Translated by Murray Rosovsky and Nina Copaken. Lanham, Md.: Rowman & Littlefield, 1999.

ISRAEL, JORDAN, AND THE PALESTINIAN TERRITORIES

5.1 The Jordan River

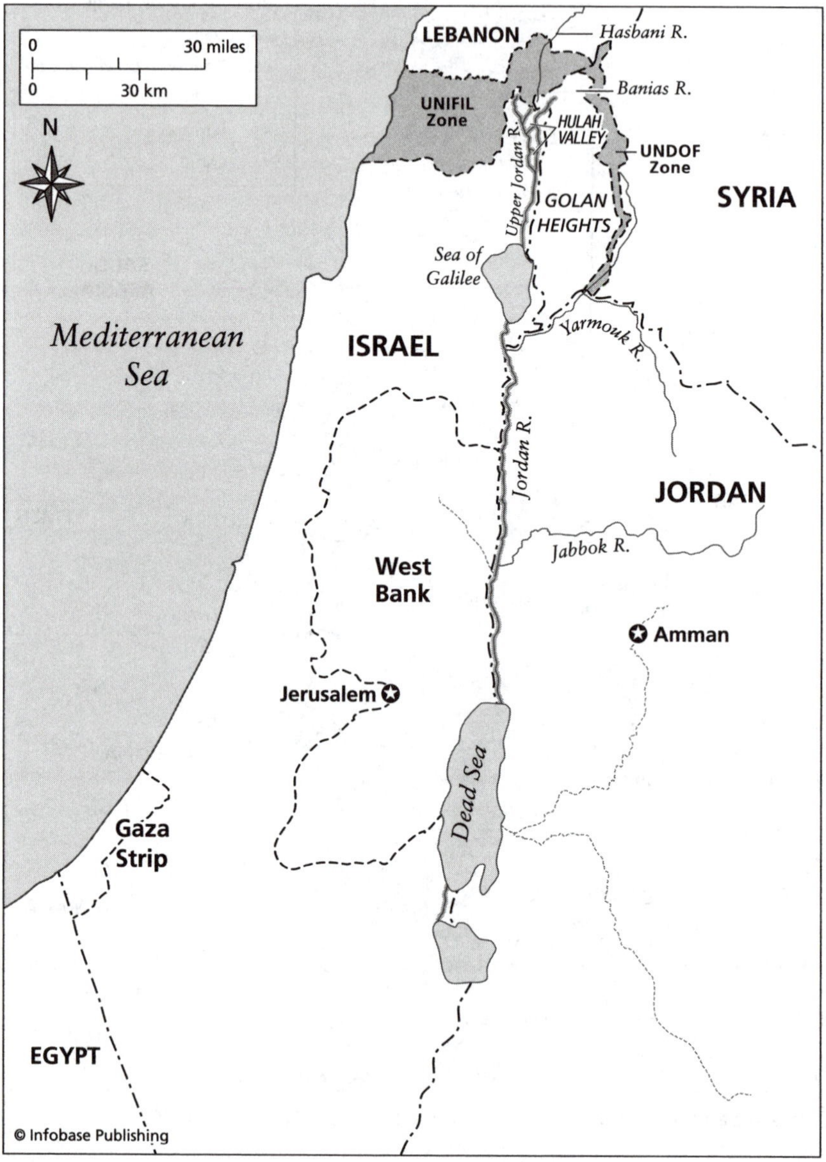

Among the smallest of the world's important and historical rivers, the Jordan runs from the Hasbani and Banias headwaters, incorporating Lake Kineret (Sea of Galilee), to the Dead Sea—a total of 206 miles.

5.2 Israeli and Palestinian Aquifers

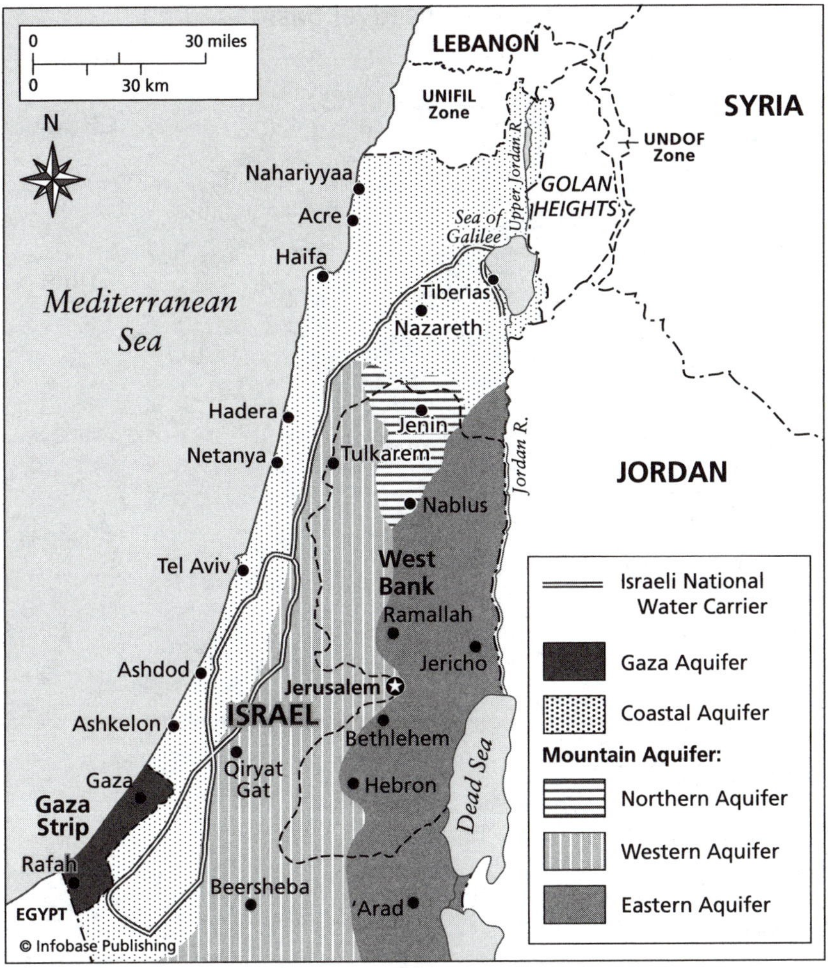

Groundwater rights may be a sticking point in future negotiations between Israel and the Palestinian territories over the issue of Palestinian statehood.

Source: Robin Clarke and Jannet King. *The Water Atlas: A Unique Visual Analysis of the World's Most Critical Resource.* New York and London: The New Press, 2004.

INDIA, PAKISTAN, AND BANGLADESH

6.1 The Indus River Basin

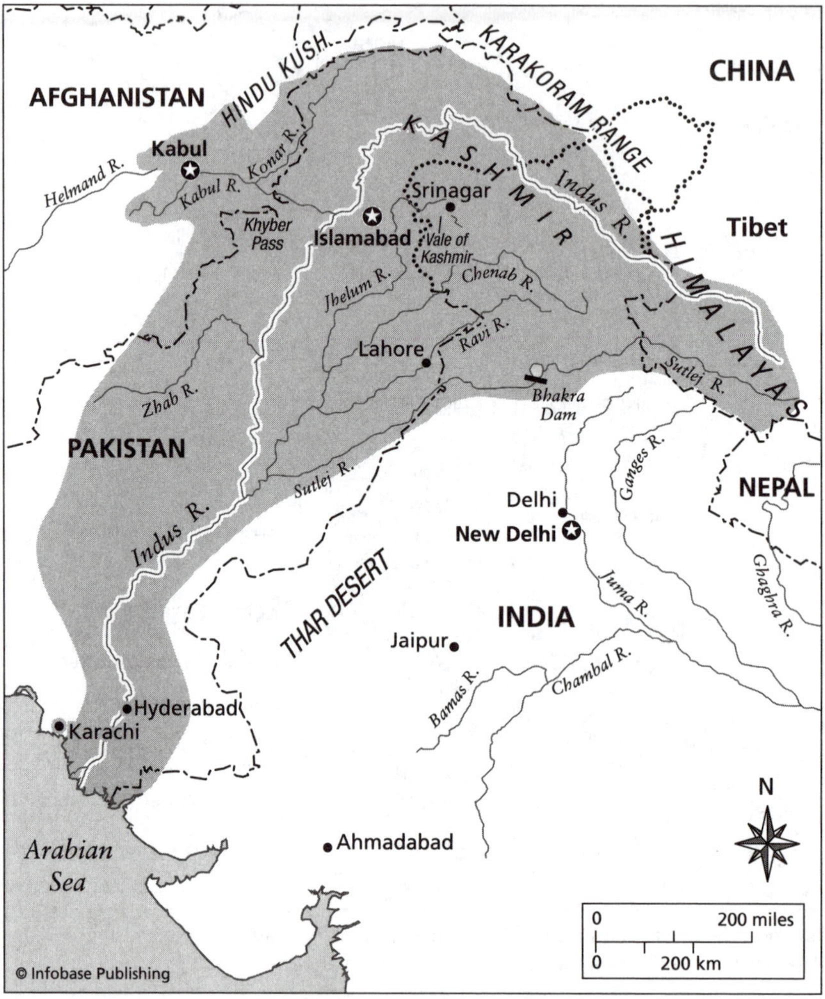

Though the Indus is most associated with Pakistan, it originates in Tibet and flows first into India. The 1960 Indus Waters Treaty resolved long-standing water rights disputes in the basin, particularly those concerning the tributary rivers.

6.2 The Ganges River Basin

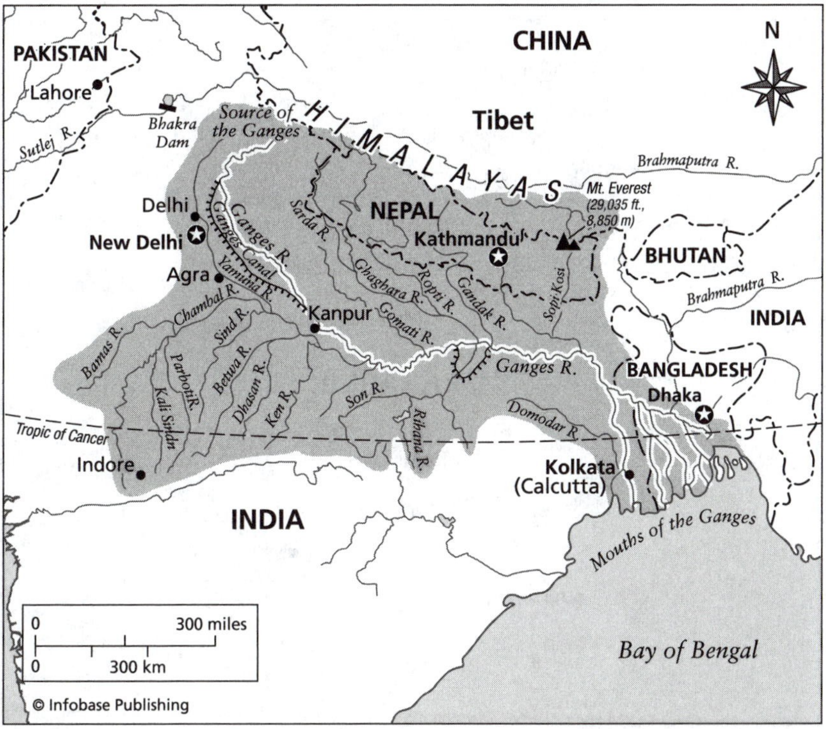

India's holiest river begins in the Himalayas and flows some 1,500 miles to the Bay of Bengal in the Indian Ocean. Recent efforts to clean up the river have yielded only limited results.

6.3 The Ganges River in Bangladesh

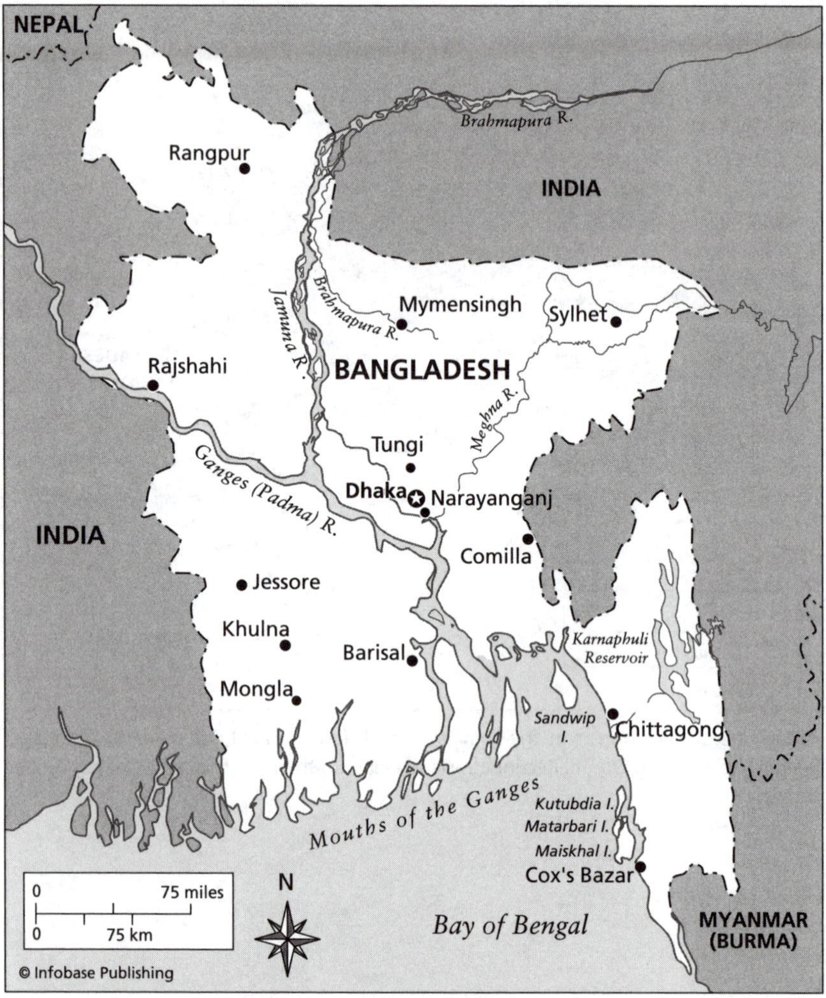

In Bangladesh, the Ganges is called the Padma River. Its delta, which is also in India, forms most of the country's coastline. After nearly 30 years of discussion and planning, Bangladesh has finally begun its water development projects on the river.

6.4 Bangladesh Groundwater Arsenic Rates

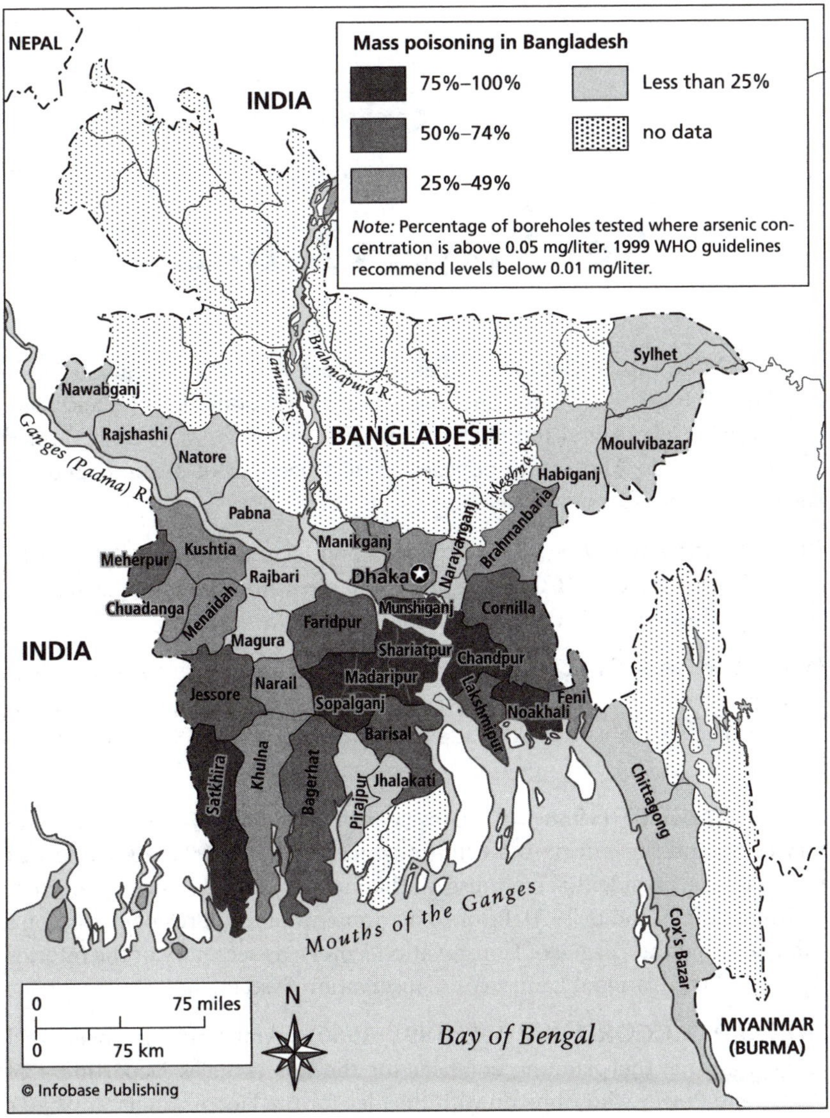

By the turn of the century, arsenic was appearing in unusually high rates in numerous regions of Bangladesh.

Source: Robin Clarke and Jannet King. *The Water Atlas: A Unique Visual Analysis of the World's Most Critical Resource.* New York and London: The New Press, 2004.

8

Key Players A to Z

BABA AMTE (1914–2008) Indian social activist, born Murlidhar Devidas Amte ("Baba" is an honorific), who became an adherent of Gandhism, worked with lepers, and founded a number of ashrams. In 1990, he joined Narmada Bachao Andolan to oppose the displacement of peasants and villagers along the Narmada River.

REUBIN ASKEW (1928–) American politician who, as the 37th governor of Florida from 1971 to 1979, promoted environmentalism, particularly the Water Resources Act.

WAYNE N. ASPINALL (1896–1983) Longtime congressional representative from Colorado who favored big dams and western water development projects such as the Colorado River Storage Act and Central Arizona Project.

BRUCE BABBITT (1938–) Governor of Arizona (1978–87) and secretary of the interior during the Clinton administration (1993–2001), he also served on the presidential commission that investigated the nuclear accident at Three Mile Island in 1979. Prior to his appointment to President Clinton's cabinet he led the League of Conservation Voters; as secretary of the interior he created the National Landscape Conservation System.

DONALD MCCORD BAKER (1891–1960) American engineer and author, he was the planning engineer for the Los Angeles Department of Water and Power who came up with the idea for the North American Water and Power Alliance in the early 1950s, the plan to harness Canada's western rivers for use in central Canada and the United States.

HUGO BANZER (1926–2002) Bolivian dictator (1971–78) and president (1997–2001) whose refusal to void the privatization law DS 21060 and promulgation of Law 2029, which strengthened a contract with a transnational consortium to take over Cochabamba's water delivery system, led to the water wars in that city.

MAUDE BARLOW (1947–) Canadian environmental activist and author who is outspoken in her opposition to the commodification of water and transnational corporations that seek to profit from the resource. She is a cofounder of the Blue Planet Project and heads the Council of Canadians and the Food & Water Watch. In 2008 she was named Senior Adviser on Water Issues with the United Nations.

GUNNAR BARNES (dates unknown) Norwegian engineer who in the 1970s developed the bamboo treadle water pump that is used extensively in Bangladesh and Nepal.

MENACHIM BEGIN (1913–1992) Israeli prime minister who served from 1977 to 1983. He signed a historic treaty with Egypt, but turned down the Egyptian offer of water via the Al Salam Canal across the Sinai to Israel's Negev Desert because its construction was linked to the question of Jerusalem's sovereignty and it would potentially lead to Israeli dependence on Egypt for water.

EUGENE BLACK (1898–1992) American investment banker who as president of the World Bank (1949–63), helped broker the 1960 Indus Waters Treaty between India and Pakistan.

HUGO BLACK (1886–1971) American politician and jurist, he was a U.S. Supreme Court justice from 1937 to 1971. In 1963 he wrote the majority opinion for *Arizona v. California,* which not only upheld Arizona's water claims but also acknowledged Native American water rights.

DAVID BROWER (1912–2000) Notable American environmentalist who was the first executive director of the Sierra Club, as well as the founder of numerous environmental organizations. He opposed many water development plans throughout the American West during the latter half of the 20th century.

RACHEL CARSON (1907–64) American marine biologist and nature writer whose books, notably *Silent Spring,* helped kick-start the modern environmental movement. She was opposed particularly to the use of pesticides, such as DDT, which not only kill more than intended but seep into ground and surface water.

JAMES EARL "JIMMY" CARTER (1924–) The 39th president of the United States, in 1977 he sought to end federal funding of water development programs by compiling a so-called hit list of 80 projects that would be cut and/or decommissioned, many of which were western projects. Some historians consider that the animosity this created contributed to his defeat for reelection in 1980.

HENRI DARCY (1803–1858) French hydrological engineer whose experiments on the flow of water through sand led to the equation that has since become known as Darcy's Law. The "darcy" is a unit of fluid permeability, or the rate at which water flows through sand or other sediment.

VICTOR HUGO DAZA (1983–2000) Bolivian worker who was the lone fatality in Cochabamba during the water wars of 2000. According to Oscar Olivera, he was an innocent bystander on his way home from work when he stopped to watch the protests and was shot and killed by an army sniper.

FLOYD DOMINY (1909–2010) American bureaucrat, he served as commissioner of the Bureau of Reclamation from 1959 to 1969. He favored big water development projects in the American West, such as the Glen Canyon Dam on the Colorado River.

FREDERICK EATON (1856–1934) American hydrological engineer and politician who became superintendent of the then privately owned Los Angeles City Water Company in 1883 and subsequently mayor of Los Angeles, from 1898 to 1900. During his term as mayor the city took over the Los Angeles City Water Company. Eaton also played a major role in the construction of the Los Angeles Aqueduct, which diverted water from the Owens River (and ultimately Owens Lake) to Los Angeles and elsewhere.

LUIS ECHEVERRÍA ÁLVAREZ (1922–) President of Mexico from 1970 to 1976. As a presidential candidate and later as president he threatened to take legal action against the United States in the World Court at the Hague over the issue of water quality from the Colorado River.

LOIS GIBBS (1951–) American environmental activist who in 1978 led the opposition to the decadeslong cover-up of toxic dumping at the Love Canal site in Niagara Falls, New York, where she lived. In 1980, she founded what eventually became known as the Center for Health, Environment, and Justice and in 1990 she received the Goldman Environmental Prize.

PETER GLEICK (1956–) American scientist and cofounder of the Pacific Institute (in 1987), he specializes in global water issues and since 1998 has published a biannual reference on the state of freshwater in the world.

GARRETT HARDIN (1915–2003) American ecologist and author of the controversial article "The Tragedy of the Commons," first published in

Science magazine in 1968, which claims that self-interested individuals will eventually destroy a common resource. His theory has been used to support the privatization of water, the argument being that only ownership with a vested financial interest will protect the resource.

JUDSON HARMON (1846–1927) U.S. attorney general from 1895 to 1897 (during the second administration of President Grover Cleveland) who, in that post, declared what became known as the Harmon Doctrine, in which a nation exercises complete sovereignty over its water resources. The doctrine was issued in relation to a dispute with Mexico over Rio Grande diversion. Harmon also served as governor of Ohio from 1909 to 1913.

CARL HAYDEN (1877–1972) U.S. senator from Arizona who introduced legislation to pay for the Central Arizona Project in 1946. Though it took 22 years, the eventual signing of the legislation transformed the project into a federal one.

HERBERT HOOVER (1874–1964) The 31st president of the United States; in 1922, as secretary of commerce in the administration of President Warren G. Harding, he presided over the negotiations that ultimately led to the Colorado River Compact, the success of which was in no small part due to Hoover's skillful negotiating; construction of Boulder Dam (renamed Hoover Dam) was begun during his presidency.

HAROLD ICKES (1874–1952) American politician who served as secretary of the interior from 1933 to 1946, during the entire administration of President Franklin D. Roosevelt and the first year of President Harry S. Truman's term. During the Great Depression he oversaw various public works programs and was responsible for the expansion of the Bureau of Reclamation (an Interior Department agency), which expanded nearly ninefold during his time in office.

RAKESH JAISWAL (1959–) Indian environmental activist and founder of the group Eco Friends, which has been involved in the movement to clean the Ganges River since the early 1990s.

MOHAMMED AYUB KHAN (1907–1974) Pakistani military commander-in-chief who came to power in a military coup in 1958 and had himself declared president. He sought to reform the country's economy by stimulating the industrial and agricultural sectors and encouraging foreign investment; in 1960 he signed the Indus Waters Treaty on behalf of Pakistan. He resigned from power in 1969.

DAVID LILIENTHAL (1899–1981) American lawyer and bureaucrat who served as chairman of the Tennessee Valley Authority from 1941 to 1946 and of the U.S. Atomic Energy Commission from 1947 to 1949. In 1951, he traveled to India and Pakistan to research a series of articles for *Collier's* magazine, in which he linked Indus water storage and distribution to the countries' overriding antagonisms. His articles helped persuade the World Bank to become involved in negotiations that eventually led to the Indus Waters Treaty.

JOSEPH B. LIPPINCOTT (1864–1942) American civil engineer who in the first years of the 20th century, while working for the U.S. Reclamation Service, assisted in the acquisition of Owens River water for the City of Los Angeles and environs.

WALTER CLAY LOWDERMILK (1888–1974) American soil conservationist who developed a comprehensive water and land management plan for the Middle East that called for diverting water from the Jordan and Yarmuk Rivers for hydroelectric power and for making the Negev Desert bloom.

ELWOOD C. MEAD (1858–1936) American civil engineer who served as chairman of the Bureau of Reclamation from 1924 to 1936. During his tenure, the bureau constructed or began construction of some of its biggest water projects, including the Hoover Dam on the Colorado River (1936), the Grand Coulee Dam on the Columbia River (1942), and the Owyhee Dam on the Owyhee River (1932). Lake Mead, the reservoir created by the Hoover Dam, is named in his honor.

MENGISTU HAILE MARIAM (1937–) Ethiopian dictator and a member of the Derg military junta that overthrew Emperor Haile Selassie in 1974; he became leader of the Derg and the country in 1977, created the office of president in 1987, and occupied it until he was overthrown in 1991. In the late 1970s, he challenged Egypt's historical water rights to the Nile, noting that the majority of Nile water originates in Ethiopia.

BENJAMIN MOEUR (1869–1937) American politician and physician, governor of Arizona from 1933 to 1937. In 1934, during an ongoing war of words between the states of Arizona and California over Colorado River water, he sent troops from the Arizona National Guard to interfere with the construction of the Parker Dam. Eventually Congress authorized the dam's construction.

JOHN MUIR (1838–1914) Scottish-born American naturalist who cofounded the Sierra Club in 1892 and was its first president. He is considered the philosophical inspiration behind the environmentalist movement.

WILLIAM MULHOLLAND (1855–1935) Irish-born American water engineer who, as head of the Los Angeles Department of Water and Power, oversaw the creation of the Los Angeles Aqueduct, which carried water from Owens River to the city of Los Angeles and elsewhere.

ORRICE ABRAM (ABE) MURDOCK, JR. (1893–1979) U.S. senator from Utah who introduced legislation for the Central Utah Project in 1946, possibly in a move to hold his seat in the upcoming election, which he lost.

JAWAHARLAL NEHRU (1889–1964) First prime minister of India, who held the post from 1947, when India gained independence, until his death in 1964. In 1960, he signed the Indus Waters Treaty on behalf of India; he also served as external affairs minister and was a founder of the Non-Aligned Movement.

RICHARD M. NIXON (1913–1994) The 37th president of the United States, he signed into law numerous environmental bills including the Clean Water Act of 1972 and the Endangered Species Act of 1973, the latter affecting some water development projects.

OSCAR OLIVERA (1955–) Resident of Cochabamba, Bolivia, and a shoe factory worker who became a leader of the workers' union, Federation of Factory Workers, and from late 1999 the main spokesperson for *Coordinadora de Defensa del Agua de la Vida* (Coalition for the Defense of Water and Life), which successfully opposed the privatization of the city's water delivery service; he also opposed transnational exploitation of Bolivian natural gas. In 2001 he received the Goldman Environmental Prize.

MEDHA PATKAR (1954–) Indian social activist who since the mid-1980s has been a central leader of the Narmada Bachao Andolan, the group that opposes the vast Narmada water development project, particularly the high dam Sardar Sarovar. She has received numerous prizes for her work, including the Right Livelihood Award (1991), the Goldman Environmental Prize (1992), and Amnesty International's Human Rights Defenders Award. She has also served on the World Commission on Dams.

VÍCTOR PAZ ESTENSSORO (1907–2001) Four-time president of Bolivia who, at the behest of the International Monetary Fund (IMF), promulgated Law DS 21060 in 1985 (the first year of his final term in office), which led to privatization of industry in Bolivia.

LEWIS A. PICK (1890–1956) American civil engineer and army officer who was responsible for the World War II construction of the Ledo Road, which ran from India to China. After the war he was head of the Missouri River Division for the Army Corps of Engineers, in which capacity he oversaw the construction of the Garrison Dam, whose reservoir forced the Three

Tribes of the Mandan, Arikara, and Hidatsa to sell their land or face inundation. He was chief of engineers (head of the Army Corps of Engineers) from 1949 to 1953.

SANDRA POSTEL (1956?–) American science writer and researcher who specializes in water and food issues; she is the founder and director of the Global Water Policy Project and was a lead researcher at the Worldwatch Institute. In 2002, *Scientific American* magazine named her one of the "Scientific American 50" in recognition of her work on water issues.

JOHN WESLEY POWELL (1834–1902) American soldier, geologist, and explorer best known for his 1869 Powell Geographic Expedition, a three-month journey down the Green and Colorado Rivers and through the Grand Canyon. He served as the second director of the U.S. Geological Survey, from 1881 to 1894. Powell identified the 100th meridian as the dividing line between the arid West and the moist East and wrote a report that discussed land characteristics and rainfall in the American West.

YITZHAK RABIN (1922–1995) Israeli water engineer and twice prime minister (1974–77, 1992–95) who promoted the Israeli–Jordan peace treaty in 1994, for which he was cowinner of the 1994 Nobel Peace Prize.

NATHANIEL REED (1933–) American environmentalist who was environmental adviser to the governor of Florida (1967–71), chairman of Florida's Department of Air and Water Pollution Control (1969–1971), and assistant secretary of interior for fish, wildlife and national parks (1971–77). Thereafter he served as advisor to seven Florida governors and as chairman of the Commission on Florida's Environmental Future; he played a major role in the preservation and revitalization of the Everglades.

MARC REISNER (1948–2000) American environmentalist and author of the now-classic *Cadillac Desert,* a history and exposé of water development in the American West, including the federal subsidies that have supported big development projects since the early 20th century and the politicians and bureaucrats who fought for them.

ARUNDHATI ROY (1961–) Indian novelist, essayist, and social activist who won the Booker Prize in 1997 for her novel *The God of Small Things* and who has been an outspoken critic of globalization and of the Narmada Dam Project and the Sardar Sarovar Dam.

ANWAR SADAT (1918–1981) Egyptian president from 1970 to 1981, he signed a historic peace treaty with Israel and offered to build the Peace Canal to pipe Nile water to Israel's Negev Desert (the offer was turned down); he also defended Egypt's historical water rights to the Nile in a war of words with Ethiopia.

ISMAIL SERAGELDIN (1944–) Egyptian technocrat who held numerous positions with the World Bank between 1972 and 1998; from 1996 to 2000 he was chairman of the Global Water Partnership, which he also founded, and from 1998 to 2000 he served as chairman of the World Commission for Water in the 21st Century. He famously predicted that the next major war would be over water.

MOSES SHERMAN (1853–1932) American developer who helped create modern Los Angeles and built a streetcar line to connect outlying areas to the city center. As one of the city's movers and shakers, he was involved in the diversion of Owens River water to San Fernando Valley, where he owned land. Sherman Oaks, California, is named for him.

IGOR SHIKLOMANOV (1939–) Russian hydrologist and director of the State Hydrology Institute in St. Petersburg, Russia, and academician of the Russian Academy of Natural Sciences on Hydrology and Water Resources. He has more than 200 publications to his credit, including books on water management problems, climate variability, and interbasin water transfers. He received the International Hydrological Prize in 2001 and the Tyler Prize in 2006.

VANDANA SHIVA (1952–) Indian environmental activist and physicist; she has been involved in different aspects of the Indian and global environmental movement since the 1970s and is a leader of the International Forum on Globalization. She is also a recognized antiglobalist and a leader in the worldwide eco-feminist movement. She has numerous publications to her credit and has received awards including the Right Livelihood Award, the Global 500 Award, the Earth Day International Award (all in 1993), and the Blue Planet Award (2007).

JAMES GUSTAVE SPETH (1942–) American environmental lawyer and dean of the Yale School of Forestry and Environmental Studies from 1999 to 2009; during the administration of President Jimmy Carter he was a member and later chairman of the Council of Environmental Quality, and thus the president's top environmental adviser.

MICHAEL STRAUS (1897–1970) American bureaucrat who served as commissioner of the U.S. Bureau of Reclamation from 1945 to 1953, during which time he promoted and oversaw extensive water projects in the American West.

MORRIS K. UDALL (1922–1998) American politician and younger brother of STEWART UDALL, he served in the U. S. House of Representatives (in his brother's vacated seat) from 1961 to 1991. While in Congress he helped shepherd the passage of various environmental and wilderness bills; he was chairman of the House Committee on Interior and Insular Affairs from 1977 to 1991. In his honor, Congress established the Morris K. Udall Scholarship and Excellence in National Environmental Policy Foundation.

STEWART UDALL (1920–2010) American politician and older brother of MORRIS UDALL, he served in the U.S. House of Representatives from 1955 to 1961 and as secretary of the interior from 1961 to 1969, during the administrations of Presidents John F. Kennedy and Lyndon Baines Johnson. His tenure as secretary of the interior was marked by the burgeoning environmental movement in the United States and the passage of the Wild and Scenic Rivers Act and creation of the Land and Water Conservation Fund.

9

Organizations and Agencies

American Rivers
URL: http://www.americanrivers.org
1101 14th Street NW, Suite 1400
Washington, DC 20005
Phone: (877) 347-7550

An environmental group whose concerns are protecting and restoring rivers in the United States, ensuring that they have sufficient water supply to remain healthy and that the water is clean. The organization also pays special attention to the effects of climate change on U.S. rivers.

American Water Resources Association
URL: http://www.awra.org/index.html
4 West Federal Street
PO Box 1626
Middleburg, VA 20118
Phone: (540) 687-8390

Founded in 1964, the association's primary goal is to encourage the development of water resources. It also supports activities that highlight research into water resources, such as the writing of papers on the subject. Additionally, the association advances the improvement of standards, codes, and recommended practices for water development.

American Water Works Association
URL: http://www.awwa.org
6666 W. Quincy Avenue
Denver, CO 80235
Phone: (800) 926-7337

Despite its name, this is actually an international organization dedicated to ensuring that drinking water supplies are safe. This is the largest and oldest organization of water professionals in the world, with more than 60,000 members including academics, environmentalists, and treatment plant operators.

Centre for Science and Environment
URL: http://www.cseindia.org
41 Tughlakabad Institutional Area
New Delhi, India 110062
Phone: (011-91-11) 299-55-124/299-55-125
Rainwater Harvesting
URL: http://www.rainwaterharvesting.org

This organization monitors and studies air, water, and soil pollution and engages in environmental education and training, media outreach, and rainwater harvesting. The last has its own dedicated Web site under the CSE umbrella (see above). The Web site examines the water crisis in India and the conflicts that have resulted from it and displays various solutions such as technologies for rural and urban rainwater harvesting.

Council of Canadians
Le Conseil des Canadiens
URL: http://www.canadians.org
700–170 Laurier Avenue West
Ottawa, Ontario K1P 5V5
Canada
Phone: (800) 387-7177

A watchdog organization whose main issues of concern are water, trade, energy, food, health care, deep integration (corporate breakdown of borders), peace, and the Blue Planet Project. Concerning water, the council seeks to reform Canada's federal water policy and tighten the Great Lakes Compact. It continues to oppose the privatization of water resources and the infringement of transnationals, including those that bottle freshwater.

Eco Friends
URL: http://www.ecofriends.org/default.htm
83 Durga Housing Society
Shivkatra Road, Lal Bangla
Kanpur 208007
India
Phone: (011-91) 979-363-3300

The primary concern of Eco Friends is to clean up the Ganges River through public awareness, legal activism, and advocacy. The organization is responsible for pressing for the successful passage of landmark antipollution and industrial remediation bills.

Ecological Society of America
URL: http://esa.org
1990 M Street NW
Washington, DC 20036
Phone: (202) 833-8773

One of the oldest environmental organizations in the United States, ESA was founded in 1915 as an educational resource and continues to provide ecological information to legislators and lobby for strong science and education expenditures. It sponsors or cosponsors three congressional briefings annually. It also publishes *Ecology* magazine.

Environment Canada/Environnement Canada
URL: http://www.ec.gc.ca/default.asp?lang=En&n=76D556B9-1
Inquiry Centre
351 St. Joseph Boulevard
Place Vincent Massey, 8th Floor
Gatineau, Québec
Canada K1A 0H3

A Canadian government agency whose mandate is to preserve and enhance the quality of the natural environment; conserve Canada's renewable resources, especially water; enforce rules relating to water boundaries; and coordinate environmental policies and programs for the Canadian government.

Friends of the Earth International
URL: http://www.foei.org
PO Box 19199
1000 gd Amsterdam
The Netherlands
Phone: (011-31-20) 622-1369

The largest grassroots environmental organization in the world, it has affiliates in 75 countries. Among the water issues the group is involved in are pollution, sanitation, and conservation.

Friends of the River Narmada
URL: http://www.narmada.org
See Web site for individual contact information.

This is an international coalition of individuals and organizations opposed to development projects on the Narmada River in India; it provides support for the organization Narmada Bachao Andolan. Its goals are to record and document the struggle and provide public outreach.

Global Water Partnership
URL: http://www.awwa.org
Drottninggattan 33
SE-111 51 Stockholm
Sweden
Phone: (011-46-8) 522-126-30

This organization was founded in 1986 by the World Bank, the United Nations Development Program, and the Swedish International Development Agency to foster integrated water resource management. It facilitates a network of regional partnerships in the Caribbean, Central Africa, Central America, Central and Eastern Europe, Central Asia and the Caucasus, China, Eastern Africa, the Mediterranean, the Pacific, South America, South Asia, Southeast Asia, Southern Africa, and West Africa with the goal of establishing water security.

Green Cross International
URL: http://gci.ch
160a, rte de Florissant
1231 Conches
Geneva
PO Box 80
Switzerland
Phone: (011-41-22) 789-1662

Cofounded by Mikhail Gorbachev in 1993, Green Cross International follows the example of the Red Cross emergency response, applying it to ecological crises and conflicts. It quickly established international affiliates, including in the United States. As of 2009 there were affiliates in 30 nations.

International Joint Commission
URL: http://www.ijc.org/en/home/main_accueil.htm
U.S. Section Office
2401 Pennsylvania Avenue NW, 4th Floor
Washington, DC 20440
Phone: (202) 736-9024
Canadian Section Office
234 Laurier Avenue West, 22nd Floor

Ottawa, Ontario K1P 6K6
Phone: (613) 947-1420
Great Lakes Regional Office
100 Ouellette Avenue, 8th Floor
Windsor, Ontario N9A 6T3
Phone: (519) 257-6714

The six-member commission was established by the 1909 Boundary Waters Treaty between the United States and Canada. The IJC is an advisory, monitoring, and investigatory agency whose concerns are water pollution, water development, and balancing the competing interests of stakeholders in the Great Lakes ecosystem.

International Rivers
URL: http://internationalrivers.org
1847 Berkeley Way
Berkeley, CA 94703
Phone: (510) 848-1155

Founded in 1985, the organization's mission is to protect rivers and defend the rights of communities that depend on them. Activities are focused in Africa, China, Latin America, South Asia, and Southeast Asia. The organization also publishes fact sheets on dams and irrigation, as well as the journal *World River Review*.

National Institutes for Water Resources
URL: http://snr.unl.edu/niwr

This is the representative organization for the various U.S. water research institutes. It serves as a facilitator of information and helps them implement the provisions of the Water Resources Research Act of 1984. Each of the 50 U.S. states and Puerto Rico, the Virgins Islands, and Guam have a Water Resources Research Institute that is usually, but not always, located at a state university. Contact information for the individual institutes can be found at the Web site listed above.

Nile Basin Society
URL: http://www.nilebasin.com
730-3 Greystone Walk Drive
Toronto, Ontario M1K 5J4
Canada

Founded in 2001, this is primarily an educational organization whose goal is to increase awareness in the West of the freshwater problems facing the countries of the Nile River Basin.

Oxfam
URL: http://www.oxfam.org.uk
Oxfam House
John Smith Drive
Cowley, Oxford OX4 2JY
United Kingdom
Phone: (011-44) (0) 1865-47-2602

An emergency, environmental, economic, and peace organization that works on numerous issues in Africa, Asia, Latin America, and the Middle East. It is well known for its work in war-damaged areas, where it helps to restore water and sanitation systems.

Pacific Institute
URL: http://www.pacinst.org
California Office
654 13th Street,
Preservation Park
Oakland, CA 94612
Phone: (510) 251-1600
Colorado Office
2260 Baseline Road, Suite 205
Boulder, CO 80302
Phone: (720) 564-0651

Founded in 1987, the Pacific Institute works to find real-world solutions to problems such as water shortages, habitat destruction, global warming, and environmental injustice. It conducts research, publishes reports, recommends solutions, and works with decision makers, advocacy groups, and the public.

Polaris Institute
URL: http://www.polarisinstitute.org
180 Metcalfe Street, Suite 500
Ottawa, Ontario
Canada K2P 1P5
Phone: (866) 346-6602 (toll free)

Founded in 1997, the Polaris Institute assists citizens' groups with strategy and planning in reversing the role of corporate power in government. Its three main areas of focus are water, trade, and energy.

Riverkeeper
URL: http://www.riverkeeper.org

828 South Broadway
Tarrytown, NY 10591
Phone: (800) 217-4837

Established in 1966 (and originally named the Hudson River Fisherman's Association), this organization is the primary clean water advocate in New York State. Its three main areas of focus are the protection of New York City's drinking water supply, restoration of the Hudson River ecosystem, and improving public access to the Hudson River.

Sierra Club
URL: http://www.sierraclub.org
85 Second Street, 2nd Floor
San Francisco, CA 94105
Phone: (415) 977-5500

Founded in 1892 by naturalist John Muir, the Sierra Club is one of the leading environmental organizations in the United States. It has a long record of fighting encroachment on and destruction of natural resources. Clean water is one of its many areas of concern.

UNESCO—Water
URL: http://www.unesco.org/water
2 United Nations Plaza, Room 900
New York, NY 10017
Phone: (212) 963-5995

Among the various programs under UNESCO's auspices are the World Water Assessment Programme, which monitors freshwater issues, develops case studies, and enhances assessment capacity and decision making at the national level, and the Internationally Shared Aquifers Resource Management Initiative, whose goal is to facilitate understanding of all the issues involved in cross-boundary aquifer management.

United States Army Corps of Engineers
URL: http://www.usace.army.mil/Pages/Default.aspx
441 G Street NW
Washington, DC 20314-1000
Phone: (202) 761-0011

The corps' mission is to provide public engineering services during war and peacetime that reduce risks from natural disasters (such as flooding) and strengthen the economy. Historically, the majority of the corps' domestic public works projects have been east of the Mississippi River.

United States Bureau of Reclamation
URL: http://www.usbr.gov
1849 C Street NW
Washington, DC 20240-0001

An agency within the Department of the Interior, the bureau was established in 1902. In the past century, it has built more than 600 dams, mostly in the American West, that have contributed to the economic growth of the western United States. It is presently a water management agency and the largest water wholesaler in the country.

United States Environmental Protection Agency
URL: http://www.epa.gov
1200 Pennsylvania Avenue NW
Washington, DC 20460
Phone: (202) 272-0167

The EPA studies environmental issues, develops and enforces regulations, provides grants, sponsors partnerships with businesses, nonprofit organizations, and state and local governments, disseminates environmental information for students and educators, and publishes technical reports and newsletters.

United States Geological Survey
URL: http://www.usgs.gov
Reston, VA

Established in 1879, the USGS publishes fact sheets and hydrological maps, among other things. It runs 11 programs related to water. The Web site offers plentiful information on water, including access to water data collected from nearly 1.5 million sites around the country.

WaterAid
URL: http://www.wateraid.org
232 Madison Avenue, Suite 1202
New York, NY 10016
Phone: (212) 683-0430

An international organization with offices in London, New York, and Australia, WaterAid works with various local organizations to provide people in the developing world with clean water and good sanitation by means of low-cost, sustainable technology.

WaterCan
URL: http://www.watercan.com
321 Chapel Street
Ottawa, Ontario
K1N 7Z2 Canada
Phone: (800) 370-5658

WaterCan works to deliver clean water and sanitation in Ethiopia, Kenya, Tanzania, and Uganda. It works with local partners in an integrative manner, with a focus on sustainability and women, who bear the burden of collecting the domestic water supply in these countries.

Water for People
URL: http://www.waterforpeople.org
6666 W. Quincy Avenue
Denver, CO 80235
Phone: (303) 734-3490
E-mail: info@waterforpeople.org

A project of the American Water Works Association, this international organization partners with local organizations in developing countries to provide safe drinking water and good sanitation.

WaterPartners International
URL: http://www.water.org
2405 Grand Boulevard, Suite 860
Box 12
Kansas City, MO 64108
Phone: (913) 312-8600

WaterPartners International uses sustainable technologies to provide safe drinking water and sanitation for people in developing countries. It has assisted more than 200 communities in eight nations. Additional offices are located in North Carolina, India, and Kenya.

Water Science and Technology Association
URL: http://www.wsta.org.bh
PO Box 20018
Manama, Bahrain
Phone: (011-973) 826-512

This organization promotes research and training programs, water conservation, antipollution programs, information exchange, and water resource development in the Arabian Gulf coast countries.

Worldwatch Institute
URL: http://www.worldwatch.org
1776 Massachusetts Avenue NW
Washington, DC 20036
Phone: (202) 452-1999

An independent research organization founded in 1974, whose main focuses are climate change, resource degradation, and population growth. Worldwatch develops solid data and effective strategies to meet these challenges.

10

Annotated Bibliography

The following annotated bibliography is divided into six sections, which include a general grouping of material and the five case study areas as they are presented in chapters 2 and 3: the United States, Canada, and Mexico; Bolivia; Egypt, Ethiopia, and Sudan; Israel, Jordan, and the Palestinian Territories; and India, Pakistan, and Bangladesh. The general section contains material on hydrology, water politics, water economics, and studies that cover more than one of those countries. The other five sections are further subdivided into books, articles, and Web documents.

GENERAL

Books

Anderson, Terry L., and Peter J. Hill, eds. *Water Marketing—The Next Generation.* Lanham, Md., and London: Rowman & Littlefield, 1997. A collection of essays that examines the possibilities, as envisioned in the late 20th century, of the free market's ability to solve the water crisis and prevent water shortages. It focuses mainly on the United States, though Australian water trading and world markets are also examined.

Ball, Philip. *Life's Matrix: A Biography of Water.* New York: Farrar, Straus and Giroux, 1999. This book looks at water in myriad ways, from the mythological to the scientific, to offer some possible solutions to the global freshwater crisis.

Barlow, Maude. *Blue Covenant: The Global Water Crisis and the Coming Battle for the Right to Water.* New York: The New Press, 2007. Maude Barlow surveys the worldwide freshwater problem from the perspective of commodification. She details the water wars between countries, municipalities, and activists opposing transnational corporations that seek financial profit from the resource.

———, and Tony Clarke. *Blue Gold: The Fight to Stop the Corporate Theft of the World's Water.* New York: The New Press, 2002. Focusing on the commodification and privatization of the second-most important resource (after air), the authors show how various multinational corporations, sometimes with the assistance of the

International Monetary Fund and the World Bank, are creating water "markets" and reaping profits by either supplying water to communities and municipalities or bottling it and selling it as a luxury item.

Bauman, Duane D., and Daniel M. Dworkin, eds. *Planning for Water Reuse.* Chicago, Ill.: Maaroufa Press, 1978. Cites different examples of early water reuse and planned reuse, including in urban areas such as Denver, Colorado. Some essays explore the potential of water reuse and issues surrounding its psychological acceptance by the public.

Biswas, Asit K. *History of Hydrology.* Amsterdam and London: North-Holland Publishing Co., 1970. Without water no life would exist on Earth, and without the ability to capture and use water for purposes other than drinking it, civilization would not exist. Beginning with hydrology in Mesopotamia, the author surveys human water use in ancient Greek, Roman, and Egyptian civilizations, the Middle Ages, the Age of Enlightenment, through the 19th century.

Caldicott, Helen, M.D. *If You Love This Planet: A Plan to Heal the Earth.* New York and London: W. W. Norton, 1992. The internationally known antinuclear activist here discusses the numerous problems that have produced and continue to produce a sick planet, devoting extensive space to water pollution and waterborne diseases.

Calhoun, Yael, series ed. *Environmental Issues: Water Pollution.* Philadelphia: Chelsea House Publishers, 2005. This collection of essays and excerpts from government agencies and nongovernmental organizations covers both saltwater and freshwater pollution and the challenges in halting and reversing it. Some of the essays focus on issues, such as forests, that provide the reader with a more holistic background into the nature of the solution to water pollution.

Chorley, Richard J., ed. *Water, Earth, and Man: A Synthesis of Hydrology, Geomorphology, and Socio-Economic Geography.* London: Methuen & Co., 1969. This collection of essays provides a complete hydrological picture of the Earth from the scientific and technical points of view. Numerous charts, tables, graphs, and mathematical formulae supplement the text.

Clarke, Robin. *Water: The International Crisis.* London: Earthscan Publications, Ltd., 1991. This is longtime water activist Clarke's early book sounding the alarm on the growing freshwater crisis. It covers water scarcity and degradation, coping with water scarcity, traditional as well as high-tech solutions to the water crisis, and scenarios for securing the future.

——— and Jannet King. *The Water Atlas: A Unique Visual Analysis of the World's Most Critical Resource.* New York and London: The New Press, 2004. A statistical gold mine illustrated with colorful, easy-to-follow charts, with comparative tables in the final chapter. The book covers every aspect of water need and use, as well as demand and abuse. The introductions to each chapter highlight the relevant points and brief text accompanies each of the charts.

Conca, Ken. *Governing Water: Contentious Transnational Politics and Global Institution Building.* Cambridge, Mass.: MIT Press, 2006. This book examines social and political roles in the global water crisis, focusing on "formal international regimes for shared rivers, networking among water experts and professionals, social movements opposing the construction of large dams," and the controversy that surrounds the commodification of water.

Dean, Robert B., and Ebba Lund, eds. *Water Reuse: Problems and Solutions.* New York: Academic Press, 1981. The authors provide a thorough examination of wastewater reuse that includes potential hazards such as bacteria, viruses, metals, asbestos, and radioactivity. They also look at many different water treatment systems (which have undoubtedly been updated since the book's publication), including their cost, management, and control. Examples of water reuse are also provided.

De Villiers, Marq. *Water: The Fate of Our Most Precious Resource.* Boston and New York: Houghton Mifflin Company, 2000. A thorough overview of the water crisis, covering the "haves" and the "have nots," that also looks at the importance of water throughout history, as well as the intra- and international politics of water acquisition and distribution.

Gleick, Peter H., ed. *Water in Crisis: A Guide to the World's Freshwater Resources.* Oxford and New York: Oxford University Press, 1993. Part I contains nine essays by noted water experts such as Igor A. Shiklomanov, Linda Nash, Alan P. Covich, Sandra Postel, Malin Falkenmark and Gunnar Lindh, Stephen C. McCaffrey, and Gleick himself. They cover such topics as freshwater resources, the relationship between water quality and health, ecosystems, agriculture, energy, economic development, politics and international law, and the water crisis in the 21st century. Part II, the bulk of the book, contains a good deal of statistical data, which though somewhat dated still provides an overview of the global situation.

———, et al. *The World's Water, 2006–2007: The Biennial Report on Freshwater Resources.* Washington, D.C.: Island Press, 2006. Produced under the auspices and by analysts of the Pacific Institute for Studies in Development, Environment, and Security, this report includes six major essays, on desalination, in-stream allocation restoration and preservation, flooding and drought, environmental justice, terrorism, and water risks for business. There are also five shorter pieces that cover topics ranging from water on Mars to the bottling of water for sale. Part III includes 22 tables that reflect some of the latest worldwide water data.

Hoekstra, Arjen Y., and Ashok K. Chapagain. *Globalization of Water: Sharing the Planet's Freshwater Resources.* Malden, Mass., and Oxford, UK: Blackwell Publishing, 2008. Trade is the engine behind globalization, and this book examines the link between free trade and water management. It may be a little too technical for the general reader, but its charts, tables, and maps make it worthwhile for anyone interested in the economics of the globalization of water.

Hunt, Constance Elizabeth. *Thirst Planet: Strategies for Sustainable Water Management.* London and New York: Zed Books, 2004. The author, a biologist and environmentalist, has served as an adviser to the World Wildlife Fund and the World Water Council. This book examines the complexity of global water issues in order to redefine the problem, thus leading to new solutions that break from the traditional thinking that hampered progress toward resolving the water crisis in the 20th century.

Hunt, Cynthia, and Robert M. Garrels. *Water: The Web of Life.* New York: W. W. Norton & Co., 1972. Covering all aspects of the hydrologic cycle, the authors reveal not just the interconnectedness of water to life, but the connection between all types of water, from the shallowest pools to the deepest oceans, to polluted waters and their negative impact.

Kandel, Robert. *Water from Heaven: The Story of Water from the Big Bang to the Rise of Civilization and Beyond.* New York: Columbia University Press, 2003. As the subtitle makes clear, this is a book about the history of water, beginning with a fine survey of the scientific knowledge and theories that led to the molecular combination of two atoms of hydrogen and one of oxygen. In the first section, Kandel looks at water's role in the evolution of life on Earth; in Part II he examines water in the modern world; and in Part III he looks at water's role in human history, and its probable role in our future.

Markham, Adam. *A Brief History of Pollution.* New York: St. Martin's Press, 1994. Covering all forms of environmental pollution—as well as how they came about—Markham discusses various forms of water pollution, including acid rain. He includes a chapter on citizen action, both personal and political.

McDonald, Bernadette, and Douglas Jehl, eds. *Whose Water Is It?: The Unquenchable Thirst of a Water-Hungry World.* Washington, D.C.: National Geographic Society, 2003. Thirteen essays by experts on the water crisis look at the topics of ownership, scarcity, conflict, and prospects. The essays examine these themes from different viewpoints and angles, thus not only highlighting the enormity of the problem but also its attendant fallout.

Overman, Michael. *Water: Solutions to a Problem of Supply and Demand.* Garden City, N.Y.: Doubleday & Co., 1969. This book contains a thorough discussion of the hydrologic cycle and how it has been interfered with. The author also discusses how water is measured before examining the effects of irrigation and dams and the use of hydroelectric power. Among the solutions he looks at are desalination, water purification, rain making, and river rerouting.

Pearce, Fred. *Keepers of the Spring: Reclaiming Our Water in an Age of Globalization.* Washington, D.C.: Island Press, 2004. Pearce first examines the various methods people have used to extract water from the ground or capture it from the earth's surface, showing how these methods have, over time, depleted supplies. He also surveys techniques currently used for capturing freshwater, which include reviving ancient methods in Cyprus, reviving wetlands, and using netting to capture fog in Chile.

Pielou, E. C. *Fresh Water.* Chicago: University of Chicago Press, 1998. Pielou's fourth book is *the* handbook on the subject of freshwater. She doesn't merely discuss the hydrologic cycle—which the first chapter is devoted to—but examines it in depth as it pertains to freshwater. This includes: the workings of surface and groundwater, the special ecological properties of wetlands, human interventions such as dams and diversions, water in the atmosphere, frozen water, and microscopic life in water.

Postel, Sandra. *Conserving Water: The Untapped Alternative.* Washington, D.C.: Worldwatch Institute, 1985. This is actually Worldwatch Paper 67 and, as the subtitle reveals, the water conservation movement was in its infancy in the mid-1980s when this was published. It is a concise overview of how things stood then and the projected changes over the next few decades. It is illuminating for its historical background information on water consumption.

———. *Last Oasis: Facing Water Scarcity.* New York and London: W. W. Norton & Co., 1992. Sounding the alarm at the end of the 20th century, this book describes both water scarcity that has worsened and the technologies to confront it that have since been further developed.

———. *Liquid Assets: The Critical Need to Safeguard Freshwater Ecosystems.* Washington, D.C.: Worldwatch Institute, 2005. This book discusses the value of healthy ecosystems, how ecosystems have become unhealthy worldwide over the past 50 years, and the problems this causes. It also outlines measures for restoring damaged ecosystems and for maintaining their health in the 21st century.

———. *Pillar of Sand: Can the Irrigation Miracle Last?* New York and London: W. W. Norton & Co., 1999. The author examines the fates of past irrigation civilizations and compares their problems and downfalls to the contemporary situation throughout the world. Among the issues Postel discusses are salination of the soil and reduced quality water due to irrigation and poor drainage and the battle for water between farmers and cities.

Postel, Sandra, and Brian Richter. *Rivers for Life: Managing Water for People and Nature.* Washington, D.C.: Island Press, 2003. An examination of the worldwide use of river water for domestic, municipal, agricultural, and industrial consumption and electrical generation and the consequences these uses are having on the rivers.

Roddick, Anita, with Brooke Shelby Biggs, eds. *Troubled Water: Saints, Sinners, Truths and Lies about the Global Water Crisis.* West Sussex, UK: Anita Roddick Books, 2004. Thirteen essays, interspersed with numerous factoids about the water crisis, discuss the battles—grassroots, regional, and international—both those fought and won and those yet to be fought. The book also covers myths about the crisis that lure many into a sense of security.

Rothfeder, Jeffrey. *Every Drop for Sale: Our Desperate Battle over Water in a World about to Run Out.* New York: Tarcher/Putnam, 2001. An early 21st-century examination of the worldwide freshwater crisis by an acclaimed business journalist, this book covers all the bases succinctly.

Simon, Paul, Dr. *Tapped Out: The Coming World Crisis in Water and What We Can Do about It.* New York: Welcome Rain, 1998, 2001. The late senator from Illinois began investigating the water crisis during his tenure on the U.S. Senate Foreign Relations Committee, focusing on the water problems of the Middle East. This book first defines the problem in various regions of the United States and around the world, then discusses methods of solving the problem, from conservation to desalination.

Soffer, Arnon. *Rivers of Fire: The Conflict over Water in the Middle East,* tr. by Murray Rosovsky and Nina Copaken. Lanham, Md.: Rowman & Littlefield, 1999. At the end of the 20th century, the author examined the potential water hot spots in the Middle East and Africa, particularly the basins of the Nile, Jordan-Yarmuk, Tigris and Euphrates, and Orantes Rivers. He also discussed groundwater problems and looked at potential solutions to them.

Staff of the International Bank for Reconstruction and Development/The World Bank. *Making the Most of Scarcity: Accountability for Better Water Management in the Middle East and North Africa.* Washington, D.C.: The International Bank for Reconstruction and Development/The World Bank, 2007. This book covers many of the latest World Bank policy proposals for what it describes as the MENA (Middle East and North Africa), stretching from Morocco in the west to Iran in the east and including Djibouti in northeast Africa (though not Sudan, Ethiopia, and Eritrea). While taking a positive attitude toward the situation, the book's overall position is reinforced by the realities of 21st-century economics.

Starke, Linda, ed. *State of the World, 2008: Innovations for a Sustainable Economy.* New York and London: W. W. Norton & Co., 2008. The annual report from the Worldwatch Institute covers many aspects of creating and maintaining a worldwide economy—or even smaller ones within the shadow of globalization—based on ecological commitment. The report wisely takes a holistic approach to the economic and environmental challenges of the 21st century. Most relevant (to the present volume) is chapter 8: "Water in a Sustainable Economy."

Stein, Richard Joseph, ed. *Water Supply.* New York: H. W. Wilson Co., 2008. As a volume of the publisher's series, The Reference Shelf, this book contains 26 articles, previously published in print and online newspapers and journals as varied as *Smithsonian,* the *Seattle Post-Intelligencer,* the *Economist,* and *Weekly Reader.* The articles are grouped into five categories covering the global freshwater supply, conflicts, water pollution, the effects of climate change on the global supply, and water management.

Stewart, Gail B. *Acid Rain.* San Diego, Calif.: Lucent Books, 1992. A good, though slightly dated, overview of the problem of acid rain in the United States. The author discusses the causes of acid rain and its effects on lakes, rivers, and forests and human, animal, and plant life. A chapter also discusses the effects of acid rain on the "built environment." Contains charts, graphs, and photographs.

Vigil, Kenneth M. *Clean Water: An Introduction to Water Quality and Water Pollution Control.* Second edition. Corvallis, Oregon: Oregon State University Press, 2003. The book discusses the hydrologic cycle, human activities that affect the cycle, how the cycle is affected, and ways to reverse the damage. Written by an environmental engineer, it is readable for the general public.

Ward, Diane Raines. *Water Wars: Drought, Flood, Folly, and the Politics of Thirst.* New York: Riverhead Books, 2002. Covering various water crises areas around the world including, but not limited to, the American West, southern Africa, the Middle East, India and Pakistan, and Australia, the author discusses the problems of population expansion, water acquisition in dry climates, and potential conflicts. The epilogue is devoted to the problems of the Everglades.

Annotated Bibliography

UNITED STATES, CANADA, AND MEXICO

Books

Barnett, Cynthia. *Mirage: Florida and the Vanishing Water of the Eastern U.S.* Ann Arbor: University of Michigan Press, 2007. Barnett proves that the water crisis in the United States is not confined to the arid West. The historically wet eastern half of the country is also experiencing the onset of a water crisis—especially the Southeast and particularly Florida, where overdevelopment and poor planning have led to reduced aquifers and polluted waterways for a population that continues to expand rapidly.

Chapelle, Francis H. *Wellsprings: A Natural History of Bottled Spring Waters.* New Brunswick, N.J.: Rutgers University Press, 2005. With the popularity of bottled water soaring and various groups beginning to oppose the bottled-water industry, this book provides an examination of the industry itself, as well as the hazards of bottled water that are often overlooked and ignored.

Colby, Bonnie G., and Katharine L. Jacobs, eds. *Arizona Water Policy: Management Innovations in an Urbanizing, Arid Region.* Washington, D.C.: RFF Press, 2007. The 15 essays in this volume give the reader a historical perspective on Arizona's water problem and discuss policy issues, the state's freshwater supply for urban and rural residents, the climate's effect on the supply, water transactions, federal intervention, recharge and recovery within the state, and tribal claims to freshwater in the state.

Duncan, David James. *My Story as Told by Water: Confessions, Druidic Rants, Reflections, Bird-Watchings, Fish-Stalkings, Visions, Songs and Prayers Refracting Light, from Living Rivers, in the Age of the Industrial Dark.* San Francisco: Sierra Club Books, 2001. In the spirit of Thoreau and Emerson, the acclaimed novelist writes a subjective book about the rivers that have been important in his life. The book is divided into three parts, and includes a section on activism.

Gibbs, Lois Marie. *Love Canal: The Story Continues . . .* Gabriola Island, B.C., and Stony Creek, Conn.: New Society Publishers, 1998. This is the 20th-anniversary revised edition of a book by one of the residents and leaders of the 1978 grassroots movement that forced national and regional leaders to come to terms with one of the worst environmental disasters in U.S. history. This edition recounts the discovery of the toxic waste site, the health problems of people who lived near the canal, the battle to force corporate accountability, the relocation of families in 1981, the "containment" of the toxins, New York State's successful lawsuit, and Gibbs's founding of the Center for Health, Environment and Justice. The book's foreword is by Ralph Nader.

Glennon, Robert. *Water Follies: Groundwater Pumping and the Fate of America's Fresh Waters.* Washington, D.C.: Island Press, 2002. From the Atlantic Ocean to the Pacific, this book covers states and localities where overdrawing of groundwater has depleted supply. No section of the country is left out, as the author takes the reader on a veritable tour from Maine, Massachusetts, and Florida to Minnesota,

Texas, Arizona, Nevada, and California. The book includes an informative chapter on the mechanics of how a river dries out.

Longo, Peter J., and David W. Yoskowitz, eds. *Water on the Great Plains: Issues and Policies.* Lubbock: Texas Tech University Press, 2002. The editors have gathered 10 essays that describe the water situation in America's heartland. Among the topics discussed are water's role on Great Plains existence, the water politics of the region, new judicial realities concerning water and water disputes, federal policy, water markets in Texas, and Native American water rights.

Marston, Ed (*High Country News*), ed. *Western Water Made Simple.* Washington, D.C.: Island Press, 1987. Winner of the 1986 George Polk Award for Environmental Reporting, this book is a compilation of four special issues of *High Country News,* comprising 27 articles originally published in the fall of 1986. The first issues, or section of the book, provide an overview of water policy in the American West, while the next three focus on the Columbia, Missouri, and Colorado Rivers.

Martin, William E., Helen M. Ingram, Nancy K. Laney, and Adrian Griffin. *Saving Water in a Desert City.* Washington, D.C.: Resources for the Future, 1984. Though a little dated, this book provides a fine overview of the water problems Tucson, Arizona, was experiencing by the mid-1980s, and those projected for the future. Its chapters run the gamut, from describing the problem and its history to the realities of various solutions, including conservation, price and demand, and politics.

McCluskey, Dorothy S., and Claire C. Bennitt. *Who Wants to Buy a Water Company?: From Private to Public Control in New Haven.* Bethel, Conn.: Rutledge Books, 1996. An account by a former Connecticut state representative and her aide about the movement of the municipal area's water supply from private hands to public control at a time when just the reverse was beginning to happen around the world.

Midkiff, Ken. *Not a Drop to Drink: America's Water Crisis (and What You Can Do).* Novato, Calif.: New World Library, 2007. Describes the water crisis in the United States as of the first decade of the 21st century, including depleted aquifers beneath the Great Plains and elsewhere and overpopulation in the Southwest straining the Colorado River basin. It also discusses regional battles stemming from these problems, and possible positive and negative outcomes depending on whether or not steps are taken to ensure a healthy supply of clean freshwater for the future.

Mulholland, Catherine. *William Mulholland and the Rise of Los Angeles.* Berkeley and Los Angeles: University of California Press, 2000. Part biography and part social history, this book is a sympathetic examination of William Mulholland's life and his activities regarding Los Angeles' acquisition of water in the early 20th century by his granddaughter. It provides different viewpoints of the Owens Valley controversy and the collapse of the St. Francis dam.

Reisner, Marc. *Cadillac Desert: The American West and Its Disappearing Water.* Revised Edition. New York: Penguin Books, 1986, 1993. A classic of the genre, Reisner provides an overview of the fight for water in the western United States. From John Wesley Powell's 19th-century observation of the scarcity of water in the region and assertion that the West could never be settled, through the 20th century's era of dam building and pork-barrel water subsidies for agribusiness and cities.

Shortle, J. S., and D. G. Abler, eds. Environmental Policies for Agricultural Pollution Control. Wallingford, U.K., and New York: CABI Publishing, 2001. Somewhat on the technical side but chapter 1: "Agriculture and Water Quality: the issues" and chapter 5: Non-point Source Pollution Control Policy in the USA" provide good explanations of what is at stake and the steps that have been taken in certain parts of the United States to correct the problem of agricultural pollution.

Sproule-Jones, Mark. *Restoration of the Great Lakes: Promises, Practices, Performances.* Vancouver, B.C., and Toronto: UBC Press, 2002. Professor Sproule-Jones examines governmental policy—on the provincial, state, and national government levels—regarding the Great Lakes environment. His analysis of government documents, along with surveys and interviews, reveals failures that outweigh the gains.

Stevens, Joseph E. *Hoover Dam: An American Adventure.* Norman: University of Oklahoma Press, 1988. A history of the construction of the world's first high dam during the Great Depression, which became a paradigm for all that followed, especially in the United States.

Symons, James M. *Plain Talk about Drinking Water: Questions and Answers about the Water You Drink.* Fourth edition. Denver, Colo.: American Water Works Association, 2001. Provides answers to 190 questions about drinking water in North America, primarily the United States. It covers topics such as health, sources, aesthetics, distribution, and testing.

Troesken, Werner. *The Great Lead Water Pipe Disaster.* Cambridge, Mass.: MIT Press, 2006. An examination of the municipal water supply system in the United States and the various illnesses, first recorded in the 19th century, that have resulted from the use of lead pipes to transport freshwater. The author focuses on health problems due to lead exposure in New York City, Boston, and Glasgow, Scotland, as well as smaller towns in New England and the United Kingdom.

Warner, Sara. *Down to the Waterline: Boundaries, Nature, and the Law in Florida.* Athens: University of Georgia Press, 2005. This book takes a look at the water controversy in Florida from a completely different angle. In the words of the author, it "examines the broad shift in the American consciousness of the natural world by tracing the development of a centuries-old concept—the ordinary high water line. . . . In Florida . . . this boundary has been particularly controversial." In essence, the high water line is the boundary that prevented human intrusion of wetlands, rivers, lakes, and oceanfront—a boundary as much under siege as the state's dwindling supplies of freshwater.

Articles

Getner, Jon. "The Future Is Drying Up." *New York Times* magazine (10/21/07). Examination of the water situation in the American West focusing on evaporation from Lake Mead, the needs and resources of Las Vegas, Arizona, Southern California, and Colorado, with interviews of key water managers in those areas.

Wickstrom, Stefanie. "Cultural Politics and the Essence of Life: Who Controls the Water?" in *Environmental Justice in Latin America: Problems, Promise, and Practice,* ed. by David V. Carruthers. Cambridge, Mass.: MIT Press, 2008, pp. 287–319. This article examines the conflicts over neoliberal policy and water privatization in Chile, Bolivia, and Mexico.

Web Documents

"About CERP: Brief Overview." Comprehensive Everglades Restoration Plan. Available online. URL: http://www.evergladesplan.org/about/about_cerp_brief.aspx. Accessed January 14, 2009. The plan overview lists background history, goals, and areas under discussion. There are also links to an in-depth review of the plan and to the Water Resources Development Act of 2000.

Arizona Water Settlements Act, 2004. Available online. URL: http://www.azwater.gov/dwr/Content/Hot_Topics/AZ_Water_Settlements/GRIC_files/Cong_Record_S437%20-108-360.pdf. Accessed October 10, 2008. The water act is reprinted in full in a PDF format.

Beck Eckardt C. "The Love Canal Tragedy." *EPA Journal.* United States Environmental Protection Agency. Available online. URL: http://www.epa.gov/history/topics/lovecanal/01.htm. Accessed October 15, 2008. Brief recap of the history of toxic dumping at Love Canal and the shock and dismay when the tragedy began to unfold in August 1978.

"Boulder Canyon Project: All American Canal System." United States Department of Interior, Bureau of Reclamation. Available online. URL: http://www.usbr.gov/dataweb/html/allamcanal.html. Accessed September 30, 2008. A good overall history of the project, it includes the general plan, the project's development, and its water and engineering data.

"CAP Subcontracting Status Report, October 5, 2009: CAP Non-Indian Municipal and Industrial Subcontracts." Central Arizona Project. Available online. URL: http://www.cap-az.com/includes/media/docs/SubcontractStatusReport-10-05-09.pdf. Accessed January 12, 2010. Lists the names of the municipalities or companies receiving Central Arizona Project water, the date of the contract, and the amount in acre-feet. The list is heavily footnoted and includes commentary.

"CERP: The Plan in Depth—Part 1." Comprehensive Everglades Restoration Plan. Available online. URL: http://www.evergladesplan.org/about/rest_plan_pt_01.aspx. Accessed January 14, 2009. Part 1 discusses CERP goals and estimated costs and lists its 13 major components.

"Chapter Five: Joint Management of the Great Lakes." The Great Lakes Atlas, U.S. Environmental Protection Agency. Available online. URL: http://www.epa.gov/glnpo/atlas/glat-ch5.html. Accessed October 30, 2008. Discusses treaties and agreements between the United States and Canada including the Boundary Waters Treaty, the International Joint Commission, and the Great Lakes Water Quality Agreements of 1972, 1978, and 1987.

Clemons, Josh. "Water-Sharing Compact Dissolves." Mississippi-Alabama Sea Grant Legal Program. Available online. URL: http://www.olemiss.edu/orgs/SGLC/

MS-AL/Water%20Log/23.3watershare.htm. Accessed January 14, 2009. Discusses the background of the Apalachicola-Chattahoochie-Flint basin water apportionment compact, how negotiations dragged past the imposed deadline, what caused them to do so, and that the probable next step will be the United States Supreme Court.

"Current Problems Facing the Region." Duke University Wetlands Center. Available online. URL: http://www.nicholas.duke.edu/wetland/current.htm. Accessed November 30, 2008. Discusses water quality, mercury contamination, invasive exotic plant species, hydrology, and land use in the Everglades.

"Cuyahoga River Fire." Ohio History Central: An Online Encyclopedia of Ohio History. Available online. URL: http://www.ohiohistorycentral.org/entry.php?rec=1642. Accessed November 1, 2008. Overview of the infamous Cuyahoga River fire in 1969, its probable causes, and the effect it had on people in Cleveland, the region, and the United States.

"District-wide Watershed Management Plan." Metropolitan North Georgia Water Planning District. Available online. URL: http://www.northgeorgiawater.com/html/253.htm. Accessed January 16, 2009. The final plan, in PDF format, for the Metropolitan North Georgia Water District, adopted in September 2003. Also included are amendments, updated in December 2006.

"Everglades Agricultural Area." Duke University Wetlands Center. Available online. URL: http://www.nicholas.duke.edu/wetland/eaa.htm. Accessed November 30, 2008. This Web site provides an overview of the agricultural area within the Everglades and two water conservation sites.

"Federal Water Pollution Control Act, as Amended by the Clean Water Act of 1977." Clean Water Act (CWA), United States Environmental Protection Agency. Available online. URL: http://www.epa.gov/npdes/pubs/cwatxt.txt. Accessed November 1, 2008. This is the full text of the amended act.

Fuller, Craig. "Central Utah Project," in *Utah History Encyclopedia* available at Utah History to Go. URL: http://historytogo.utah.gov/utah_chapters/utah_today/centralutahproject.html. Accessed October 15, 2008. Reference article discussing the history of the Central Utah Project.

Gelt, Joe. "Sharing Colorado River Water: History, Public Policy and the Colorado River Compact." Water Resources Research Center, University of Arizona. Available online. URL: http://ag.arizona.edu/AZWATER/arroyo/101comm.html. Accessed October 5, 2008. As the subtitle explains, this article examines the history of appropriation of Colorado River water. It includes sections on Native American rights to the river's water, depletion projections, and water marketing.

"Great Lakes Tonnage Tumbles Due to Lower Water Levels." *Platts Coal Outlook* (1/14/08). Article in a coal industry newsletter that examines one effect of lower water levels in the Great Lakes.

"Helping Solve Georgia's Water Problems—The USGS Cooperative Water Program." United States Geological Survey. Available online. URL: http://pubs.usgs.gov/fs/2006/3032. Accessed December 17, 2008. Fact sheet that details the USGS water program in Georgia as of 2006. Among other things it discusses coastal hydrology, groundwater supply, and the metro-Atlanta situation.

Lydersen, Kari. "Bottled Water at Issue in Great Lakes: Conservation and Commerce Clash." *Washington Post* (9/29/08). Available online. URL: http://www.washingtonpost.com/wp-dyn/content/article/2008/09/28/AR2008092802997.html?nav=emailpage. Accessed June 7, 2009. Reports on what is known as the bottle-water loophole of the Great Lakes Compact, which allows for freshwater to be extracted from the Great Lakes basin if it is done so in small bottles.

"North American Water and Power Alliance (NAWAPA)." San José State University Department of Economics. Available online. URL: http://www.sjsu.edu/faculty/watkins/NAWAPA.htm. Accessed October 22, 2008. Article with map and charts outlining the projected costs and revenues for the proposed U.S.-Canada water and power system.

"Review of Aquifer Storage and Recovery in the Floridian Aquifer System of Southern Florida." United States Geological Survey. Available online. URL: http://pubs.usgs.gov/fs/2004/3128. Accessed December 17, 2008. A USGS fact sheet that covers the technique of well-drilling used for freshwater storage and recovery in the southern portion of the Floridian Aquifer. The report is supplemented by maps and charts.

Shelton, Stacey. "Georgia Loses Round in Fight over Lanier Water." *Atlanta Journal-Constitution* (1/12/09). Available online. URL: http://www.ajc.com/services/content/metro/stories/2009/01/12/lanier_water_fight.html. Accessed January 17, 2009. Article discusses U.S. Supreme Court denial of a request by the state of Georgia to review a lower court decision to invalidate a prior agreement that would have apportioned more water from Lake Lanier to metropolitan Atlanta.

———. "Lake Lanier No Longer Rising." *Atlanta Journal-Constitution* (1/15/09). Available online. URL: http://www.ajc.com/services/content/metro/stories/2009/01/15/lake_lanier_drought.html?cxtype=rss&cxsvc=7&cxcat=13. Accessed January 16, 2009. Mentions how high the reservoir's water level was in early January 2009 and how far below normal it stood.

Tortajada, Cecilia. "Water Management in Mexico City Metropolitan Area." *Water Resources Development* 22, no. 2, (June 2006): pp. 353–376. Available online. URL: http://www.adb.org/Documents/Books/AWDO/2007/br03.pdf. Accessed January 21, 2009. This article is a fine analysis of the current water problems of Mexico City and its environs. It is supplemented with two maps and six statistical tables.

"Water Supply and Water Conservation Management Plan." Draft. Metropolitan North Georgia Water Planning District. Available online. URL: http://www.northgeorgiawater.com/files/2008-12-12_WaterSupply_Conservation_Public_Comment_DRAFT.pdf. Accessed January 16, 2009. This is the December 2008 draft of the water conservation plan for north Georgia. The plan includes sections detailing the existing freshwater supply, demand forecasts, water conservation, reuse, policy recommendations for the state and region, public education on the issues, an implementation plan, and a plan for future evaluation.

BOLIVIA

Books

Olivera, Oscar, with Tom Lewis. *Cochabamba!: Water War in Bolivia.* Cambridge, Mass.: South End Press, 2004. This book details the history of the most infamous incident of globalized water privatization so far in the world, when Bechtel, through a Bolivian subsidiary, held the water rights in Cochabamba. The book chronicles not only the people's fight to drive out the water company, but the solidarity that spread to include a gas war. Oscar Olivera was one of the leaders in the struggle against water privatization in Cochabamba. The activist Vandana Shiva wrote the foreword.

Articles

Forero, Juan. "As Andean Glaciers Sink, Water Worries Grow." *New York Times* (11/29/02). The author details climate change in the Andean region as it affects the glaciers in Bolivia and Venezuela.

Web Documents

Alcázar, José Luis. "Pilcomayo River to Be Saved from Ruin." Tierramérica. Available online. URL: http://www.tierramerica.net/2005/0521/iacentos.shtml. Accessed June 2, 2009. This article discusses the steps to be taken by Bolivia, Argentina, and Paraguay, beginning in 2008, to clean up the Pilcomayo River. It provides a population breakdown by country of the river basin, as well as a brief history of mining in the region, dating back to colonial times. It also discusses the European Union's input into the clean-up plan.

"Bolivia: Ministry Announces 'Water Belonging to All and for All' Declaration." IRC. Available online. URL: http://www.irc.nl/page/38528. Accessed June 1, 2009. This is a news bulletin highlighting Bolivia's first national water meeting and the water philosophy ratified at the meeting.

"Cochabamba Water Dispute Settled." Bechtel (1/19/06). Available online. URL: http://www.bechtel.com/2006-01-19.html. Accessed February 14, 2009. A news brief from the Bechtel Corporation outlining the terms of the settlement between the Bolivian government and Bechtel's subsidiary, Aguas del Tunari, concerning the Cochabamba water dispute.

"Law 2878." Global Legal Information Network. Available online. URL: http://www.glin.gov/view.action?glinID=123807. Accessed June 2, 2009. This Web page provides an abstract of Bolivian Law 2878, promulgated in 2004, which sought to rectify disparities in previous Bolivian water law and reaffirm traditional water rights.

Martinez, Elizabeth, and Arnoldo Garcia. "What Is Neoliberalism?: A Brief Definition for Activists." CorpWatch. Available online. URL: http://www.corpwatch.org/article.php?id=376. Accessed February 3, 2009. This brief article describes the term *neoliberalism* and how it is used. The authors list five main points of neoliberal policy.

"Secretive World Bank Tribunal Bans Public and Media Participation in Bechtel Lawsuit over Access to Water." The Center for International Environmental Law (2/12/03). Available online. URL: http://www.ciel.org/Ifi/Bechtel_Lawsuit_12Feb03.html. Accessed February 14, 2009. This article reports on the decision by the International Center for the Settlement of Investment Disputes to bar the public from participating in or even witnessing the hearing between the Bolivian government and Bechtel subsidiary Aguas del Tunari. The article briefly outlines the history of the dispute and contains links to other Web sites for more contextual information.

"Timeline: Cochabamba Water Revolt." Frontline World. Available online. URL: http://www.pbs.org/frontlineworld/stories/bolivia/timeline.html. Accessed February 13, 2009. This time line of the Cochabamba water dispute covers the period September 1998 to April 23, 2002.

EGYPT, ETHIOPIA, AND SUDAN

Books

Collins, Robert O. *The Nile.* New Haven, Conn.: Yale University Press, 2002. The author's experience traveling along the Nile is the background for this well-written study of the river's history, politics, and development projects. The book includes maps and interesting photographs.

Little, Tom. *High Dam at Aswan: The Subjugation of the Nile.* New York: John Day Co., 1965. Possibly the earliest book to disclose the environmental and social problems that would result from the Aswan High Dam in Egypt, it was written and published before the dam was completed. Among the points the author makes are the submergence of marshland and the displacement of at least 100,000 people by the giant reservoir—now called Lake Nasser.

Web Documents

"Applied Training Project." Nile Basin Initiative. Available online. URL: http://atp.nilebasin.org/index.php?option=com_content&task=view&id=14&Itemid=27. Accessed March 6, 2009. The home page of the Applied Training Project of the Nile Basin Initiative, it contains a menu that provides links to project background information, resources, etc.

Askouri, Ali. "A Culture Drowned: Sudan Dam Will Submerge Historically Rich Area, Destroy Nile Communities." *World Rivers Review* 19, no. 2 (April 2004). Available online. URL: http://internationalrivers.org/files/WRR.V19.N2.pdf. Accessed February 23, 2009. Askouri points out how the planned Merowe Dam in northern Sudan will displace thousands of people and create social instability. The dam and reservoir, he further contends, will negatively affect all of Sudan through the loss of cultural treasures.

"Confidence Building and Stakeholders Involvement." Nile Basin Initiative. Available online. URL: http://cbsi.nilebasin.org/index.php?option=com_content&task=vi

ew&id=13&Itemid=27. Accessed March 6, 2009. The home page of the Confidence Building and Stakeholders Involvement Project of the Nile Basin Initiative, it contains a menu that provides links to project background information, resources, etc.

"Dams in Sudan." Miraya FM (8/21/08). Available online. URL: http://www.mirayafm.org/reports/reports/_200808214531/. Accessed February 23, 2009. This Web page lists and briefly describes the major Nile Basin dams in Sudan. The descriptions tell which branch they are on, when they were constructed, and what their main purpose is (e.g., hydropower or irrigation).

"Eastern Nile Subsidiary Action Program." Nile Basin Initiative. Available online. URL: http://ensap.nilebasin.org/. Accessed March 7, 2009. The home page of one of two investment programs of the Nile Basin Initiative, it contains a menu that provides links to project background information, resources, etc.

"Efficient Water Use for Agricultural Production." Nile Basin Initiative. Available online. URL: http://ewuap.nilebasin.org/. Accessed March 6, 2009. The home page of one of the shared vision programs of the Nile Basin Initiative, it contains a menu that provides links to project background information, resources, etc.

"Key Achievements, 1998–2006." Nile Basin Initiative. Available online. URL: http://www.nilebasin.org/index.php?option=com_content&task=view&id=13&Itemid=42. Accessed March 6, 2009. This site lists 26 major Nile Basin Initiative events in chronological order, providing the location and important outcome.

Lirri, Evelyn. "East Africa: Regional Countries to Jointly Manage River Kagera." *Monitor* (Kampala) (9/9/08). Available online. URL: http://allafrica.com/stories/200809090249.html. Accessed February 22, 2009. This brief article discusses the ongoing process to reach an agreement among the Nile equatorial states for management of the Kagera River, a Nile tributary. The four countries involved are Burundi, Rwanda, Uganda, and Tanzania.

"Nile Equatorial Lakes Subsidiary Action Program." Nile Basin Initiative. Available online. URL: http://nelsap.nilebasin.org/index.php?option=com_content&task=view&id=20 &Itemid=85. Accessed March 7, 2009. The home page of one of two investment programs of the Nile Basin Initiative, it contains a menu that provides links to project background information, resources, etc.

"Operational Structure." Nile Basin Initiative. Available online. URL: http://www.nilebasin.org/index.php?option=com_content&task=view&id=30&Itemid=77. Accessed March 6, 2009. This Web page provides general and particular details on the organization of the Nile Basin Initiative.

"Regional Power Trade Project." Nile Basin Initiative. Available online. URL: http://rpt.nilebasin.org/index.php?option=com_content&task=view&id=18&Itemid=99. Accessed March 6, 2009. The home page of the Regional Power Trade Project of the Nile Basin Initiative, it contains a menu that provides links to project background information, resources, etc.

"Sequence of Major Events of the Nile Basin Initiative Process." Available online. URL: http://www.africanwater.org/nbihistory.htm. Accessed March 6, 2009. A Web page that provides an annotated time line of the Nile Basin Initiative, in tabular form. The time line runs from 1992 to April 17–24, 2002.

"Shared Vision Coordination Project." Nile Basin Initiative. Available online. URL: http://svpcp.nilebasin.org/index.php?option=com_content&task=view&id=12&Itemid=29. Accessed March 6, 2009. The home page of the Shared Vision Coordination Project of the Nile Basin Initiative, it contains a menu that provides links to project background information, resources, etc.

"Sinai Development Projects." Sinai Liberation ... 26 Years. Available online. URL: http://www.sis.gov.eg/VR/sinia/html/esinia10.htm. Accessed March 5, 2009. A concise listing of seven major development projects in the Sinai, including a breakdown of the North Sinai Agricultural Development Project.

"Socioeconomic Development and Benefits Sharing Project." Nile Basin Initiative. Available online. URL: http://sdbs.nilebasin.org/index.php?option=com_content&task=view&id=16&Itemid=39. Accessed March 6, 2009. The home page of the Socioeconomic Development and Benefits Sharing Project of the Nile Basin Initiative, it contains a menu that provides links to project background information, resources, etc.

"Toshka Project—Mubarak Pumping Station/Sheikh Zayed Canal, Egypt." Water-Technology.net. Available online. URL: http://www.water-technology.net/projects/mubarak. Accessed March 4, 2009. This Web page gives the vital statistics of the Mubarak Pumping Station and discusses the background of the station and the Zayed Canal and their roles in reclaiming 500,000 acres of the desert.

"Trans-boundary Environmental Action Project." Nile Basin Initiative. Available online. URL: http://nteap.nilebasin.org/index.php?option=com_content&task=view&id=50&Itemid=72. Accessed March 6, 2009. The home page of the Trans-boundary Environmental Action Project of the Nile Basin Initiative, it contains a menu that provides links to project background information, resources, etc.

"Water Resources Planning and Management Project." Nile Basin Initiative. Available online. URL: http://wrpmp.nilebasin.org/index.php?option=com_content&task=view&id=12&Itemid=50. Accessed March 6, 2009. The home page of the Water Resources Planning and Management Project of the Nile Basin Initiative, it contains a menu that provides links to project background information, resources, etc.

Zarley, Kermit. "Extending Egypt's Al Salam Canal to the Palestinian State." (1/24/05). Available online. URL: http://www.kermitzarley.com/pdf/alsalamcanal.pdf. Accessed March 4, 2009. This article explores ways of making more water available to the Palestinian territories from Egypt.

ISRAEL, JORDAN, AND THE PALESTINIAN TERRITORIES

Books

Amery, Hussein A., and Aaron T. Wolf, eds. *Water in the Middle East: A Geography of Peace.* Austin: University of Texas Press, 2000. The 10 essays in this book address the Middle East's water needs from the perspective of the region's geography. In addition to the editors themselves, contributors include Peter Beaumont, Steve

Lonergan, Frederick C. Hof, Paul A. Kay and Bruce Mitchell, Nurit N. Kliot, Gwyn Rowley, and John Kolars.

Haddadin, Munther J., ed. *Water Resources in Jordan: Evolving Policies for Development, the Environment, and Conflict Resolution.* Washington, D.C.: RFF Press, 2006. Published by the Resources for the Future Group, this is "the first comprehensive, multidisciplinary book to address water policy in Jordan." The 12 essays cover such areas as environment; economics, including the role of trade, cultural and social issues; sanitation; legislation; and regional cooperation.

Sosland, Jeffery K. *Cooperating Rivals: The Riparian Politics of the Jordan River Basin.* Albany, N.Y.: SUNY Press, 2007. The author covers the slow process toward water cooperation in the Jordan River Valley from the early 20th century when Jordan first gained its independence, through the creation of Israel, the gradual acceptance of the Palestinian Authority, and the 1990s treaty that was a major step toward peaceful cooperation regarding the region's water resources.

Wolf, Aaron T. *Hydropolitics along the Jordan River: Scarce Water and Its Impact on the Arab-Israeli Conflict.* Tokyo: United Nations University Press, 1995. This book begins with an examination of the hydrography of the Jordan River Basin, goes on to reiterate the recent history of water conflict in the region, and explores cooperative methods for solving the entwined political and hydrological problems in the Jordan River Basin.

Articles

Mizroch, Amir. "Ministry: Desalination Can't Meet Water Needs." *Jerusalem Post* (5/14/08). A news article that quotes the Israeli Environmental Protection Ministry's chief scientist, who claims that Israel's current (as of mid-2008) plans for water conservation and desalination are not comprehensive enough. Among the measures listed in the article are a better public relations campaign, a campaign to instruct local leaders on building a better infrastructure, and regular maintenance of the existing and future infrastructure.

Web Documents

Beyth, Michael. "The Red Sea and the Mediterranean-Dead Sea Canals Project." Israel Ministry of Foreign Affairs. (8/10/02). Available online. URL: http://www.mfa.gov.il/MFA/MFAArchive/2000_2009/2002/8/The%20Red%20Sea%20and%20the%20Mediterranean%20Dead%20Sea%20canals. Accessed March 30, 2009. At the time of writing, the author was the chief scientist of Israel's Ministry of National Infrastructures. The article discusses the goals, the various alignments (such as the canals, pumping stations, reservoirs, and desalination plants), and the environmental effects of the Med–Dead and the Red–Dead canals—especially the latter.

"Dead Sea–Red Sea Canal Could Cause Quakes—Official." Reuters/Planet Ark (7/27/05). Available online. URL: http://www.planetark.com/dailynewsstory.cfm/newsid/31801/newsDate/27-Jul-2005/story.htm. Accessed June 3, 2009. Egyptians' fears regarding construction of the Dead Sea–Red Sea Canal are highlighted, primarily the possibility of increased seismic activity and that it would provide water to cool Israel's nuclear power plant.

Kress, Rory. "World Bank Promotes Dead Sea Canal." *Jerusalem Post* (7/25/07). Available online. URL: http://www.jpost.com/servlet/Satellite?cid=1185379003420&pagename=JPost%2FJPArticle%2FShowFull. Accessed June 3, 2009. This article discusses a then upcoming meeting (held in August 2007) called by the World Bank to discuss the Dead Sea–Red Sea Canal. It also quotes opponents of the canal, mainly environmentalists.

"Mountain and Coastal Aquifers." Israel-Palestine Water Issues. Available online. URL: http://mapsomething.com/demo/waterusage/hydrology.php. Accessed March 25, 2009. This Web site covers not only the aquifers beneath Israeli and Palestinian territory but also watershed sources (including Lebanon, Syria, and Jordan) and water infrastructure. Hydrological maps are included.

"Seawater Desalination Projects: The Challenge and the Options to Meet the Water Shortage," Jewish Virtual Library. (March 1999). Available online. URL: http://www.jewishvirtuallibrary.org/jsource/History/desal.html. Accessed March 28, 2009. This comparatively brief article, whose ultimate source is the Israeli Foreign Ministry, outlines the Israeli desalination project as it stood at the turn of the century, including its initial and main phases.

INDIA, PAKISTAN, AND BANGLADESH

Books

Daniélou, Alain. *The Gods of India.* New York: Inner Traditions International, 1985. Originally published as *Hindu Polytheism,* this book discusses various aspects of the Hindu gods, including symbolism, incarnations, and divine power. It includes sections on the sacredness of the Ganges as it relates to mythology.

Leslie, Jaques. *Deep Water: The Epic Struggle over Dams, Displaced People, and the Environment.* New York: Farrar, Straus and Giroux, 2005. In the dam-building era that marked the mid to late 20th century, dams were seen as unquestionably beneficial. Little thought appeared to be given to their effects on the people whose towns and villages were to be submerged or the environment. This book examines these social and natural aspects of dam-building, and the controversies that have erupted. The first section covers India, followed by South Africa and Australia.

Michel, Aloys Arthur. *The Indus Rivers: A Study of the Effects of Partition.* New Haven, Conn.: Yale University Press, 1967. Although somewhat dated, this book provides good information on the water tensions caused by the partition of India and Pakistan. It describes the 12 years of problems and temporary solutions leading up to the 1960 Indus Waters Treaty. An appendix provides the treaty in full.

Ray, Binayak. *Water: The Looming Crisis in India.* Lanham, Md.: Lexington Books, 2008. The author examines India's internal and regional (including China) freshwater policies both internally and regionally and finds them lacking. He offers a new approach to the crisis that includes more involvement by the stakeholders, particularly municipalities.

Shiva, Vandana. *Water Wars: Privatization, Pollution, and Profit.* Cambridge, Mass.: South End Press, 2002. Though the author's focus is on India's water problems, she also looks at various ways in which water rights around the world have become consolidated into fewer and fewer hands, including through damming, mining, and the international water trade. She also examines the roles of the World Bank and the World Trade Organization in the decline of international water rights, as well as the major corporations on the scene.

Articles

"Bangladesh." *The World Almanac and Book of Facts, 2007.* New York: World Almanac Books, 2007. This article is in the "Nations of the World" section. It provides the country's vital statistics, plus a brief history.

Hammer, Joshua. "A Prayer for the Ganges: Across India, Environmentalists Battle a Tide of Troubles to Clean Up a River Revered as the Source of Life." *Smithsonian* (November 2007). As the author describes his trip down the Ganges, he details the various ways in which the river is polluted and the seemingly impossible clean-up attempts. He meets and interviews activists who have struggled with the problem for decades.

Web Documents

"About the Ganges River." Available online. URL: http://web.bryant.edu/~langlois/ecology/gangesmap.htm. Accessed April 7, 2009. This is a brief, bullet-point fact page, with a map, about the Ganges River. The site includes pages that cover the river's history, its problems, some solutions to those problems, and the economy of the river versus its environment.

Badiwala, Mitesh. "The Narmada Sagar and Sardar Sarovar Dam Projects: Social, Environmental, and Economic Outcomes." Indian Dam Inquiry. Available online. URL: http://www.geocities.com/CollegePark/Library/9175/inquiry2.htm. Accessed April 11, 2009. An informative essay about the Narmada Project in which the author discusses the history of the Narmada Sagar and Sardar Sarovar dam projects, as well as their locations and descriptions, social implications, and environmental and economic implications.

Cullet, Phillipe. "The Sardar Sarovar Dam Project: An Overview." Available online. URL: https://eprints.soas.ac.uk/2985/1/a0704.pdf. Accessed April 13, 2009. This 40-page article discusses the history of the Sardar Sarovar Dam, places it in the context of big dams, and discusses the opposition to the project.

"Extracts from 'Sardar Sarovar' the Report of the Independent Review (Morse Committee) Chapter 17: The Findings: Resettlement and Rehabilitation." Available online. URL: http://narmada.aidindia.org/content/view/52/1. Accessed April 13, 2009. These extracts from the World Bank's Morse Committee are from the sections titled "Resettlement and Rehabilitation," "Environment," and "The Bank." They detail why the World Bank pulled out of the Sardar Sarovar Dam project.

"Final Order and Decision of the Tribunal." Available online. URL: http://www.sscac.gov.in/NWDT.pdf. Accessed June 3, 2009. This is a full-text, PDF file of the 1979 decision of the Narmada Waters Disputes Tribunal that, among other things, decided the various state allotments of Narmada River water.

"Govt. Begins Process on Ganges Barrage." South Asian Media Net (4/10/09). Available online. URL: http://www.southasianmedia.net/cnn.cfm?id=571021&category=Development&Country=BANGLADESH. Accessed April 15, 2009. This article is about Bangladesh finally beginning construction on the Ganges Barrage Project after nearly 30 years. (Note: This Web site may be unsafe; a warning was provided on April 19, 2009.)

"Indus Waters Treaty: A History." Henry L. Stimson Center. Available online. URL: http://www.stimson.org/?SN=SA20020116301. Accessed April 6, 2009. A good article that describes the events leading up to the treaty and the involvement of the World Bank in getting India and Pakistan to restart negotiations and conclude the treaty.

"River Ganges." The Water Page. Available online. URL: http://www.africanwater.org/ganges.htm#Dams and the Farakka Barrage. Accessed April 7, 2009. This is an interesting overview of the Ganges River that includes sections on the delta, the river's religious and economic significance, dams, pollution, and the Ganges River dolphins.

Schneider, Ann-Kathrin. "Dam Boom in Himalayas Will Create Mountains of Risk." *World Rivers Review* (3/2/09). Available online. URL: http://internationalrivers.org/en/node/3924. Accessed April 15, 2009. The author discusses the current trend of dam building in the Himalayan Mountains during a time of climate change.

Chronology

CA. 6000 B.C.E.

- The earliest dams are constructed on the Tigris and Euphrates Rivers.

CA. 3000 B.C.E.

- Nilometers are first used to record the Nile River's fluctuations.

CA. 2950–2750 B.C.E.

- The Egyptians dam the Nile River south of Cairo at what is now known as the Wadi el-Garawi. It is theorized that the dam, called the Sadd el-Kafara (Dam of the Pagans), failed during the first flood season.

CA. 2750 B.C.E.

- The earliest supply and drainage systems are constructed in the Indus Valley.

CA. 2500 B.C.E.

- The earliest known water war occurs in Mesopotamia between the city-states of Lagash and Umma.

CA. 2200 B.C.E.

- Dams are constructed in Persia.
- A canal is constructed on the island of Crete to convey springwater to the palace at Knossos.
- Massive waterworks are constructed in China. Down through medieval times and into the early modern period these are attributed to the legendary engineer and emperor Yu the Great.

CA. 1750 B.C.E.

- King Hammurabi of Babylon institutes his famous code of laws, which contain regulations pertaining to irrigation.

CA. 1319–1304 B.C.E.

- Egyptians construct a dam on the Orontes River that is still in use.

CA. 1300 B.C.E.

- Extensive irrigation and drainage systems are constructed in and around the city-state of Nippur in Mesopotamia (present-day Iraq).

CA. 1200 B.C.E.

- Water tunnels, called *sinnōrs,* are in use in Palestine to protect municipal water supplies, derived from local springs, during times of war.

CA. 1000 B.C.E.

- Earliest *qanāts* (underground channels to carry either springwater or groundwater) are built. There is scholarly dispute as to whether they originated in Persia or Armenia.

CA. 950 B.C.E.

- The Siloam channel is built during the reign of King Solomon to carry water from the Gihon Spring to the city of Jerusalem.

CA. 700 B.C.E.

- King Hezekiah improves and protects the conveyance of Jerusalem's water by constructing the Siloam Sinnōr.

CA. 703–690 B.C.E.

- King Sennacherib of Assyria orders the construction of massive waterworks involving weirs, dams, and canals to provide water to his palace at Khosabad and the capital at Ninevah and for irrigation.

689 B.C.E.

- King Sennacherib orders the damming of the Euphrates, causing the flooding of the defeated city of Babylon.

EARLY FIFTH CENTURY B.C.E.

- During the reign of Darius I, Persians introduce the *qanāt* system to Egypt; the *qanāt* eventually spreads as far east as northern India.

LATE FOURTH CENTURY B.C.E.

- The first mention of a rain gauge is found in the *Arthashastra,* written by the Indian minister Kautilya.

Chronology

CA. 27–17 B.C.E.

- Vituvius writes his famous treatise *De architectura* (*On Architecture*). Book eight of the work covers the topic of water: how to find it, how to capture it, and how different soils hold water and the quality of the water found in each type.

CA. 50 C.E.

- Hero (or Heron) of Alexandria calculates stream flow in his book *Dioptra.* It is not until the 17th century, however, that his principle is put into general use.

64

- Lucius Annaeus Seneca writes *Quaetiones naturales.* Among the subjects it deals with are physical geography and meteorology.

97

- Sextus Julius Frontinus is appointed curator aquarum of Rome. Rome has more than 260 miles of aqueducts that carry water whose quality ranges from excellent to poor—the latter is used outside the city, probably for irrigation.

CA. 98

- Julius Sextus Frontius writes *De aquis urbis Romae, libri II* (*On the Water of the City of Rome, book II*).

1247

- The earliest known rain gauges appear in China.

1441

- First mention of rain gauges in Korea; this type remained in use until the first decade of the 20th century.

CA. 1490

- Leonardo da Vinci makes notes for his never completed *Treatise on Water.* Among the 15 books, he plans to include studies of the sea, rivers, groundwater, surface water (other than rivers), canals, machines turned by water, and raising water.

CA. 1575

- The artisan and scientist Bernard Palissy begins public lectures on natural history in Paris, based on his observations and experimentation. These are published five years later under the title *Discours admirables.* Among Palissy's discourses is a discussion of the hydrologic cycle.

CA. 1610

- The Italian physician Santorio Santorio constructs the earliest known meter for measuring water velocity.

1687

- Edmond Halley publishes the first of four papers in the *Philosophical Transactions of the Royal Society* on his experiments into evaporation. The other papers are published in 1691, 1694, and 1715.

LATE EIGHTEENTH CENTURY

- Rain gauges that not only capture rain but record the amount of rainfall are developed in Britain and elsewhere.

1854

- The British construct the Haridwar Dam on the Upper Ganges River to divert Himalayan snowmelt for irrigation.

1855

- The British admiral William Allen proposes an ambitious system of canals to connect the Mediterranean, Dead, and Red Seas.

1862

- Construction of the Delta barrage on the Nile delta is completed, but the diversion dam is never put into operation, as it is discovered that its foundation is too weak.

1869

- John Wesley Powell leads a 10-man, three-month expedition to explore the Green and Colorado Rivers. His exploration will also take him through the Grand Canyon.

1871

- The Treaty of Washington is signed by Canada and the United States. While Great Lakes navigation rights are the main concerns of the treaty, it also settles the boundary between the two countries on the lakes.

1892

- William T. Love proposes to build a canal to connect the upper and lower Niagara River, which are separated by Niagara Falls. The financial panic of 1893 and congressional fiat cause the plan to be abandoned, leaving a ditch of more than 3,000 feet long.

Chronology

1895

- John Wesley Powell publishes *Canyons of the Colorado,* the record of his various trips of exploration down the Colorado River. It is subsequently retitled *The Exploration of the Colorado River and Its Canyons.*

1899

- The U.S. Congress passes the Rivers and Harbors Act, the nation's first environmental law and the precursor to the Clean Water Act.

1902

- The Aswan Low Dam is built on the Nile River at Aswan in Egypt; it will eventually be replaced by a high dam. Further north the Assyut Barrage is constructed and in the delta the Zifta Barrage is built. The latter is the first modern working barrage in the Nile delta.
- The Zionist leader, Theodor Herzl, proposes a canal to connect the Mediterranean and Dead Seas.

1905

- Canada and the United States create the International Waterways Commission to advise both governments on Great Lakes water levels and flow as these relate to the generation of hydropower.

1909

- The Boundary Waters Treaty is signed by the United States and Canada; the treaty creates the International Joint Commission (IJC) to address problems such as pollution, which are not part of the International Waterways Commission's mandate. The IJC begins studying Great Lakes' pollution problems in 1912.
- The Isna and Nag Hammadi Barrages are built north of the Aswan Low Dam. These barrages, like the Assyut further north and those in the delta, channel water from the Nile River for irrigation.

1913

- ***November 5:*** The Los Angeles Aqueduct is dedicated. It diverts Owens Valley water to Los Angeles and the San Fernando Valley and eventually drains Owens Lake.

1920s

- The abandoned ditch in the city of Niagara Falls that is the remnant of the failed Love Canal, by which it is still known, becomes a municipal dumpsite. Eventually the U.S. Army and a chemical company also dump waste at the site.

1922

- ***November 24:*** The Colorado River Compact is signed by representatives from six of the seven states that have been negotiating their allotments of the river's water: Colorado, Wyoming, Utah, New Mexico, Nevada, and California. Arizona is the holdout.

1926

- The Sinnar Dam is constructed on the Blue Nile in Sudan. Its waters irrigate as much as 60 percent of Sudan's agricultural land.

1928

- A disastrous flood strikes Florida and sets the stage for future flood control projects that will involve draining the Everglades.
- ***December 21:*** The U.S. federal government approves the Boulder Canyon project, which includes Boulder Dam and the All-American Canal. The project also finalizes state allotments of Colorado River water.

1929

- Egypt and Great Britain (on behalf of Sudan) sign an agreement that allots 48 billion cubic meters (12.67 trillion gallons) of Nile water to Egypt and 4 billion cubic meters (1.06 trillion gallons) to Sudan.
- Great Britain, on behalf of what are now the nations of Sudan, Kenya, Tanganyika, and Uganda, signs another water agreement with Egypt that effectively gives Egypt consensual power over all water development projects on the Nile River and the equatorial lakes.

1934

- Arizona governor Benjamin Moeur sends troops from the Arizona National Guard to halt construction of the Parker Dam in what becomes known as the Great Colorado River War. No one is injured and no shots are fired. The incident is mainly a political ploy.

1935

- ***September 30:*** President Franklin D. Roosevelt dedicates Boulder Dam on the Colorado River, which at the time of construction is the world's largest concrete structure and generates more hydroelectric power than any other dam. Boulder Dam becomes Hoover Dam in 1947.

1937

- The Jabal Awlia Dam is built on the White Nile south of the Sudanese capital of Khartoum.

Chronology

1939

- The soil conservationist Walter Clay Lowdermilk tours the Jordan Basin and develops a comprehensive plan for water storage and usage. One of the components of his plan is a canal linking the Mediterranean and Dead Seas.

1942

- The All-American Canal, the world's largest irrigation canal, is completed.
- The Hooker Chemical Company begins using the ditch called Love Canal as a dump for toxic waste, though it doesn't purchase the property until 1947.

1944

- ***February 3:*** Arizona ratifies the Colorado River Compact.

1945

- The Grand Coulee Dam is built on the Columbia River in Washington. It is larger than the Hoover Dam and generates more hydroelectricity.

1946

- Separate legislation is introduced in the U.S. Senate to fund the Central Arizona Project and the Central Utah Project.

1947

- ***August 14:*** Largely Muslim Pakistan (consisting of West Pakistan and East Bengal, later renamed East Pakistan) is partitioned from largely Hindu India. The new political boundaries create more water problems, as India is by-and-large the upriver nation on both the Ganges and the Indus Rivers.

1948

- The Presa Morelos Dam is constructed on the Colorado River in Mexico.
- ***April 1:*** India shuts off water supplies to Pakistan from the Ferozepore headworks on the Sutlej River. More than 5 percent of Pakistan's agricultural land, which has already been sown, as well as the city of Lahore are without water.
- ***May 4:*** The Inter-Dominion Agreement between India and Pakistan ends the water conflict begun a month earlier, though it is not a permanent solution.
- ***October 22:*** The upper basin states of Colorado, New Mexico, Utah, and Wyoming, as well as Arizona, sign the Upper Colorado River Storage Compact. The compact eventually leads to the Colorado River Storage Project, which in turn leads to the Central Utah Project. The states begin drawing their full allotments of Colorado River water.

1950

- The Roseires Dam, a hydropower dam on the Blue Nile in Sudan, is completed. By 2008, it is responsible for almost 50 percent of Sudan's combined power output.

1951

- The Adfina and Mohammed Ali barrages are constructed in the Nile Delta, the latter at the site of the failed 19th-century Delta barrage.
- On a visit to the South Asian subcontinent, the former Tennessee Valley Authority and Atomic Energy Commission chairman David Lilienthal recognizes the need for a permanent solution to the Indus River Basin problem. He subsequently convinces World Bank chairman Eugene Black to involve his institution in the negotiations.

1952

- In the ongoing water dispute between Arizona and California, this time over whether water from Arizona's Salt River Project should be charged against its Colorado River allotment, California threatens to block the Central Arizona Project. In retaliation, Arizona sues California and the case eventually goes to the U.S. Supreme Court.
- Soon after the military coup that topples Egypt's King Faruq, the generals in power begin discussing the construction of a high dam at Aswan to replace the dam that has been in operation for 50 years. The political entanglement related to the dam's financing will lead to the Suez Crisis in 1956.

1953

- The Hooker Chemical Company closes the Love Canal as a toxic dumping site, fills the ditch with dirt, and sells the site to the City of Niagara Falls, New York, for one dollar. The area eventually becomes a residential neighborhood with a school.
- Lake Lanier, named for Georgia poet Sydney Lanier (1842–1881), is created by damming the Chattahoochee River near Atlanta.

1954

- Sudan and Egypt reach a compromise on the Jonglei Canal, which will divert water from the Sudd Swamps in southern Sudan before it evaporates. However, it will be 20 years before the plan is fully approved, while the project itself has yet to be completed due to civil war in southern Sudan and other internal strife.

1955

- Congress passes and President Dwight D. Eisenhower signs into law the Air Pollution Control Act. This is followed by the Clean Air Act of 1963 and the Air Quality Act of 1967.

Chronology

1957

- The St. Lawrence Seaway, a joint Canadian–U.S. effort, is completed. It allows oceanbound vessels to navigate the Great Lakes.

1959

- Egypt and Sudan sign an agreement that paves the way for the former's construction of the Aswan High Dam—part of the reservoir of which will lie in Sudan and eventually displace thousands of Nubians. As per the agreement, which abrogates the one signed in 1929, Egypt's water allotment increases to 55.5 billion cubic meters (14.65 trillion gallons), while Sudan's increases to 18.5 billion cubic meters (4.88 trillion gallons).

1960s

- Large stretches of Lake Erie are considered "dead," or lacking marine life, due to eutrophication caused by phosphate and nitrate dumping in the lake.

1960

- ***January 9–11:*** Construction of the Aswan High Dam in Egypt begins with financing and technical assistance from the Soviet Union. The dam's first stage will be completed in 1964.
- ***September 19:*** President Mohammed Ayub Khan of Pakistan and Prime Minister Jawaharlal Nehru of India sign the Indus Waters Treaty. By the terms of the treaty, Pakistan controls the Indus Basin's western tributaries and India its eastern tributaries. The treaty also calls for payments to Pakistan to make up for water it would otherwise have. The treaty remains in effect.

1964

- Israel's National Water Carrier begins diverting water from Lake Kinneret. This will be a cause of future hostilities between Israel and Syria that culminate in the Six-Day War.
- Syria begins construction of its own water diversion project, upriver from Israel.
- The Khashm Al-Gerba Dam on the Atbara River, west of the border with Eritrea, is completed.

1965

- ***March 17:*** Israeli tanks shell Syrian water diversion equipment from inside the Israeli border.
- ***May 13:*** Israeli tanks again fire on Syrian diversion equipment. Syria ceases work at the shelled site on May 22, but begins work at another site.

1966

- ***July 14:*** The Israeli Air Force bombs the Syrian diversion dam under construction at Mukheiba. Eleven months later, at the end of the Six-Day War, Israel occupies the Golan Heights, where part of the dam was to be constructed.

1969

- The Cuyahoga River, a tributary of Lake Erie, catches on fire at Cleveland, Ohio. The fire burns for about half an hour and causes approximately $50,000 in damage. The fire focuses national attention on the problems of water pollution in the United States.
- Congress passes and President Richard Nixon signs into law the National Environmental Policy Act, one of the first U.S. laws to address environmental problems from a holistic perspective.

1970

- ***July 21:*** The Aswan High Dam project is completed, and the reservoir created by the dam, which extends into Sudan, is named Lake Nasser after Egypt's president, who was the driving force behind the project.

1971

- Mexico promulgates its first Federal Water Law. It is succeeded by the Federal Water Law of 1992, which in turn is amended in December 2003.
- ***December 16:*** East Pakistan gains independence from Pakistan and is renamed Bangladesh. The new country now faces the challenges of asserting its riparian rights on the Ganges River.

1972

- ***April 15:*** The United States and Canada sign the first Great Lakes Water Quality Agreement (GLWQA), which deals with phosphate and nitrate dumping in Lake Erie and Lake Ontario.
- Congress passes and President Richard Nixon signs into law the Clean Water Act, formally known as the Water Pollution Control Act. It becomes law 17 years after the first air pollution act.

1973

- Construction begins on the Central Arizona Project, the largest aqueduct system in the United States. It is completed in 1993.
- The United States agrees to construct a desalination plant at Yuma, Arizona, to improve the quality of Colorado River water received by Mexico.

Chronology

1974

- Congress passes and President Gerald Ford signs into law the Safe Drinking Water Act.

1975

- The Farakka barrage in the Indian state of West Bengal near the border with Bangladesh is completed.

1976

- Lake Nasser achieves its highest capacity for the first time. By 1988 the reservoir will be at about one-third capacity due to drought.

1977

- The Kagera Basin Agreement is signed by the nations of Burundi, Rwanda, and Tanzania; Uganda becomes a signatory nation in 1981. The agreement effectually abrogates a 1929 treaty between Egypt and Great Britain, as the colonial ruler of the equatorial states, and a 1934 treaty Great Britain signed with Belgium barring the client states from damming the Kagera River.
- Egyptian President Anwar Sadat offers to deliver water to Israel's Negev Desert via a canal to be built across the Sinai Desert. The canal is to be named Al Salaam Canal (Peace Canal). The Israelis refuse the offer for political reasons.

1978

- ***February 16:*** Ethiopian president Mengistu Haile Mariam challenges Egypt's historical water rights to the Nile River, the majority of whose waters originate in Ethiopia.
- ***May:*** Egyptian president Anwar Sadat responds to Mengistu's challenge by threatening war if Egypt is denied its water rights.
- ***July:*** Following heavy rainfall, chemical leaching at the Love Canal site begins. Children are burned, and miscarriages and at least five birth defects are attributed to the toxic waste rising to the earth's surface. The subsequent national uproar leads Congress to create the Superfund Program to clean up toxic waste sites throughout the United States.
- ***November 22:*** Canada and the United States sign the second Great Lakes Water Quality Agreement. More comprehensive than its predecessor, it covers all five lakes, their tributaries and connecting rivers, prohibits toxic chemical and nutrient (phosphate and nitrate) dumping, sets down stricter standards, provides for municipal and industrial abatement programs, and strengthens the IJC's monitoring mandate.

1979

- Construction of the Narmada Dam Project, a massive hydroelectric and irrigation plan, begins.

1980s

- It is estimated that over the course of the decade more than 15,000 people in Punjab—one of the states that was divided between India and Pakistan during the partition—die in water conflicts.

1980

- Bangladesh makes preliminary plans for the Ganges Barrage Project, the purpose of which is to increase navigation on the Ganges and its tributaries in Bangladesh and reduce the river's salinity to expand agricultural production and strengthen the region's fishing industries. It is planned that the project will also increase the flow of freshwater and save the tidal forest located in the Ganges delta known as the Sundarbans, which is a World Heritage Site.

1983

- The Mexican Supreme Court declares groundwater a national property.

1985

- ***February 11:*** In an unprecedented move, the eight governors and two premiers of the Great Lakes basin states and provinces sign the Great Lakes Charter, an agreement to work together on the state and provincial level to clean up the basin and monitor and preserve it in the future.
- The World Bank approves a $450 million loan for construction of India's Sardar Sarovar Dam, reservoir, and hydropower facilities.
- India launches Phase 1 of the Ganga Action Plan to restore the health of the Ganges River; most deem it a failure by the early 1990s.

1986

- Narmada Bachao Andolan (Save Narmada Movement) is founded to oppose the Narmada Dam Project, in particular the Sardar Sarovar Dam.

1989

- The Mexican government establishes Comisión Nacionel del Agua (CONAGUA), the National Water Commission.

1990

- In an ongoing tristate water conflict between Georgia, Alabama, and Florida, Alabama files suit against the U.S. Army Corps of Engineers, charging the Corps with favoring Georgia's water projects to Alabama's detriment.

Chronology

1992

- The Council of Ministers of Water Affairs of the Nile riparian countries (NILE-COM) embark on an initiative to promote intrabasin development and cooperation.

1993

- ***April 18:*** Former Soviet president Mikhail Gorbachev helps establish Green Cross International, which becomes an influential ecology organization.
- The World Bank withdraws its support for the Sardar Sarovar project, citing environmental concerns and concerns about India's ability to successfully relocate displaced people. The Indian state of Gujarat vows to pick up the financial slack.

1994

- ***January 1:*** The North American Free Trade Agreement (NAFTA) comes into force, creating a trade block consisting of Canada, the United States, and Mexico. In the 21st century the agreement has far-ranging implications regarding the commodification of water.
- ***July 25:*** Israel's prime minister Yitzhak Rabin and Jordan's King Hussein are present at the signing of the Israeli–Jordan Treaty of Peace, a good deal of which has to do with the freshwater resources in the Jordan River Basin.
- An Indian law requires industrial polluters along the Ganges River, in particular tanneries, to do preliminary clean-up of wastewater, but there is no oversight.
- The Egyptian government begins the Northern Sinai Agricultural Development Project (NSADP) to develop approximately 415,000 acres of land for farming along the Mediterranean coast of the Sinai Desert in northeast Egypt. By doing this, it is hoped the region's population will increase from approximately 50,000 to 500,000.

1995

- The Nile River Basin Action Plan (NRBAP) is endorsed by NILE-COM, which also requests coordinating assistance from the World Bank.

1996

- A mine accident in Bolivia causes increased and unavoidable pollution of the Pilcomayo River. This has repercussions for the downriver nations of Argentina and Paraguay.
- ***December 12:*** India and Bangladesh sign a 30-year treaty dealing with water sharing at the Farakka barrage.

1997

- An Australian multistate commission begins correcting ecological abuses of the Murray-Darling River system, the nation's largest freshwater system, by establishing caps on withdrawals.
- The Green Cross steps in to mediate a solution to the ecological crisis of the Pilcomayo River, which threatens to become a military crisis. Bolivia acknowledges its responsibility in the disaster, while Argentina and Paraguay admit their own riparian responsibilities; a clean-up plan is gradually developed.
- The Egyptian government devises the Toshka, or New Valley, project, an irrigation plan to make the desert west of Lake Nasser bloom. The plan is to irrigate more than 600,000 acres of desert.
- ***June:*** After a second request by NILE-COM, the World Bank decides to offer coordinating assistance for the Nile River Basin Action Plan.
- ***October 26:*** Egyptian president Hosni Mubarak dedicates the Al Salam Canal (Peace Canal), which transports Nile water under the Suez Canal to the Sinai Desert.

1998

- The Republic of South Africa passes the National Water Act, under which water resources are treated as ecosystems.
- ***July–September:*** Flooding in Bangladesh directly affects about half the country, with about one-quarter of the country severely affected. More than 1,300 die as result of the flooding and disease and approximately 31 million are left homeless for at least some period of time. The flooding also destroys some 16,000 kilometers (10,000 miles) of roads.

1999

- The city of Atlanta, Georgia, signs a 20-year deal with United Water Resources, a subsidiary of Suez Lyonnaise, under which the company will run the city's water system at a cost of $22 million per year. The contract is considered the largest water privatization deal in U.S. history, but it falls apart within a few years amid acrimony on both sides.
- Four Indian rivers—the Ganges, Godavari, Sabarmati, and Tapti—register 100 percent violations of World Health Organization pollution standards.
- The Indian Supreme Court rules that the height of the Sardar Sarovar Dam on the Narmada River can be increased from 80 meters (262 feet) to 88 meters (289 feet).
- ***February 22:*** The Nile Basin Initiative is formally established in Dar es Salaam, Tanzania, to deal with regional water problems. The representative nations include Egypt, Ethiopia, Sudan, Tanzania, Rwanda, Burundi, the Democratic Republic of the Congo, Uganda, Kenya, and (as a nonvoting member) Eritrea.

- ***May:*** The first shared vision projects planning meeting of the Nile Basin Initiative is held in Sodere, Ethiopia.
- ***June:*** The Nile Basin Initiative Secretariat (NILE-SEC) is established in Entebbe, Uganda.
- ***September 3:*** The Bolivian government signs a 40-year contract with Aguas del Tunari, a subsidiary of an international consortium, to manage the water in Cochabamba, Bolivia's third-largest city.
- ***October 29:*** The Bolivian government passes Law 2029, which eliminates legal blockades to water privatization.
- ***November 1:*** Aguas del Tunari takes over operation of Cochabamba's water supply and distribution.
- ***November 12:*** *Coordinadora de Defensa del Agua y de la Vida* (Coalition in Defense of Water and Life) is formed to oppose water privatization in Cochabamba. The organization's main spokesperson is Oscar Olivera. Another protest leader is Evo Morales, leader of the *cocaleros* (coca growers), and later president of Bolivia.

2000

- ***February 4:*** Coordinadora de Defensa del Agua y de la Vida organizes massive protests in Cochabamba against water privatization.
- ***February 6:*** The Bolivian government agrees in principle to a freeze on water rate increases in Cochabamba, but within six weeks Cochabambinos realize the government has no intention of carrying out the agreement.
- ***March 22:*** In a referendum sponsored by Coordinadora de Defensa del Agua y de la Vida, 96 percent of approximately 50,000 voters disapprove of the Aguas del Tunari contract.
- ***April 4:*** First day of the final phase of the Cochabamba water war. Blockades are set up and manned by Cochabambinos; the government sends in the army to assist the police. Some 20,000 people have gathered in the city's main plaza.
- ***April 6:*** Oscar Olivera and other Coordinadora leaders are arrested and briefly detained.
- ***April 8:*** President Hugo Banzer of Bolivia declares a 90-day state of siege, which is legal under Bolivian law. Following the declaration, violence spreads beyond Cochabamba; the capital of La Paz is among the places where civil unrest occurs.
- In Cochabamba, an army sniper kills 17-year-old Hugo Daza, the water war's only fatality in the city.
- ***April 10:*** The Bolivian government and the Coordinadora reach an agreement under which the contract with Aguas del Tunari is cancelled and Law 2029 is repealed.
- The Indian Supreme Court again allows for an increase in the height of the Sardar Sarovar Dam, from 289 to 295 feet (88 to 90 meters).

- The U.S. Congress passes and President Bill Clinton signs into law the Water Resources Development Act. From this law comes the Comprehensive Everglades Restoration Plan.
- It is estimated that 25 percent of Pakistan's total irrigated land has been damaged to some degree by salt.
- In seven Bangladeshi states, mostly in the delta region, 75 percent or more of test boreholes show higher concentrations of arsenic than the standard set by the World Health Organization of .05 milligrams per liter of water. In another six states, located mostly in the delta or along the river, from 50 percent to 74 percent of the test boreholes reveal concentrations of arsenic higher than the organization's standard. It is theorized that the arsenic leaching is caused by the lowering of the water table due to the over-drawing of groundwater.

2001

- ***July:*** U.S. president George W. Bush makes the observation that Canada's water resources ought to be treated like its energy reserves—as a commodity bound by NAFTA rules.

2002

- ***February:*** The consortium that owns Aguas del Tunari, led by major stakeholder Bechtel, charges that Bolivia's cancellation of the Aguas del Tunari contract is a violation of a trade agreement it holds with the Netherlands (where another Bechtel water subsidiary is incorporated) and takes its case to the International Centre for the Settlement of Investment Disputes (ICSID), the arbitration branch of the World Bank. Bechtel seeks $25 million in reparations.
- The Narmada Control Authority approves another height increase for the Sardar Sarovar Dam, from 90 to 95 meters (295 to 312 feet).

2003

- Water company Suez Environnement, which traces its existence to the construction company that built the Suez Canal, serves some 100 million people worldwide in more than 130 countries, but the world's largest water company is Veolia Water, part of Veolia Environnement (formerly Vivendi Environnement) and itself an umbrella company created by its parent company, the entertainment conglomerate Vivendi.

2004

- The Narmada Control Authority approves another height increase for the Sardar Sarovar Dam, from 95 to 110 meters (312 to 361 feet).

- The Mexican government signs water agreements with indigenous people concerning the diversion of the Cutzamala River, water pollution, crop destruction, and increased costs. Further agreements are signed the following year, but the government is slow to live up to its end of the deal.
- A second Bolivian water war occurs in and around La Paz and El Alto over a 30-year water and sewage contract the Bolivian government had signed with Aguas de Illimani, a subsidiary of the French Transnational Suez. The contract is eventually canceled.
- The Living Murray Initiative is established as Australian riparian state and federal governments pledge (Aus.) $500 million over five years to recapture as much as 500 million cubic feet (500 gigaliters) of water annually for the Murray-Darling River system.
- Another disastrous flood occurs in Bangladesh; it results in nearly 1,000 deaths and $7 billion in property damage.

2005

- ***March:*** The Mubarak Pumping Station goes online to pump reservoir water into the Sheikh Zayed Canal for the Toshka Project. Within four years, the station is pumping 14.5 million cubic meters (3.83 billion gallons) of water per day from Lake Nasser. Completion of the entire project is expected to take another 15 years.
- ***May 9:*** Israel, Jordan, and the Palestinian Authority sign an agreement for a new feasibility study regarding the construction of the Red Sea–Dead Sea Canal.
- A report issued by Environnement Canada warns of a national water crisis caused by pollution and overextraction.
- ***December 22:*** Evo Morales is officially declared the winner of Bolivia's presidential election, with slightly more than 53 percent of the vote.

2006

- ***January:*** Bechtel announces that the case of *Aguas del Tunari v. Republic of Bolivia* is settled by mutual agreement of the parties: the consortium and its subsidiary are absolved of all blame and the Bolivian government is free from indemnity.
- The Narmada Control Authority approves another height change for the Sardar Sarovar Dam (and the fifth increase overall) to 121.92 meters (approximately 400 feet)—more than 50 percent higher than the initial construction figure. It is estimated that at least 35,000 more families will be displaced as a result.

2008

- ***January:*** Lower water levels lead to decreased freight transport, especially of coal, on the Great Lakes.

- ***June:*** It is estimated that Israel is recycling 60 percent of its wastewater.
- A U.S. federal appeals court rules that Georgia's plan to divert more water from Lake Lanier during a continuing drought would require congressional approval.
- More than 10 years after the Bolivian mining disaster that set off an ecological, and nearly military, crisis among the riparian nations of the Pilcomayo River, the clean-up of the river begins.
- Indian Prime Minister Manmohan Singh declares the Ganges a national river.

2009

- The U.S. Supreme Court denies Georgia's request to review a federal appeals court decision concerning the state's plan to divert more water from Lake Lanier.
- ***February:*** China's news agency Xinhua reports that Chinese scientists have determined that Himalayan glaciers are melting at a troubling rate.
- ***April:*** Bangladesh begins construction on the Ganges Barrage Project, whose planning began in 1980.
- ***April 22–27:*** Hartford, Connecticut, area residents are advised to boil their water after copepods and rotifers are discovered in the system. The organisms themselves are harmless, but water officials believe that their passing through the filters may signal that harmful microscopic bacteria may also have passed through the filters.
- ***May 5:*** The U.S. Army Corps of Engineers announces plans to increase the outflow of Lake Lanier in Georgia after months of heavy rainfall. The decision means less storage for Georgia and more available water for the downriver states of Alabama and Florida.
- ***July:*** Preliminary damage and loss figures due to water shortage in five California counties of San Joaquin Valley total approximately $1.4 billion.
- ***September:*** Following months of drought, water is rationed in Mexico City.

2010

- ***February:*** Mumbai's water shortage is characterized as the worst in the city's history. A sporadic monsoon season forces authorities to cut supply by 30 percent leaving approximately 20 million residents to face acute water shortage.
- ***April:*** Researchers at Carnegie Mellon University announce they have devised a tool that better estimates direct and indirect industrial water usage.
- ***April:*** Facing their most serious drought in half a century, the nations of the lower Mekong River Basin—Vietnam, Cambodia, Laos, Thailand, and Myanmar—blame China's large dams for exacerbating the rainfall shortage by diverting too much water headed downstream.

Glossary

acid rain the result of atmospheric water vapor mixing with air pollutants such as sulfur or nitrogen compounds.

acre-foot the amount of water required to cover an acre of land one foot deep; a unit of measurement used in the western United States, it equals 325,851 gallons.

Aguas del Tunari Bolivian subsidiary of an international consortium that briefly held the water distribution contract for the city of Cochabamba, Bolivia.

alluvium any form of matter, such as sand, silt, or gravel, that is deposited by moving water.

anoxia severe lack of oxygen; in bodies of water this results in "dead" zones.

aquicludes rocks that are too dense for water to penetrate.

aquifer an area of underground water amid permeable rock; essentially an aquifer is an underground lake.

arsenic an element found naturally in the environment, high concentrations are poisonous.

assimilative capacity the ability of a body of water to cleanse itself.

Aswan High Dam replacement dam for the original Aswan Dam on the Nile River in Egypt. Completed in 1970, its purpose was flood control, water storage, and the generation of hydroelectricity.

barrage a dam that diverts water into canals instead of storing it in a reservoir.

bioassay a lab test that evaluates municipal and industrial effluents.

blackwater water that contains sewage.

blending the mixing of desalinated water with groundwater that does not meet the minimum standards for potability to produce water that meets those standards.

brackish water water that contains higher concentrations of dissolved salts and other substances than normally occurring freshwater, but a lesser amount than seawater. It generally falls in the range of 500–30,000 parts per million. Thus, at its high end, brackish water is moderately saline.

brine water containing dissolved salts at a concentration greater than 50,000 parts per million.

Bureau of Reclamation an agency of the Department of the Interior responsible for dam and canal construction in the American West and general water resource management; originally named the United States Reclamation Service.

Clean Water Act the federal law that covers water pollution in the United States, it was signed by President Richard M. Nixon in 1972.

Cochabamba Bolivian city that was the site of a water war in 2000.

Colorado River Compact 1922 agreement between the seven riparian states on the Colorado River that divided up its waters. The states are Colorado, Utah, Wyoming, New Mexico, Nevada, California, and Arizona (ratified only in 1944). Since Mexico also has riparian rights with respect to the Colorado and was mentioned in a section of the compact, it was later added.

confined aquifer groundwater trapped by rock and/or sediment layers.

Coordinadora de Defensa del Agua y de la Vida translated as Coalition in Defense of Water and Life, the Bolivian ad hoc organization headed by Oscar Olivera that opposed the sale of the Cochabamba water company to AGUAS DEL TUNARI.

cubic foot a U.S. water industry term of measurement equal to 7.48 gallons.

cubic meter a measurement of water that is equal to 1,000 liters, or 264 gallons.

cusec a measurement of water flow that refers to one cubic foot per second.

Cuyahoga River a tributary river of Lake Erie that runs through Cleveland, Ohio. In 1969, it caught fire in a notorious incident that briefly made the river and the city infamous.

delta the fan-shaped area of alluvial deposits at the mouth of a river.

desalination the process of removing salts from water either by distillation or reverse osmosis.

discharge the volume of water that flows past a given area in a given time.

distillation the desalination process of boiling saline water and retaining, then cooling the steam to produce pure freshwater. The salts and other contaminants are left behind in a brine.

drip irrigation the process by which water is conveyed to plants through tiny holes or nozzles in pipes or tubing.

ecosystem the complete community of organisms and environment.

effluent anything that flows out, but usually associated with farm runoff or treated wastewater.

eutrophication the process by which elevated levels of nitrates and phosphates in a body of water led to increased algal growth, which then depletes the oxygen from the water.

evaporation the process by which water converts into vapor. This occurs naturally from sunlight and is an important component of the hydrologic cycle.

evapotranspiration a catchall word that includes the processes of evaporation, transpiration, and sublimation.

fallowing a water conservation program under which farmers are paid to fallow their land. In some areas the farmers then lease their water rights to municipalities.

floodplain forest a forest area near a river that relies on regular flooding of the river for its survival.

freshwater water in which dissolved salts are less than 500 parts per million.

Ganges Barrage Project Bangladeshi water project whose planning began in 1980, though construction did not begin until 2009. Its purpose is to reduce the salinity of the Ganges River in Bangladesh and increase river navigation and agricultural and fishing production.

glacier a huge ice mass on land. Glaciers are sources of renewable freshwater and feed some rivers, but are shrinking worldwide due to climate change.

godwater a western U.S. term for rainwater.

gray water water from sinks, showers, washing machines, and the like that can be treated and recycled.

Great Lakes Charter a 1985 agreement signed by the governors of eight U.S. states and premiers of two Canadian provinces of the Great Lakes Basin that draws on the 1978 Great Lakes Water Quality Agreement, but also confirms the ecological unity of the basin regardless of political boundaries.

Great Lakes Compact the most recent treaty between Canada and the United States concerning Great Lakes waters, it lessened but did not eliminate fears of water being diverted out of the basin.

Great Lakes Water Quality Agreements two treaties between the United States and Canada signed in 1972 and 1978 that covered phosphate and nitrate dumping in Lake Erie and Lake Ontario (1972) and later expanded to cover all toxins in all five lakes, their tributaries, and the connecting rivers.

green revolution the agricultural movement, begun after World War II, but which gained momentum in the 1960s, to increase output and alleviate hunger in developing nations. In some areas it placed more stress on the local water supply.

groundwater water that is found underground, for example in aquifers.

groundwater overdrafting depletion of aquifers by overpumping.

hard water water with higher levels of minerals such as calcium and magnesium.

high dam a dam that is taller than 165 feet. The first high dam was Hoover Dam in the United States; four other notable high dams are Grand Coulee Dam in the United States, Aswan High Dam in Egypt, Three Gorges Dam in China, and Sardar Sarovar Dam in India.

Hoover Dam the world's first high dam, originally called Boulder Dam because it was constructed in the Boulder Canyon on the Colorado River in the United States. It was designed for hydroelectric power generation and water storage.

hydraulic imperative a political theory that regards Israeli territorial acquisitions as the fulfillment of a desire for water and to control water sources.

hydrologic cycle the cycle of water on Earth whereby surface water evapotranspiration produces atmospheric clouds, which then produce rain, hail, or snow. The precipitation that falls percolates through the ground to become groundwater, refills surface bodies of water (some of which makes its way to the sea), or is used by plants and animals, including humans. It eventually finds its way back into the atmosphere.

hydrology the study of water circulation on and below the surface of the Earth and in the atmosphere.

hydronationalism a theory whose proponents claim Israel must retain control of the West Bank and the Golan Heights to maintain water security.

hydropower technically any form of power that is produced by moving water, including waterwheels and hydraulic power. The term is now generally used to mean electricity produced by flowing water that moves a turbine which powers a generator. Most hydropower is produced by dams and their reservoirs: the higher the dam, the larger the reservoir, and the more electricity is produced.

hypertrophication eutrophication that has expanded quickly; also called galloping eutrophication.

Indus Waters Treaty treaty signed in 1960 between Pakistan and India that provides for equitable distribution of water from the Indus River and those of its tributaries that flow through both countries.

irrigation a general term for supplying water to crops.

Lake Mead the reservoir created by the Hoover Dam and named for Elwood Mead, who was head of the Bureau of Reclamation during the dam's construction.

Lake Nasser reservoir created by the Aswan High Dam, it extends southward into Sudan; named for Egyptian president Gamal Abdel Nasser, under whose rule the dam was built.

law of priority water law doctrine that evolved out of riparian rights; it protected the water rights of entrepreneurs during the Industrial Revolution.

leach a process by which something is removed by percolating, or trickling, liquid; it usually refers to the removal of components in soil by water.

leachate also called landfill leachate, this comes about when water (e.g., rainwater) has passed through solid waste and picked up various pollutants. The leachate can in turn pollute groundwater.

Love Canal an uncompleted canal in the city of Niagara Falls in upstate New York that became infamous in the late 1970s when it was discovered that for decades it had been used as a toxic waste dump.

Med–Dead Canal a proposed canal that would run from the Mediterranean Sea to the Dead Sea; it would provide hydropower, some potable (desalinated) water, and raise the level of the Dead Sea. The idea has been overshadowed by one to create the Red–Dead Canal.

mesotrophic the medium between eutrophic and ogliotrophic.

Narmada Bachao Andolan organization founded in 1986 to oppose construction of the Narmada Basin Project, specifically the Sardar Sarovar Dam. Originally led by Medha Patkar and Baba Amte, the organization was a recipient of a Swedish Right Livelihood Award in 1991.

National Water Carrier Israel's system of diverting Jordan River water to the Negev Desert and elsewhere; it began operating in 1964.

Nile Basin Initiative a riparian partnership for Nile Basin water management launched in 1999, it involves the nations of Egypt, Sudan, Ethiopia, Uganda, Kenya, Tanzania, Burundi, Rwanda, and the Democratic Republic of the Congo as full partners and Eritrea (a nonriparian basin state) as a nonvoting observer.

nitrates generally either of two compounds, sodium nitrate or potassium nitrate, used in fertilizers.

nonpoint source pollution water pollution that is caused by runoff from adjacent land.

oligotrophic low in nutrients; when a body of water is oligotrophic, plant growth is inhibited.

outfall the drainage point of a body of water, such as the mouth of a river.

overtopping occurs when a reservoir's water flows over the top of the dam, holding it back.

phosphates phosphatic compounds used in fertilizers (and formerly detergents).

point source pollution water pollution that is caused by discharge from pipes (or outfalls) directly into a body of water; also called direct pollution.

potable safe to drink.

prior appropriation a legal concept for allocating water developed in the western United States. It is essentially a first-come, first-served rule whose main requirement is that the water have beneficial use to the rights' holder. Under this concept the first rights' holder is entitled to his or her full amount (from a river or lake) before the second rights' holder can draw water, etc.

qana ancient Persian horizontal wells; they are still in use.

Red–Dead Canal a canal that would run from the Red Sea to the Dead Sea, it would have much the same purpose as the Med–Dead Canal, which it has succeeded, though its benefits would be spread among Israelis, Jordanians, and Palestinians.

renewable water groundwater and surface water that is recharged due to precipitation. If the majority of a country's renewable water is within its boundaries, it is considered water independent.

residence time the average length of time water molecules remain in a body of water or in the atmosphere. For a body of water it is calculated by dividing the water volume by the rate at which the water leaves it.

reverse osmosis a desalination process whereby saline or brackish water is forced through a semipermeable membrane to remove salts and other contaminants.

riparian rights developed from English common law, these are the rights of landowners whose property abuts or includes a body of water, usually freshwater.

run-of-the-river project a method of producing hydropower that uses a river's flow instead of a dam; used to produce small amounts of electricity.

Safe Drinking Water Act federal law that defines the maximum contaminant levels for microorganisms and both organic and inorganic substances, it was signed into law in 1974 by President Gerald R. Ford.

safe yield the amount that can be withdrawn from a groundwater source without negatively affecting that source. Safe yields are not hard and fast amounts. Also called perennial yield.

saline water water in which dissolved salts are found in concentrations of 30,000 to 50,000 parts per million. Seawater is in this category.

Sardar Sarovar Dam controversial high dam under construction on the Narmada River in western India.

SEMAPA *Servicios Municipales de Agua Potable y Alcantarillado* (Municipal Water and Sanitation Services), the water supplier for the city of Cochabamba, Bolivia, before and after the city's water war.

sublimation evaporation of snow and ice.

subsidence physical depression or sinking of the Earth; it is sometimes caused by a lowering of the water table due to overdrawing. Central Mexico City is the best-known case of subsidence, and Florida sinkholes are also examples of the phenomenon.

surface water freshwater found on the Earth's surface, such as in lakes, rivers, ponds, glaciers, and oceans.

sweet water another term for freshwater.

transpiration the process by which plants and trees take up water through their roots and release it through the stomata (minute pores) of their leaves.

turbid water water that is muddy or in which particles, natural or otherwise, have been stirred up; water may become turbid after a storm.

unconfined aquifer an aquifer whose water reaches the water table.

U.S. Army Corps of Engineers branch of the U.S. Army that provides engineering services during war and peacetime; domestically, it is primarily responsible for public engineering works east of the Mississippi River.

usufructuary right the right to use but not own a resource, such as water.

wadi a riverbed that is dry except during the rainy season; the term is used throughout the Middle East and North Africa.

warabandi a three-step irrigation method that conserves water. It is commonly used in the Punjab region in both India and Pakistan.

wastewater water that has been contaminated with industrial or domestic pollutants.

waterborne diseases diseases caused by parasites in freshwater, they are a major killer of children in developing countries. They include bilharzia, cholera, guinea-worm disease, polio, and typhoid.

water footprint the total amount of freshwater used by an individual, a group, a municipality, state, nation, or region; it is expressed in annual total water use.

waterlog occurs when water is added to land faster than it can drain out.

watershed an area that is drained by an interconnected group of rivers, streams, lakes, and bays.

water table the highest level of groundwater in a given area.

wetlands land areas that are either covered with shallow water or have very moist soil, such as swamps, marshes, and bogs.

xeriscape a method of landscaping that conserves water by using native grasses and plants. It is useful particularly in arid and semiarid climates.

Index

Page numbers in **boldface** indicate major treatment of a subject. Page numbers followed by *c* indicate chronology entries. Page numbers followed by *f* indicate figures. Page numbers followed by *g* indicate glossary entries. Page numbers followed by *m* indicate maps.

D

Index

Q

R

X

Y

Z